CAMBRIDGE LIBRARY COLLECTION

Books of enduring scholarly value

Life Sciences

Until the nineteenth century, the various subjects now known as the life sciences were regarded either as arcane studies which had little impact on ordinary daily life, or as a genteel hobby for the leisured classes. The increasing academic rigour and systematisation brought to the study of botany, zoology and other disciplines, and their adoption in university curricula, are reflected in the books reissued in this series.

Himalayan Journals

Sir Joseph Hooker (1817–1911) was one of the greatest British botanists and explorers of the nineteenth century. He succeeded his father, Sir William Jackson Hooker, as Director of the Royal Botanic Gardens, Kew, and was a close friend and supporter of Charles Darwin. His journey to the Himalayas and India was undertaken between 1847 and 1851 to collect plants for Kew, and his account, published in 1854, was dedicated to Darwin. Hooker collected some 7,000 species in India and Nepal, and carried out surveys and made maps which proved of economic and military importance to the British. He was arrested by the Rajah of Sikkim, but the British authorities secured his release by threatening to invade, and annexing part of the small kingdom. Volume 1 begins at his arrival in Calcutta, and follows his travels northward to Sikkim and Nepal via Bangalore and Darjeeling, and then on to Tibet.

Himalayan Journals

Or, Notes of a Naturalist in Bengal, the Sikkim and Nepal Himalayas, the Khasia Mountains, &c

VOLUME 1

JOSEPH DALTON HOOKER

CAMBRIDGE UNIVERSITY PRESS

Cambridge, New York, Melbourne, Madrid, Cape Town,
Singapore, São Paolo, Delhi, Tokyo, Mexico City

Published in the United States of America by Cambridge University Press, New York

www.cambridge.org
Information on this title: www.cambridge.org/9781108029353

This edition first published 1854
This digitally printed version 2011

ISBN 978-1-108-02935-3 Paperback

The original edition of this book contains a number of colour plates, which cannot be printed cost-effectively in the current state of technology. The colour scans will, however, be incorporated in the online version of this reissue, and in printed copies when this becomes feasible while maintaining affordable prices.

Pl. II.

W. Taylor Esqr. B.C.S. del.t John Murray, Albemarle Street, 1854. W. L. Walton, lith.

Kinchin-junga from Mr. Hodgson's Bungalow,

HIMALAYAN JOURNALS;

OR,

NOTES OF A NATURALIST

IN BENGAL, THE SIKKIM AND NEPAL HIMALAYAS,
THE KHASIA MOUNTAINS, &c.

BY JOSEPH DALTON HOOKER, M.D., R.N., F.R.S.

WITH MAPS AND ILLUSTRATIONS.

IN TWO VOLUMES.—VOL. I.

LONDON:
JOHN MURRAY, ALBEMARLE STREET.
1854.

TO

CHARLES DARWIN, F.R.S., &c.

These Volumes are Dedicated,

BY HIS AFFECTIONATE FRIEND,

J. D. HOOKER.

KEW, *Jan. 12th*, 1854.

PREFACE.

HAVING accompanied Sir James Ross on his voyage of discovery to the Antarctic regions, where botany was my chief pursuit, on my return I earnestly desired to add to my acquaintance with the natural history of the temperate zones, more knowledge of that of the tropics than I had hitherto had the opportunity of acquiring. My choice lay between India and the Andes, and I decided upon the former, being principally influenced by Dr. Falconer, who promised me every assistance which his position as Superintendent of the H. E. I. C. Botanic Garden at Calcutta, would enable him to give. He also drew my attention to the fact that we were ignorant even of the geography of the central and eastern parts of these mountains, while all to the north was involved in a mystery equally attractive to the traveller and the naturalist.

On hearing of the kind interest taken by Baron Humboldt in my proposed travels, and at the request of my father (Sir William Hooker), the Earl of Carlisle (then Chief Commissioner of Woods and Forests) undertook to represent to Her Majesty's Government the expediency of securing my collections for the Royal Gardens at Kew; and owing to the generous exertions of that nobleman, and of the late Earl of Auckland (then First Lord of the Admiralty), my journey assumed the character of a Government mission, £400 per annum being granted by the Treasury for two years.

I did not contemplate proceeding beyond the Himalaya and Tibet, when Lord Auckland desired that I should afterwards visit Borneo, for the purpose of reporting on the capabilities of Labuan, with reference to the cultivation of cotton, tobacco, sugar, indigo, spices, gutta-percha, &c. To this end a commission in the navy (to which service I was already attached) was given me, such instructions were drawn up as might facilitate my movements in the East, and a suitable sum of money was placed at my disposal.

Soon after leaving England, my plans became, from various causes, altered. The Earl of Auckland * was dead; the interest in Borneo had in a great measure subsided; H. M. S. "Mæander," to which I had been attached for service in Labuan, had left the Archipelago; reports of the unhealthy nature of the coast had excited

* It is with a melancholy satisfaction that I here record the intentions of that enlightened nobleman. The idea of turning to public account what was intended as a scientific voyage, occurred to his lordship when considering my application for official leave to proceed to India; and from the hour of my accepting the Borneo commission with which he honoured me, he displayed the most active zeal in promoting its fulfilment. He communicated to me his views as to the direction in which I should pursue my researches, furnished me with official and other information, and provided me with introductions of the most essential use.

alarm; and the results of my researches in the Himalaya had proved of more interest and advantage than had been anticipated. It was hence thought expedient to cancel the Borneo appointment, and to prolong my services for a third year in India; for which purpose a grant of £300 (originally intended for defraying the expense of collecting only, in Borneo) was transferred as salary for the additional year to be spent in the Himalaya.

The portion of the Himalaya best worth exploring, was selected for me both by Lord Auckland and Dr. Falconer, who independently recommended Sikkim, as being ground untrodden by traveller or naturalist. Its ruler was, moreover, all but a dependant of the British government, and it was supposed, would therefore be glad to facilitate my researches.

No part of the snowy Himalaya eastward of the north-west extremity of the British possessions had been visited since Turner's embassy to Tibet in 1789; and hence it was highly important to explore scientifically a part of the chain which, from its central position, might be presumed to be typical of the whole range. The possibility of visiting Tibet, and of ascertaining particulars respecting the great mountain Chumulari,* which was only known from Turner's account, were additional inducements to a student of physical geography;

* My earliest recollections in reading are of "Turner's Travels in Tibet," and of "Cook's Voyages." The account of Lama worship and of Chumulari in the one, and of Kerguelen's Land in the other, always took a strong hold on my fancy. It is, therefore, singular that Kerguelen's Land should have been the first strange

but it was not then known that Kinchinjunga, the loftiest known mountain on the globe, was situated on my route, and formed a principal feature in the physical geography of Sikkim.

My passage to Egypt was provided by the Admiralty in H. M. steam-vessel "Sidon," destined to convey the Marquis of Dalhousie, Governor-General of India, thus far on his way. On his arrival in Egypt, his Lordship did me the honour of desiring me to consider myself in the position of one of his suite, for the remainder of the voyage, which was performed in the "Moozuffer," a steam frigate belonging to the Indian Navy. My obligations to this nobleman had commenced before leaving England, by his promising me every facility he could command; and he thus took the earliest opportunity of affording it, by giving me such a position near himself as ensured me the best reception everywhere; no other introduction being needed. His Lordship procured my admission into Sikkim, and honoured me throughout my travels with the kindest encouragement.

During the passage out, some days were spent in Egypt, at Aden, Ceylon, and Madras. I have not thought it necessary to give here the observations made in those well-known countries; they are detailed in a series of letters published in the "London Journal of Botany,"

country I ever visited (now fourteen years ago), and that in the first King's ship which has touched there since Cook's voyage, and whilst following the track of that illustrious navigator in south polar discovery. At a later period I have been nearly the first European who has approached Chumulari since Turner's embassy.

as written for my private friends. Arriving at Calcutta in January, I passed the remainder of the cold season in making myself acquainted with the vegetation of the plains and hills of Western Bengal, south of the Ganges, by a journey across the mountains of Birbhoom and Behar to the Soane valley, and thence over the Vindhya range to the Ganges, at Mirzapore, whence I descended that stream to Bhaugulpore; and leaving my boat, struck north to the Sikkim Himalaya. This excursion is detailed in the "London Journal of Botany," and the Asiatic Society of Bengal honoured me by printing the meteorological observations made during its progress.

During the two years' residence in Sikkim which succeeded, I was laid under obligations of no ordinary nature to Brian H. Hodgson, Esq., B. C. S., for many years Resident at the Nepal Court; whose guest I became for several months. Mr. Hodgson's high position as a man of science requires no mention here; but the difficulties he overcame, and the sacrifices he made, in attaining that position, are known to few. He entered the wilds of Nepal when very young, and in indifferent health; and finding time to spare, cast about for the best method of employing it: he had no one to recommend or direct a pursuit, no example to follow, no rival to equal or surpass; he had never been acquainted with a scientific man, and knew nothing of science except the name. The natural history of men and animals, in its most comprehensive sense, attracted his attention; he sent to Europe for books, and commenced the study of ethnology and zoology. His labours have now extended over upwards of twenty-five years' residence in the Himalaya. During this period he has seldom had a staff of less than from ten to twenty persons (often many more), of various tongues and races, employed as translators and collectors, artists, shooters, and stuffers. By unceasing exertions and a princely liberality, Mr. Hodgson has unveiled the mysteries of the Boodhist religion, chronicled the affinities, languages, customs, and faiths of the Himalayan tribes; and completed a natural history of the animals and birds of these regions. His collections of specimens are immense, and are illustrated by drawings and descriptions taken from life, with remarks on the anatomy,* habits, and localities of the animals themselves. Twenty volumes of the Journals, and the Museum of the Asiatic Society of Bengal, teem with the proofs of his indefatigable zeal; and throughout the cabinets of the bird and quadruped departments of our national museum, Mr. Hodgson's name stands pre-eminent. A seat in the Institute of France, and the cross of the Legion of Honour, prove the estimation in which his Boodhist studies are held on the continent of Europe. To be welcomed to the Himalaya by such a person, and to be allowed the most unreserved intercourse, and the advantage of all his information and library, exercised a material influence on the progress I made in my studies, and on my travels. When I add that many of the subjects

In this department he availed himself of the services of Dr. Campbell, who was also attached to the Residency at Nepal, as surgeon and assistant political agent.

treated of in these volumes were discussed between us, it will be evident that it is impossible for me to divest much of the information thus insensibly obtained, of the appearance of being the fruits of my own research.

Dr. Campbell, the Superintendent of Dorjiling, is likewise the Governor-General's agent, or medium of communication between the British Government and the Sikkim Rajah; and as such, invested with many discretionary powers. In the course of this narrative, I shall give a sketch of the rise, progress, and prospects of the Sanatarium, or Health-station of Dorjiling, and of the anomalous position held by the Sikkim Rajah. The latter circumstance led indirectly to the detention of Dr. Campbell (who joined me in one of my journeys) and myself, by a faction of the Sikkim court, for the purpose of obtaining from the Indian Government a more favourable treaty than that then existing. This mode of enforcing a request by *douce violence* and detention, is common with the turbulent tribes east of Nepal, but was in this instance aggravated by violence towards my fellow-prisoner, through the ill will of the persons who executed the orders of their superiors, and who had been punished by Dr. Campbell for crimes committed against both the British and Nepalese governments. The circumstances of this outrage were misunderstood at the time; its instigators were supposed to be Chinese; its perpetrators Tibetans; and we the offenders were assumed to have thrust ourselves into the country, without authority from our own government, and contrary to the will of the Sikkim Rajah; who was imagined to be a tributary of China, and protected by that nation, and to be under no obligation to the East Indian government.

With regard to the obligations I owe to Dr. Campbell, I confine myself to saying that his whole aim was to promote my comfort, and to secure my success, in all possible ways. Every object I had in view was as sedulously cared for by him as by myself: I am indebted to his influence with Jung Bahadoor* for the permission to traverse his dominions, and to visit the Tibetan passes of Nepal. His prudence and patience in negotiating with the Sikkim court, enabled me to pursue my investigations in that country. My journal is largely indebted to his varied and extensive knowledge of the people and productions of these regions.

In all numerical calculations connected with my observations, I received most essential aid from John Muller, Esq., Accountant of the Calcutta Mint, and from his brother, Charles Muller, Esq., of Patna, both ardent amateurs in scientific pursuits, and who employed themselves in making meteorological observations at Dorjiling, where they were recruiting constitutions impaired by the performance of arduous duties in the climate of the plains. I cannot sufficiently thank these gentlemen for

* It was in Nepal that Dr. Campbell gained the friendship of Jung Bahadoor, the most remarkable proof of which is the acceding to his request, and granting me leave to visit the eastern parts of his dominions; no European that I am aware of, having been allowed, either before or since, to travel anywhere except to and from the plains of India and valley of Katmandu, in which the capital city and British residency are situated.

the handsome manner in which they volunteered me their assistance in these laborious operations. Mr. J. Muller resided at Dorjiling during eighteen months of my stay in Sikkim, over the whole of which period his generous zeal in my service never relaxed; he assisted me in the reduction of many hundreds of my observations for latitude, time, and elevation, besides adjusting and rating my instruments; and I can recal no more pleasant days than those thus spent with these hospitable friends.

Thanks to Dr. Falconer's indefatigable exertions, such of my collections as reached Calcutta were forwarded to England in excellent order; and they were temporarily deposited in Kew Gardens until their destination should be determined. On my return home, my scientific friends interested themselves in procuring from the Government such aid as might enable me to devote the necessary time to the arrangement, naming, and distributing of my collections, the publication of my manuscripts, &c. I am in this most deeply indebted to the disinterested and generous exertions of Mr. L. Horner, Sir Charles Lyell, Dr. Lindley, Professor E. Forbes, and many others; and most especially to the Presidents of the Royal Society (the Earl of Rosse), of the Linnean (Mr. R. Brown), and Geological (Mr. Hopkins), who in their official capacities memorialized in person the Chief Commissioner of Woods and Forests on this subject; Sir William Hooker at the same time bringing it under the notice of the First Lord of the Treasury.

The result was a grant of £400 annually for three years.

Dr. T. Thomson joined me in Dorjiling in the end of 1849, after the completion of his arduous journeys in the North-West Himalaya and Tibet, and we spent the year 1850 in travelling and collecting, returning to England together in 1851. Having obtained permission from the Indian Government to distribute his botanical collections, which equal my own in extent and value, we were advised by all our botanical friends to incorporate, and thus to distribute them. The whole constitute an Herbarium of from 6000 to 7000 species of Indian plants, including an immense number of duplicates; and it is now in process of being arranged and named, by Dr. Thomson and myself, preparatory to its distribution amongst sixty of the principal public and private herbaria in Europe, India, and the United States of America.

For the information of future travellers, I may state that the total expense of my Indian journey, including outfit, three years and a half travelling, and the sending of my collections to Calcutta, was under £2000 (of which £1200 were defrayed by government), but would have come to much more, had I not enjoyed the great advantages I have detailed. This sum does not include the purchase of books and instruments, with which I supplied myself, and which cost about £200, nor the freight of the collections to England, which was paid by Government. Owing to the kind services of Mr. J. C. Melvill, Secretary of the India House, many small parcels of seeds, &c., were conveyed to England, free of cost; and I have to record my great obligations and sincere thanks to the Peninsular and Oriental Steam Navigation Company, for conveying, without charge, all small parcels of books, instruments and specimens, addressed to or by myself.

It remains to say something of the illustrations of this work. The maps are from surveys of my own, made chiefly with my own instruments, but partly with some valuable ones for the use of which I am indebted to my friend Captain H. Thuillier, Deputy Surveyor-General of India, who placed at my disposal the resources of the magnificent establishment under his control, and to whose innumerable good offices I am very greatly beholden.

The landscapes, &c. have been prepared chiefly from my own drawings, and will, I hope, be found to be tolerably faithful representations of the scenes. I have always endeavoured to overcome that tendency to exaggerate heights, and increase the angle of slopes, which is I believe the besetting sin, not of amateurs only, but of our most accomplished artists. As, however, I did not use instruments in projecting the outlines, I do not pretend to have wholly avoided this snare; nor, I regret to say, has the lithographer, in all cases, been content to abide by his copy. My drawings will be considered tame compared with most mountain landscapes, though the subjects comprise some of the grandest scenes in nature. Considering how conventional the treatment of such subjects is, and how unanimous artists seem to be as to the propriety of exaggerating those features which should predominate in the landscape, it may fairly be doubted whether the total effect of steepness and elevation, especially in a mountain view, can, on a small scale, be conveyed by a strict adherence to truth. I need hardly add, that if such is attainable, it is only by those who have a power of colouring that few pretend to. In the list of plates and woodcuts I have mentioned the obligations I am under to several friends for the use of drawings, &c.

With regard to the spelling of native names, after much anxious discussion I have adopted that which assimilates most to the English pronunciation. For great assistance in this, for a careful revision of the sheets as they passed through the press, and for numerous valuable suggestions throughout, I am indebted to my fellow-traveller, Dr. Thomas Thomson.

CONTENTS.

CHAPTER I.

CHAPTER II.

CHAPTER III.

CHAPTER IV.

CHAPTER V.

CHAPTER VI.

CHAPTER VII.

CHAPTER VIII.

CHAPTER IX.

CHAPTER X.

CHAPTER XI.

CHAPTER XII.

CHAPTER XIII.

CHAPTER XIV.

CHAPTER XV.

CHAPTER XVI.

CHAPTER XVII.

LIST OF ILLUSTRATIONS.

LITHOGRAPHIC VIEWS.

WOOD ENGRAVINGS.

MAPS

1. A GENERAL MAP OF LOWER AND EASTERN BENGAL, WITH THE HIMALAYA AND ADJACENT PROVINCES OF TIBET.

The Tibetan portion of this map is to a great extent conjectural, and is intended to convey a general idea of the arrangement of the mountains, according to the information collected by Dr. Campbell and myself, and to show the position of the principal groups of snowed peaks between the Yaru-tsampu and the plains of India, and their relations to the water-shed of the Himalaya.

The positions and direction of the minor spurs of the mountain ranges of Central India and Behar are also, to a great extent, conjectural. It is particularly requisite to observe, that the only object of this map is to give a better general idea of the physical geography of South-eastern Tibet and Central India, from the materials at my command, and hence to afford a better guide to the understanding of some of the points I have attempted to explain in these volumes, than is obtainable from any map with which I am acquainted.

Above the map is a view of the Sikkim Himalaya, from Nango to Donkia, as seen from Dorjiling. On the right are four views of celebrated mountains, as seen from great distances :—

1. CHUMULARI, FROM TONGLO.
2. KINCHINJUNGA, FROM EAST NEPAL.
3. DITTO FROM BHOMTSO IN TIBET.
4. THE GHASSA MOUNTAINS, TIBET, FROM BHOMTSO IN TIBET.

On the left is a survey of the moraines, &c., in the Yangma valley, as described in vol. i. p. 231-238.

I beg to return my acknowledgments to Mr. Petermann for the skill and care which he has devoted to the construction of this map. The scale is approximate only, and perhaps very erroneous.

2. GENERAL MAP OF SIKKIM, &C., FROM A SURVEY BY THE AUTHOR.

On the cover of this work is a Sikkim chait of the ordinary construction, with a pole, to which is attached a long narrow banner or strip of cotton cloth, inscribed with Tibetan characters.

On the back is a copy of the sacred sentence, "Om mani padmi om," in the Uchen character of Tibet.

DHURMA RAJAH'S SEAL.

ERRATA.—VOL. I.

Page 141, line 2, *for* "Looties" *read* "Cookies."
,, 165, ,, 4 from bottom, *erase* "7000 feet."
,, 187, ,, 10, *for* "700" *read* "7000."
,, 389, heading, *for* "Rajah of Cooch" *read* "Rajah of Jeel."
Chap. XVII., heading, *for* "Behar" *read* "Pigoree."

HIMALAYAN JOURNALS.

CHAPTER I.

Sunderbunds vegetation—Calcutta Botanic Garden—Leave for Burdwan—Rajah's gardens and menagerie—Coal-beds, geology, and plants of—Lac insect and plant—Camels—Kunker—Cowage—Effloresced soda on soil—Glass, manufacture of—Atmospheric vapours—Temperature, &c —Mahowa oil and spirits —Maddaobund—Jains—Ascent of Paras-nath—Vegetation of that mountain.

I LEFT England on the 11th of November, 1847, and performed the voyage to India under circumstances which have been detailed in the Introduction. On the 12th of January, 1848, the "Moozuffer" was steaming amongst the low swampy islands of the Sunderbunds. These exhibit no tropical luxuriance, and are, in this respect, exceedingly disappointing. A low vegetation covers them, chiefly made up of a dwarf-palm (*Phœnix paludosa*) and small mangroves, with a few scattered trees on the higher bank that runs along the water's edge, consisting of fan-palm, toddy-palm, and *Terminalia*. Every now and then, the paddles of the steamer tossed up the large fruits of *Nipa fruticans*, a low stemless palm that grows in the tidal waters of the Indian ocean, and bears a large head of nuts. It is a plant of no interest to the common observer, but of much to the geologist, from the nuts of a similar plant abounding in the tertiary formations at the mouth of the Thames, and having floated about there in as great profusion as here, till buried deep in the silt and mud that now forms the island of Sheppey.*

Higher up, the river Hoogly is entered, and large trees, with villages and cultivation, replace the sandy spits and marshy jungles of the great Gangetic delta. A few miles below Calcutta, the scenery becomes beautiful, beginning with the Botanic Garden, once the residence of Roxburgh and Wallich, and now of Falconer,—classical ground to the naturalist. Opposite are the gardens of Sir Lawrence Peel; unrivalled in India for their beauty and cultivation, and fairly entitled to be called the Chatsworth of Bengal. A little higher up, Calcutta opened out, with the batteries of Fort William in the foreground, thundering forth a salute, and in a few minutes more all other thoughts were absorbed in watching the splendour of the arrangements made for the reception of the Governor-General of India.

During my short stay in Calcutta, I was principally occupied in preparing for an excursion with Mr. Williams of the Geological Survey, who was about to move his camp from the Damooda valley coal-fields, near Burdwan, to Beejaghur on the banks of the Soane, where coal was reported to exist, in the immediate vicinity of water-carriage, the great desideratum of the Burdwan fields.

My time was spent partly at Government-House, and partly at Sir Lawrence Peel's residence. The former I was kindly invited to consider as my Indian home, an honour which I appreciate the more highly, as the invitation was accompanied with the assurance that I should

* Bowerbank "On the Fossil Fruits and Seeds of the Isle of Sheppey," and Lyell's "Elements of Geology," 3rd ed. p. 201.

have entire freedom to follow my own pursuits; and the advantages which such a position afforded me, were, I need not say, of no ordinary kind.

At the Botanic Gardens I received every assistance from Dr. McLelland,* who was very busy, superintending the publication of the botanical papers and drawings of his friend, the late Dr. Griffith, for which native artists were preparing copies on lithographic paper.

Of the Gardens themselves it is exceedingly difficult to speak; the changes had been so very great, and from a state with which I had no acquaintance. There had been a great want of judgment in the alterations made since Dr. Wallich's time, when they were celebrated as the most beautiful gardens in the east, and were the great object of attraction to strangers and townspeople. I found instead an unsightly wilderness, without shade (the first requirement of every tropical garden) or other beauties than some isolated grand trees, which had survived the indiscriminate destruction of the useful and ornamental which had attended the well-meant but ill-judged attempt to render a garden a botanical class-book. It is impossible to praise too highly Dr. Griffith's abilities and acquirements as a botanist, his perseverance and success as a traveller, or his matchless industry in the field and in the closet; and it is not wonderful, that, with so many and varied talents, he should have wanted the eye of a landscape-gardener, or the education of a horticulturist. I should, however, be wanting in my duty to his predecessor, and to his no less illustrious successor, were these remarks withheld, proceeding, as they do, from an unbiassed observer, who had the honour of standing in an equally friendly relation to all parties. Before leaving India, I saw great improvements,

* Dr. Falconer's *locum tenens*, then in temporary charge of the establishment.

but many years must elapse before the gardens can resume their once proud pre-eminence.

I was surprised to find the Botanical Gardens looked upon by many of the Indian public, and even by some of the better informed official men, as rather an extravagant establishment, more ornamental than useful. These persons seemed astonished to learn that its name was renowned throughout Europe, and that during the first twenty years especially of Dr. Wallich's superintendence, it had contributed more useful and ornamental tropical plants to the public and private gardens of the world than any other establishment before or since.* I speak from a personal knowledge of the contents of our English gardens, and our colonial ones at the Cape, and in Australia, and from an inspection of the ponderous volumes of distribution lists, to which Dr. Falconer is daily adding. The botanical public of Europe and India is no less indebted than the horticultural to the liberality of the Hon. East India Company, and to the energy of the several eminent men who have carried their views into execution.†

* As an illustration of this, I may refer to a Report presented to the government of Bengal, from which it appears that between January, 1836, and December, 1840, 189,932 plants were distributed gratis to nearly 2000 different gardens.

† I here allude to the great Indian herbarium, chiefly formed by the staff of the Botanic Gardens under the direction of Dr. Wallich, and distributed in 1829 to the principal museums of Europe. This is the most valuable contribution of the kind ever made to science, and it is a lasting memorial of the princely liberality of the enlightened men who ruled the counsels of India in those days. No botanical work of importance has been published since 1829, without recording its sense of the obligation, and I was once commissioned by a foreign government, to purchase for its national museum, at whatever cost, one set of these collections, which was brought to the hammer on the death of its possessor. I have heard it remarked that the expense attending the distribution was enormous, and I have reason to know that this erroneous impression has had an unfavourable influence upon the destination of scarcely less valuable collections, which have for years been lying untouched in the cellars of the India House. I may add that officers who have exposed their lives and impaired their health in forming similar ones at the

The Indian government, itself, has already profited largely by these gardens, directly and indirectly, and might have done so still more, had its efforts been better seconded either by the European or native population of the country. Amongst its greatest triumphs may be considered the introduction of the tea-plant from China, a fact I allude to, as many of my English readers may not be aware that the establishment of the tea-trade in the Himalaya and Assam is almost entirely the work of the superintendents of the gardens of Calcutta and Seharunpore.

From no one did I receive more kindness than from Sir James Colvile, President of the Asiatic Society, who not only took care that I should be provided with every comfort, but presented me with a completely equipped palkee, which, for strength and excellence of construction, was everything that a traveller could desire. Often *en route* did I mentally thank him when I saw other palkees breaking down, and travellers bewailing the loss of those forgotten necessaries, with which his kind attention had furnished me.

I left Calcutta to join Mr. Williams' camp on the 28th of January, driving to Hoogly on the river of that name, and thence following the grand trunk-road westward towards Burdwan. The novelty of palkee-travelling at first renders it pleasant; the neatness with which every thing is packed, the good humour of the bearers, their merry pace, and the many more comforts enjoyed than could be expected in a conveyance *horsed by men*, the warmth when the sliding doors are shut, and the breeze when they are open, are all fully appreciated on first

orders and expense of the Indian government, are at home, and thrown upon their own resources, or the assistance of their scientific brethren, for the means of publishing and distributing the fruits of their labours.

starting, but soon the novelty wears off, and the discomforts are so numerous, that it is pronounced, at best, a barbarous conveyance. The greedy cry and gestures of the bearers, when, on changing, they break a fitful sleep by poking a torch in your face, and vociferating "Buck-sheesh, Sahib;" their discontent at the most liberal largesse, and the sluggishness of the next set who want bribes, put the traveller out of patience with the natives. The dust when the slides are open, and the stifling heat when shut during a shower, are conclusive against the vehicle, and on getting out with aching bones and giddy head at the journey's end, I shook the dust from my person, and wished never to see a palkee again.

On the following morning I was passing through the straggling villages close to Burdwan, consisting of native hovels by the road side, with mangos and figs planted near them, and palms waving over their roofs. Crossing the nearly dry bed of the Damooda, I was set down at Mr. M'Intosh's (the magistrate of the district), and never more thoroughly enjoyed a hearty welcome and a breakfast.

In the evening we visited the Rajah of Burdwan's palace and pleasure-grounds, where I had the first glimpse of oriental gardening: the roads were generally raised, running through rice fields, now dry and hard, and bordered with trees of Jack, Bamboo, *Melia*, *Casuarina*, &c. Tanks were the prominent features: chains of them, full of Indian water-lilies, being fringed with rows of the fan-palm, and occasionally the Indian date. Close to the house was a rather good menagerie, where I saw, amongst other animals, a pair of kangaroos in high health and condition, the female with young in her pouch. Before dark I was again in my palkee, and hurrying onwards. The night was cool and clear, very different from the damp and foggy atmosphere I had left at Calcutta. On the following morning I was travelling over a flat and apparently rising country, along an excellent road, with groves of bamboos and stunted trees on either hand, few villages or palms, a sterile soil, with stunted grass and but little cultivation; altogether a country as unlike what I had expected to find in India as well might be. All around was a dead flat or table-land, out of which a few conical hills rose in the west, about 1000 feet high, covered with a low forest of dusky green or yellow, from the prevalence of bamboo. The lark was singing merrily at sunrise, and the accessories of a fresh air and dewy grass more reminded me of some moorland in the north of England than of the torrid regions of the east.

At 10 P.M. I arrived at Mr. Williams' camp, at Taldangah, a dawk station near the western limit of the coal basin of the Damooda valley. His operations being finished, he was prepared to start, having kindly waited a couple of days for my arrival.

Early on the morning of the last day of January, a motley group of natives were busy striking the tents, and loading the bullocks, bullock-carts and elephants: these proceeded on the march, occupying in straggling groups nearly three miles of road, whilst we remained to breakfast with Mr. F Watkins, Superintendent of the East India Coal and Coke Company, who were working the seams.

The coal crops out at the surface; but the shafts worked are sunk through thick beds of alluvium. The age of these coal-fields is quite unknown, and I regret to say that my examination of their fossil plants throws no material light on the subject. Upwards of thirty species of fossil plants have been procured from them, and of these the majority are referred by Dr. McLelland* to the inferior oolite epoch of England, from the prevalence of species of *Zamia*, *Glossopteris*, and *Tæniopteris*. Some of these genera, together with *Vertebraria* (a very remarkable Indian fossil), are also recognised in the coal-fields of Sind and of Australia. I cannot, however, think that botanical evidence of such a nature is sufficient to warrant a satisfactory reference of these Indian coal-fields to the same epoch as those of England or of Australia; in the first place the outlines of the fronds of ferns and their nervation are frail characters if employed alone for the determination of existing genera, and much more so of fossil fragments: in the second place recent ferns are so widely distributed, that an inspection of the majority affords little clue to the region or locality they come from: and in the third place, considering the wide difference in latitude and longitude of Yorkshire, India, and Australia, the natural conclusion is that they could not have supported a similar vegetation at the same epoch. In fact, finding similar fossil plants at places widely different in latitude, and hence in climate, is, in the present state of our knowledge, rather an argument against than for their having existed cotemporaneously. The *Cycadeæ* especially, whose fossil remains afford so much ground for geological speculations, are far from yielding such precise data as is supposed. Species of the order are found in Mexico, South Africa, Australia, and India, some inhabiting the hottest and dampest, and others the driest climates on the surface of the globe; and it appears to me rash to argue much from the presence of the order in the coal of Yorkshire and India, when we reflect that the geologist of some future epoch may find as good reasons for referring the present Cape, Australian, or Mexican

* Reports of the Geological Survey of India. Calcutta, 1850.

Flora to the same period as that of the Lias and Oolites, when the *Cycadeæ* now living in the former countries shall be fossilised.

Specific identity of their contained fossils may be considered as fair evidence of the cotemporaneous origin of beds, but amongst the many collections of fossil plants that I have examined, there is hardly a specimen, belonging to any epoch, sufficiently perfect to warrant the assumption that the species to which it belonged can be again recognised. The botanical evidences which geologists too often accept as proofs of specific identity are such as no botanist would attach any importance to in the investigation of existing plants. The faintest traces assumed to be of vegetable origin are habitually made into genera and species by naturalists ignorant of the structure, affinities and distribution of living plants, and of such materials the bulk of so-called systems of fossil plants is composed.

A number of women were here employed in making gunpowder, grinding the usual materials on a stone, with the addition of water from the Hookah; a custom for which they have an obstinate prejudice. The charcoal here used is made from an *Acacia:* the Seiks, I believe, employ *Justicia Adhatoda*, which is also in use all over India: at Aden the Arabs prefer the *Calotropis*, probably because it is most easily procured. The grain of all these plants is open, whereas in England, closer-grained and more woody trees, especially willows, are preferred.

The jungle I found to consist chiefly of thorny bushes, Jujube of two species, an *Acacia* and *Butea frondosa*, the twigs of the latter often covered with lurid red tears of Lac, which is here collected in abundance. As it occurs on the plants and is collected by the natives it is called Stick-lac, but after preparation Shell-lac. In Mirzapore, a species of

Celtis yields it, and the Peepul very commonly in various parts of India. The elaboration of this dye, whether by the same species of insect, or by many from plants so widely different in habit and characters, is a very curious fact; since none have red juice, but some have milky and others limpid.

After breakfast, Mr. Williams and I started on an elephant, following the camp to Gyra, twelve miles distant. The docility of these animals is an old story, but it loses so much in the telling, that their gentleness, obedience, and sagacity seemed as strange to me as if I had never heard or read of these attributes. The swinging motion, under a hot sun, is very oppressive, but compensated for by being so high above the dust. The Mahout, or driver, guides by poking his great toes under either ear, enforcing obedience with an iron goad, with which he hammers the animal's head with quite as much force as would break a cocoa-nut, or drives it through his thick skin down to the quick. A most disagreeable sight it is, to see the blood and yellow fat oozing out in the broiling sun from these great punctures! Our elephant was an excellent one, when he did not take obstinate fits, and so docile as to pick up pieces of stone when desired, and with a jerk of the trunk throw them over his head for the rider to catch, thus saving the trouble of dismounting to geologise!

Of sights on the road, unfrequented though this noble line is, there were plenty for a stranger; chiefly pilgrims to Juggernath, most on foot, and a few in carts or pony gigs of rude construction. The vehicles from the upper country are distinguished by a far superior build, their horses are caparisoned with jingling bells, and the wheels and other parts are bound with brass. The kindness of the people towards animals, and in some cases towards their suffering relations, is very remarkable, and may in part have given origin to the prevalent idea that they are less cruel and stern than the majority of mankind; but that the "mild" Hindoo, however gentle on occasion, is cruel and vindictive to his brother man and to animals, when his indolent temper is roused or his avarice stimulated, no one can doubt who reads the accounts of Thuggee, Dacoitee, and poisoning, and witnesses the cruelty with which beasts of burthen are treated. A child carrying a bird, kid, or lamb, is not an uncommon sight, and a woman with a dog in her arms is still more frequently seen. Occasionally too, a group will bear an old man to see Juggernath before he dies, or a poor creature with elephantiasis, who hopes to be allowed to hurry himself to his paradise, in preference to lingering in helpless inactivity, and at last crawling up to the second heaven only. The costumes are as various as the religious castes, and the many countries to which the travellers belong. Next in wealth to the merchants, the most thriving-looking wanderer is the bearer of Ganges' holy water, who drives a profitable trade, his gains increasing as his load lightens, for the further he wanders from the sacred stream, the more he gets for the contents of his jar.

Of merchandise we passed very little, the Ganges being still the high road between north-west India and Bengal. Occasionally a string of camels was seen, but, owing to the damp climate, these are rare, and unknown east of the meridian of Calcutta. A little cotton, clumsily packed in ragged bags, dirty, and deteriorating every day, even at this dry season, proves in how bad a state it must arrive at the market during the rains, when the low wagons are dragged through the streams.

The roads here are all mended with a curious stone,

called Kunker, which is a nodular concretionary deposit of limestone, abundantly imbedded in the alluvial soil of a great part of India.* It resembles a coarse gravel, each pebble being often as large as a walnut, and tuberculated on the surface : it binds admirably, and forms excellent roads, but pulverises into a most disagreeable impalpable dust.

A few miles beyond Taldangah we passed from the sandstone, in which the coal lies, to a very barren country of gneiss and granite rocks, upon which the former rests; the country still rising, more hills appear, and towering far above all is Paras-nath, the culminant point, and a mountain whose botany I was most anxious to explore.

The vegetation of this part of the country is very poor, no good-sized trees are to be seen, all is a low stunted jungle. The grasses were few, and dried up, except in the beds of the rivulets. On the low jungly hills the same plants appear, with a few figs, bamboo in great abundance, several handsome *Acanthaceæ*; a few *Asclepiadeæ* climbing up the bushes; and the Cowage plant, now with over-ripe pods, by shaking which, in passing, there often falls such a shower of its irritating microscopic hairs, as to make the skin tingle for an hour.

On the 1st of February, we moved on to Gyra, another insignificant village. The air was cool, and the atmosphere clear. The temperature, at three in the morning, was 65°, with no dew, the grass only 61°. As the sun rose, Paras-nath appeared against the clear grey sky, in the form of a beautiful broad cone, with a rugged peak, of a deeper grey than the sky. It is a remarkably handsome mountain, sufficiently lofty to be imposing, rising out of an elevated country, the slope of which, upward to the base of the mountain, though imperceptible, is really considerable; and

* Often occurring in strata, like flints.

it is surrounded by lesser hills of just sufficient elevation to set it off. The atmosphere, too, of these regions is peculiarly favourable for views: it is very dry at this season; but still the hills are clearly defined, without the harsh outlines so characteristic of a moist air. The skies are bright, the sun powerful; and there is an almost imperceptible haze that seems to soften the landscape, and keep every object in true perspective.

Our route led towards the picturesque hills and vallies in front. The rocks were all hornblende and micaceous schist, cut through by trap-dykes, while great crumbling masses (or bosses) of quartz protruded through the soil. The stratified rocks were often exposed, pitched up at various inclinations: they were frequently white with effloresced salts, which entering largely into the composition tended to hasten their decomposition, and being obnoxious to vegetation, rendered the sterile soil more hungry still. There was little cultivation, and that little of the most wretched kind; even rice-fields were few and scattered; there was no corn, or gram (*Ervum Lens*), no Castor-oil, no Poppy, Cotton, Safflower, or other crops of the richer soils that flank the Ganges and Hoogly; a very little Sugar-cane, Dhal (*Cajana*), Mustard, Linseed, and Rape, the latter three cultivated for their oil. Hardly a Palm was to be seen; and it was seldom that the cottages could boast of a Banana, Tamarind, Orange, Cocoa-nut or Date. The Mahowa (*Bassia latifolia*) and Mango were the commonest trees. There being no Kunker in the soil here, the roads were mended with angular quartz, much to the elephants' annoyance.

We dismounted where some very micaceous stratified rock cropped out, powdered with a saline efflorescence.*

* An impure carbonate of soda. This earth is thrown into clay vessels with water, which after dissolving the soda, is allowed to evaporate, when the remainder is collected, and found to contain so much silica, as to be capable of being fused

Jujubes (*Zizyphus*) prevailed, with the *Carissa carandas* (in fruit), a shrub belonging to the usually poisonous family of Dog-banes (*Apocyneæ*); its berries make good tarts, and the plant itself forms tolerable hedges.

The country around Fitcoree is rather pretty, the hills covered with bamboo and brushwood, and as usual, rising rather suddenly from the elevated plains. The jungle affords shelter to a few bears and tigers, jackals in abundance, and occasionally foxes; the birds seen are chiefly pigeons. Insects are very scarce; those of the locust tribe being most prevalent, indicative of a dry climate.

The temperature at 3 A.M. was 65°; at 3 P.M. 82°; and at 10 P.M., 68°, from which there was no great variation during the whole time we spent at these elevations. The clouds were rare, and always light and high, except a little fleecy spot of vapour condensed close to the summit of Paras-nath. Though the nights were clear and starlight, no dew was deposited, owing to the great dryness of the air. On one occasion, this drought was so great during the passage of a hot wind, that at night I observed the wet-bulb thermometer to stand 20½° below the temperature of the air, which was 66°; this indicated a dew-point of 11½°, or 54½° below the air, and a saturation-point of 0·146; there being only 0·102 grains of vapour per cubic foot of air, which latter was loaded with dust. The little moisture suspended in the atmosphere is often seen to be condensed in a thin belt of vapour, at a considerable distance above the dry surface of the earth, thus intercepting the

into glass. Dr. Royle mentions this curious fact (Essay on the Arts and Manufactures of India, read before the Society of Arts, February 18, 1852), in illustration of the probably early epoch at which the natives of British India were acquainted with the art of making glass. More complicated processes are employed, and have been from a very early period, in other parts of the continent.

radiation of heat from the latter to the clear sky above. Such strata may be observed, crossing the hills in ribbon-like masses, though not so clearly on this elevated region as on the plains bounding the lower course of the Soane, where the vapour is more dense, the hills more scattered, and the whole atmosphere more humid. During the ten days I spent amongst the hills I saw but one cloudy sunrise, whereas below, whether at Calcutta, or on the banks of the Soane, the sun always rose behind a dense fog-bank.

At 9½ A.M. the black-bulb thermometer rose in the sun to 130°. The morning observation before 10 or 11 A.M. always gives a higher result than at noon, though the sun's declination is so considerably less, and in the hottest part of the day it is lower still (3½ P. M. 109°), an effect no doubt due to the vapours raised by the sun, and which equally interfere with the photometer observations. The N.W. winds invariably rise at about 9 A.M. and blow with increasing strength till sunset; they are due to the rarefaction of the air over the heated ground, and being loaded with dust, the temperature of the atmosphere is hence raised by the heated particles. The increased temperature of the afternoon is therefore not so much due to the accumulation of caloric from the sun's rays, as to the passage of a heated current of air derived from the much hotter regions to the westward. It would be interesting to know how far this N.W. diurnal tide extends; also the rate at which it gathers moisture in its progress over the damp regions of the Sunderbunds. Its excessive dryness in N.W. India approaches that of the African and Australian deserts; and I shall give an abstract of my own observations, both in the vallies of the Soane and Ganges, and on the elevated plateaus of Behar and of Mirzapore.*

* See Appendix A.

On the 2nd of February we proceeded to Tofe-Choney, the hills increasing in height to nearly 1000 feet, and the country becoming more picturesque. We passed some tanks covered with *Villarsia*, and frequented by flocks of white egrets. The existence of artificial tanks so near a lofty mountain, from whose sides innumerable water-courses descend, indicates the great natural dryness of the country during one season of the year. The hills and vallies were richer than I expected, though far from luxuriant. A fine *Nauclea* is a common shady tree, and *Bignonia indica*, now leafless, but with immense pods hanging from the branches. *Acanthaceæ* is the prevalent natural order, consisting of gay-flowered *Eranthemums*, *Ruellias*, *Barlerias*, and such hothouse favourites.*

This being the most convenient station whence to ascend Paras-nath, we started at 6 A.M. for the village of Maddao-bund, at the north base of the mountain, or opposite side from that on which the grand trunk-road runs. After following the latter for a few miles to the west, we took a path through beautifully wooded plains, with scattered trees of the Mahowa (*Bassia latifolia*), resembling good oaks: the natives distil a kind of arrack from its fleshy flowers, which are also eaten raw. The seeds, too, yield a concrete oil, by expression, which is used for lamps and occasionally for frying.

Some villages at the west base of the mountain occupy a better soil, and are surrounded with richer cultivation; palms, mangos, and the tamarind, the first and last rare

* Other plants gathered here, and very typical of the Flora of this dry region, were *Linum trigynum*, *Feronia elephantum*, *Ægle marmelos*, *Helicteres Asoca*, *Abrus precatorius*, *Flemingia*; various *Desmodia*, *Rhynchosiæ*, *Glycine*, and *Grislea tomentosa* very abundant, *Conocarpus latifolius*, *Loranthus longiflorus*, and another species; *Phyllanthus Emblica*, various *Convolvuli*, *Cuscuta*, and several herbaceous *Compositæ*.

features in this part of Bengal, appeared to be common, with fields of rice and broad acres of flax and rape, through the latter of which the blue *Orobanche indica* swarmed. The short route to Maddaobund, through narrow rocky vallies, was impracticable for the elephants, and we had to make a very considerable détour, only reaching that village at 2 P.M. All the hill people we observed were a fine-looking athletic race; they disclaimed the tiger being a neighbour, which every palkee-bearer along the road declares to carry off the torch-bearers, torch and all. Bears they said were scarce, and all other wild animals, but a natural jealousy of Europeans often leads the natives to deny the existence of what they know to be an attraction to the proverbially sporting Englishman.

OLD TAMARIND TREES.

The site of Maddaobund, elevated 1230 feet, in a clearance of the forest, and the appearance of the snow-white domes and bannerets of its temples through the fine trees by which it is surrounded, are very beautiful. Though several hundred feet above any point we had hitherto reached, the situation is so sheltered that the tamarind, peepul, and banyan trees are superb. A fine specimen of the latter stands at the entrance to the village, not a broad-headed tree, as is usual in the prime of its existence, but a mass of trunks irregularly throwing out immense branches in a most picturesque manner; the original trunk is apparently gone, and the principal mass of root stems is fenced in. This, with two magnificent tamarinds, forms a grand clump. The ascent of the mountain is immediately from the village up a pathway worn by the feet of many a pilgrim from the most remote parts of India.

Paras-nath is a mountain of peculiar sanctity, to which circumstance is to be attributed the flourishing state of Maddaobund. The name is that of the twenty-third incarnation of Jinna (Sanscrit "Conqueror"), who was born at Benares, lived one hundred years, and was buried on this mountain, which is the eastern metropolis of Jain worship, as Mount Aboo is the western (where are their libraries and most splendid temples). The origin of the Jain sect is obscure, though its rise appears to correspond with the wreck of Boodhism throughout India in the eleventh century. The Jains form in some sort a transition-sect between Boodhists and Hindoos, differing from the former in acknowledging castes, and from both in their worship of Paras-nath's foot, instead of that of Munja-gosha of the Boodhs, or Vishnoo's of the Hindoos. As a sect of Boodhists their religion is considered pure, and free from the obscenities so conspicuous in Hindoo worship; whilst, in fact, perhaps the reverse is the case; but the symbols are fewer, and indeed almost confined to the feet of Paras-nath, and the priests jealously conceal their esoteric doctrines.

The temples, though small, are well built, and carefully kept. No persuasion could induce the Brahmins to allow us to proceed beyond the vestibule without taking off our shoes, to which we were not inclined to consent. The bazaar was for so small a village large, and crowded to excess with natives of all castes, colours, and provinces of India, very many from the extreme W. and N. W., Rajpootana, the Madras Presidency, and Central India. Numbers had come in good cars, well attended, and appeared men of wealth and consequence; while the quantities of conveyances of all sorts standing about, rather reminded me of an election, than of anything I had seen in India.

The natives of the place were a more Negro-looking race than the Bengalees to whom I had previously been accustomed; and the curiosity and astonishment they displayed at seeing (probably many of them for the first time) a party of Englishmen, were sufficiently amusing. Our coolies with provisions not having come up, and it being two o'clock in the afternoon, I having had no break-fast, and being ignorant of the exclusively Jain population of the village, sent my servant to the bazaar, for some fowls and eggs; but he was mobbed for asking for these articles, and parched rice, beaten flat, with some coarse sugar, was all I could obtain; together with sweetmeats so odiously flavoured with various herbs, and sullied with such impurities, that we quickly made them over to the elephants.

Not being able to ascend the mountain and return in one day, Mr. Williams and his party went back to the road, leaving Mr. Haddon and myself, who took up our quarters under a tamarind-tree.

In the evening a very gaudy poojah was performed. The car, filled with idols, was covered with gilding and silk, and drawn by noble bulls, festooned and garlanded. A procession was formed in front; and it opened into an avenue, up and down which gaily dressed dancing-boys paced or danced, shaking castanets, the attendant worshippers singing in discordant voices, beating tom-toms, cymbals, &c. Images (of Boodh apparently) abounded on the car, in front of which a child was placed. The throng of natives was very great and perfectly orderly, indeed, sufficiently apathetic: they were remarkably civil in explaining what they understood of their own worship.

At 2 P.M., the thermometer was only 65°, though the day was fine, a strong haze obstructing the sun's rays; at 6 P.M., 58°; at 9 P.M., 56°, and the grass cooled to 49°. Still there was no dew, though the night was starlight.

Having provided doolies, or little bamboo chairs slung on four men's shoulders, in which I put my papers and boxes, we next morning commenced the ascent; at first through woods of the common trees, with large clumps of bamboo, over slaty rocks of gneiss, much inclined and sloping away from the mountain. The view from a ridge 500 feet high was superb, of the village, and its white domes half buried in the forest below, the latter of which continued in sight for many miles to the northward. Descending to a valley some ferns were met with, and a more luxuriant vegetation, especially of *Urticeæ*. Wild bananas formed a beautiful, and to me novel feature in the woods.

The conical hills of the white ants were very abundant. The structure appears to me not an independent one, but the débris of clumps of bamboos, or of the trunks of large trees, which these insects have destroyed. As they work up a tree from the ground, they coat the bark with particles of sand glued together, carrying up this artificial sheath or covered way as they ascend. A clump of bamboos is thus speedily killed; when the dead stems fall away, leaving the mass of stumps coated with sand, which the action of the weather soon fashions into a cone of earthy matter.

Ascending again, the path strikes up the hill, through a thick forest of Sal (*Vateria robusta*) and other trees, spanned with cables of scandent *Bauhinia* stems. At about 3000 feet above the sea, the vegetation becomes more luxuriant, and by a little stream I collected five species of ferns and some mosses,—all in a dry state, however. Still higher, *Clematis*, *Thalictrum*, and an increased number of grasses are seen; with bushes of *Verbenaceæ* and *Compositæ*. The white ant apparently does not enter this cooler region. At 3500 feet the vegetation again changes, the trees all become gnarled and scattered; and as the dampness also increases, more mosses and ferns appear. We emerged from the forest at the foot of the great ridge of rocky peaks, stretching E. and W. three or four miles. Abundance of a species of berberry and an *Osbeckia* marked the change in the vegetation most decidedly, and were frequent over the whole summit, with coarse grasses, and various bushes.

At noon we reached the saddle of the crest (alt. 4230 feet), where was a small temple, one of five or six which occupy various prominences of the ridge. The wind, N. W., was cold, the temp. 56°. The view was beautiful, but the atmosphere too hazy: to the north were ranges of low wooded hills, and the course of the Barakah and Adji rivers; to the south lay a flatter country, with lower ranges, and the Damooda river, its all but waterless bed snowy-white from the exposed granite blocks with which its course is strewn. East and west the several sharp ridges of the mountain itself are seen; the western considerably the highest. Immediately below, the mountain flanks appear clothed with impenetrable forest, here and there interrupted by rocky eminences; while to the north the grand trunk road shoots across the plains, like a white thread, as straight as an arrow, spanning here and there the beds of the mountain torrents.

On the south side the vegetation was more luxuriant than on the north, though, from the heat of the sun, the reverse might have been expected. This is owing partly to the curve taken by the ridge being open to the south, and partly to the winds from that quarter being the moist ones. Accordingly, trees which I had left 3000 feet below in the north ascent, here ascended to near the summit, such as figs and bananas. A short-stemmed palm (*Phœnix*) was tolerably abundant, and a small tree (*Pterospermum*) on which a species of grass grew epiphytically; forming a curious feature in the landscape.

The situation of the principal temple is very fine, below the saddle in a hollow facing the south, surrounded by jungles of plantain and banyan. It is small, and contains little worthy of notice but the sculptured feet of Paras-nath, and some marble Boodh idols; cross-legged figures with crisp hair and the Brahminical cord. These, a leper covered with ashes in the vestibule, and an officiating priest, were all we saw. Pilgrims were seen on various parts of the mountain in very considerable numbers, passing from one temple to another, and generally leaving a few grains of dry rice at each; the rich and lame were carried in chairs, the poorer walked.

The culminant rocks are very dry, but in the rains may possess many curious plants; a fine *Kalanchoe* was common, with the berberry, a beautiful *Indigofera,* and various other shrubs; a *Bolbophyllum* grew on the rocks, with a small *Begonia,* and some ferns. There were no birds, and very few insects, a beautiful small *Pontia* being the only butterfly. The striped squirrel was very busy amongst the rocks; and I saw a few mice, and the traces of bears.

At 3 P.M., the temperature was 54°, and the air deliciously cool and pleasant. I tried to reach the western peak (perhaps 300 feet above the saddle), by keeping along the ridge, but was cut off by precipices, and ere I could retrace my steps it was time to descend. This I was glad to do in a doolie, and I was carried to the bottom, with only one short rest, in an hour and three quarters. The descent was very steep the whole way, partly down steps of sharp rock, where one of the men cut his foot severely. The pathway at the bottom was lined for nearly a quarter of a mile with sick, halt, maimed, lame, and blind beggars, awaiting our descent. It was truly a fearful sight, especially the lepers, and numerous unhappy victims to elephantiasis.

Though the botany of Paras-nath proved interesting, its elevation was not accompanied by such a change from the flora of its base as I had expected. This is no doubt due to its dry climate and sterile soil; characters which it shares with the extensive elevated area of which it forms a part, and upon which I could not detect above 300 species of plants during my journey. Yet, that the atmosphere at the summit is more damp as well as cooler than at the base, is proved as well by the observations as by the vegetation;* and in some respects, as the increased

* Of plants eminently typical of a moister atmosphere, I may mention the genera *Bolbophyllum, Begonia, Æginetia, Disporum, Roxburghia, Panax, Eugenia,*

proportion of ferns, additional epiphytal orchideous plants, *Begonias*, and other species showed, its top supported a more tropical flora than its base.

Myrsine, Shorea, Millettia, ferns, mosses, and foliaceous lichens; which appeared in strange association with such dry-climate genera as *Kalanchoe, Pterospermum,* and the dwarf-palm, *Phœnix.* Add to this list the *Berberis asiatica, Clematis nutans, Thalictrum glyphocarpum*, 27 grasses, *Cardamine,* &c., and the mountain top presents a mixture of the plants of a damp hot, a dry hot, and of a temperate climate, in fairly balanced proportions. The prime elements of a tropical flora were however wholly wanting on Paras-nath, where are neither Peppers, *Pothos, Arum,* tall or climbing palms, tree-ferns, *Guttiferæ,* vines, or laurels.

CHAPTER II.

Doomree—Vegetation of table-land—Lieutenant Beadle—Birds—Hot springs of Soorujkoond—Plants near them—Shells in them—Cholera-tree—Olibanum—Palms, form of—Dunwah Pass—Trees, native and planted—Wild peacock—Poppy fields—Geography and geology of Behar and Central India—Toddy-palm—Ground, temperature of—Barroon—Temperature of plants—Lizard—Cross the Soane—Sand, ripple-marks on—Kymore hills—Ground, temperature of—Limestone—Rotas fort and palace—Nitrate of lime—Change of climate—Lime stalagmites, enclosing leaves—Fall of Soane—Spiders, &c.—Scenery and natural history of upper Soane valley—*Hardwickia binata*—Bhel fruit—Dust-storm—Alligator—Catechu—*Cochlospermum*—Leaf-bellows—Scorpions—Tortoises—Florican—Limestone spheres—Coles—Tiger-hunt—Robbery.

In the evening we returned to our tamarind tree, and the next morning regained the trunk road, following it to the dawk bungalow of Doomree. On the way I found the *Cæsalpinia paniculata*, a magnificent climber, festooning the trees with its dark glossy foliage and gorgeous racemes of orange blossoms. Receding from the mountain, the country again became barren: at Doomree the hills were of crystalline rocks, chiefly quartz and gneiss; no palms or large trees of any kind appeared. The spear-grass abounded, and a detestable nuisance it was, its long awns and husked seed working through trowsers and stockings.

Balanites was not uncommon, forming a low thorny bush, with *Ægle marmelos* and *Feronia elephantum*. Having rested the tired elephant, we pushed on in the evening to the next stage, Baghoda, arriving there at 3 A.M., and after a few hours' rest, I walked to the bungalow of Lieutenant Beadle, the surveyor of roads, sixteen miles further.

The country around Baghoda is still very barren, but improves considerably in going westward, the ground becoming hilly, and the road winding through prettily wooded vallies, and rising gradually to 1446 feet. *Nauclea cordifolia*, a tree resembling a young sycamore, is very common; with the Semul (*Bombax*), a very striking tree from its buttressed trunk and gaudy scarlet flowers, swarming with birds, which feed from its honeyed blossoms.

At 10 A.M. the sun became uncomfortably hot, the thermometer being 77°, and the black-bulb thermometer 137° I had lost my hat, and possessed no substitute but a silken nightcap; so I had to tie a handkerchief over my head, to the astonishment of the passers-by. Holding my head down, I had little source of amusement but reading the foot-marks on the road; and these were strangely diversified to an English eye. Those of the elephant, camel, buffalo and bullock, horse, ass, pony, dog, goat, sheep and kid, lizard, wild-cat and pigeon, with men, women, and children's feet, naked and shod, were all recognisable.

It was noon ere I arrived at Lieutenant Beadle's, at Belcuppee (alt. 1219 feet), glad enough of the hearty welcome I received, being very hot, dusty, and hungry. The country about his bungalow is very pretty, from the number of wooded hills and large trees, especially of banyan and peepul, noble oak-like Mahowa (*Bassia*), *Nauclea*, Mango, and *Ficus infectoria*. These are all scattered, however, and do not form forest, such as in a stunted form clothes the hills, consisting of *Diospyros*, *Terminalia*, *Gmelina*, *Nauclea parvifolia*, *Buchanania*, &c. The rocks are still hornblende-schist and granite, with a covering of alluvium, full of quartz pebbles. Insects and birds are numerous, the latter consisting of jays, crows, doves, sparrows, and maina (*Pastor*); also the *Phœnicophaus tristis* ("Mahoka" of the natives), with a note like that of the English cuckoo, as heard late in the season.

I remained two days with Lieutenant Beadle, enjoying in his society several excursions to the hot springs, &c. These springs (called Soorujkoond) are situated close to the road, near the mouth of a valley, in a remarkably pretty spot. They are, of course, objects of worship; and a ruined temple stands close behind them, with three very conspicuous trees—a peepul, a banyan, and a white, thick-stemmed, leafless *Sterculia*, whose branches bore dense clusters of greenish fœtid flowers. The hot springs are four in number, and rise in as many ruined brick tanks about two yards across. Another tank, fed by a cold spring, about twice that size, flows between two of the hot, only two or three paces distant from one of the latter on either hand. All burst through the gneiss rocks, meet in one stream after a few yards, and are conducted by bricked canals to a pool of cold water, about eighty yards off.

The temperatures of the hot springs were respectively 169°, 170°, 173°, and 190°; of the cold, 84° at 4 P.M., and 75° at 7 A.M. the following morning. The hottest is the middle of the five. The water of the cold spring is sweet but not good, and emits gaseous bubbles; it was covered with a green floating *Conferva*. Of the four hot springs, the most copious is about three feet deep, bubbles constantly, boils eggs, and though brilliantly clear, has an exceedingly nauseous taste. This and the other warm ones cover the bricks and surrounding rocks with a thick incrustation of salts.

Confervæ abound in the warm stream from the springs,

and two species, one ochreous brown, and the other green, occur on the margins of the tanks themselves, and in the hottest water; the brown is the best Salamander, and forms a belt in deeper water than the green; both appear in broad luxuriant strata, wherever the temp. is cooled down to 168°, and as low as 90°. Of flowering plants, three showed in an eminent degree a constitution capable of resisting the heat, if not a predilection for it; these were all *Cyperaceæ*, a *Cyperus* and an *Eleocharis*, having their roots in water of 100°, and where they are probably exposed to greater heat, and a *Fimbristylis* at 98°; all were very luxuriant. From the edges of the four hot springs I gathered sixteen species of flowering plants, and from the cold tank five, which did not grow in the hot. A water-beetle, *Colymbetes* (?) and *Notonecta*, abounded in water at 112°, with quantities of dead shells; frogs were very lively, with live shells, at 90°, and with various other water beetles. Having no means of detecting the salts of this water, I bottled some for future analysis.*

On the following day I botanized in the neighbourhood, with but poor success. An oblique-leaved fig climbs the other trees, and generally strangles them: two epiphytal *Orchideæ* also occur on the latter, *Vanda Roxburghii* and an *Oberonia*. Dodders (*Cuscuta*) of two species, and *Cassytha*, swarm over and conceal the bushes with their yellow thread-like stems.

I left Belcuppee on the 8th of February, following Mr. Williams' camp. The morning was clear and cold, the temperature only 56°. We crossed the nearly dry broad bed of the Burkutta river, a noble stream during the rains, carrying along huge boulders of granite and gneiss. Near this I passed the Cholera-tree, a famous peepul by

* For an account of the *Confervæ*, and of the mineral constituents of the waters, &c. see Appendix B.

the road side, so called from a detachment of infantry having been attacked and decimated at the spot by that fell disease; it is covered with inscriptions and votive tokens in the shape of rags, &c. We continued to ascend to 1360 feet, where I came upon a small forest of the Indian Olibanum (*Boswellia thurifera*), conspicuous from its pale bark, and spreading curved branches, leafy at their tips; its general appearance is a good deal like that of the mountain ash. The gum, celebrated throughout the East, was flowing abundantly from the trunk, very fragrant and transparent. The ground was dry, sterile, and rocky; kunker, the curious formation mentioned at p. 12, appears in the alluvium, which I had not elsewhere seen at this elevation.

Descending to the village of Burshoot, we lost sight of the *Boswellia*, and came upon a magnificent tope of mango, banyan, and peepul, so far superior to anything hitherto met with, that we were glad to choose such a pleasant halting-place for breakfast. There are a few lofty fan-palms here too, great rarities in this soil and elevation: one, about eighty feet high, towered above some wretched hovels, displaying the curious proportions of this tribe of palms: first, a short cone, tapering to one-third the height of the stem, the trunk then swelling to two-thirds, and again tapering to the crown. Beyond this, the country again ascends to Burree (alt. 1169 feet), another dawk bungalow, a barren place, which we left on the following morning.

So little was there to observe, that I again amused myself by watching footsteps, the precision of which in the sandy soil was curious. Looking down from the elephant, I was interested by seeing them all in *relief*, instead of *depressed*, the slanting rays of the sun in front producing this kind of mirage. Before us rose no more of those wooded hills that had been our companions for the last 120 miles, the absence of which was a sign of the nearly approaching termination of the great hilly plateau we had been traversing for that distance.

Chorparun, at the top of the Dunwah pass, is situated on an extended barren flat, 1320 feet above the sea, and from it the descent from the table-land to the level of the Soane valley, a little above that of the Ganges at Patna, is very sudden. The road is carried zizgag down a rugged hill of gneiss, with a descent of nearly 1000 feet in six miles, of which 600 are exceedingly steep. The pass is well wooded, with abundance of bamboo, *Bombax*, *Cassia*, *Acacia*, and *Butea*, with *Calotropis*, the purple Mudar, a very handsome road-side plant, which I had not seen before, but which, with the *Argemone Mexicana*, was to be a companion for hundreds of miles farther. All the views in the pass are very picturesque, though wanting in good foliage, such as *Ficus* would afford, of which I did not see one tree. Indeed the rarity of the genus (except *F. infectoria*) in the native woods of these hills, is very remarkable. The banyan and peepul always appear to be planted, as do the tamarind and mango.

Dunwah, at the foot of the pass, is 620 feet above the sea, and nearly 1000 below the mean level of the highland I had been traversing. Every thing bears here a better aspect; the woods at the foot of the hills afforded many plants; the bamboo (*B. stricta*) is green instead of yellow and white; a little castor-oil is cultivated, and the Indian date (low and stunted) appears about the cottages.

In the woods I heard and saw the wild peacock for the first time. Its voice is not to be distinguished from that of the tame bird in England, a curious instance of the perpetuation of character under widely different circumstances, for the crow of the wild jungle-fowl does not rival that of the farm-yard cock.

In the evening we left Dunwah for Barah (alt. 480 feet), passing over very barren soil, covered with low jungle, the original woods having apparently been cut for fuel. Our elephant, a timid animal, came on a drove of camels in the dark by the road-side, and in his alarm insisted on doing battle, tearing through the thorny jungle, regardless of the mahout, and still more of me: the uproar raised by the camel-drivers was ridiculous, and the danger to my barometer imminent.

We proceeded on the 11th of February to Sheergotty, where Mr. Williams and his camp were awaiting our arrival. Wherever cultivation appeared the crops were tolerably luxuriant, but a great deal of the country yielded scarcely half-a-dozen kinds of plants to any ten square yards of ground. The most prevalent were *Carissa carandas*, *Olax scandens*, two *Zizyphi*, and the ever-present *Acacia Catechu*. The climate is, however, warmer and much moister, for I here observed dew to be formed, which I afterwards found to be usual on the low grounds. That its presence is due to the increased amount of vapour in the atmosphere I shall prove: the amount of radiation, as shown by the cooling of the earth and vegetation, being the same in the elevated plain and lower levels.*

The good soil was very richly cultivated with poppy (which I had not seen before), sugar-cane, wheat, barley, mustard, rape, and flax. At a distance a field of poppies looks like a green lake, studded with white water-lilies. The houses, too, are better, and have tiled roofs; while, in such situations, the road is lined with trees.

* See Appendix, C.

A retrospect of the ground passed over is unsatisfactory, as far as botany is concerned, except as showing how potent are the effects of a dry soil and climate during one season of the year upon a vegetation which has no desert types. During the rains probably many more species would be obtained, for of annuals I scarcely found twenty. At that season, however, the jungles of Behar and Birbhoom, though far from tropically luxuriant, are singularly unhealthy.

In a geographical point of view the range of hills between Burdwan and the Soane is interesting, as being the north-east continuation of a chain which crosses the broadest part of the peninsula of India, from the Gulf of Cambay to the junction of the Ganges and Hoogly at Rajmahal. This range runs south of the Soane and Kymore, which it meets I believe at Omerkuntuk;* the granite of this and the sandstone of the other, being there both overlaid with trap. Further west again, the ranges separate, the southern still betraying a nucleus of granite, forming the Satpur range, which divides the valley of the Taptee from that of the Nerbudda. The Paras-nath range is, though the most difficult of definition, the longer of the two parallel ranges; the Vindhya continued as the Kymore, terminating abruptly at the Fort of Chunar on the Ganges. The general and geological features of the two, especially along their eastern course, are very different. This consists of metamorphic gneiss, in various highly inclined beds, through which granite hills protrude, the loftiest of which is Paras-nath. The north-east Vindhya (called Kymore), on the other hand, consists of nearly horizontal beds of sandstone, overlying inclined beds of non-fossiliferous limestone. Between the latter and the Paras-nath

* A lofty mountain said to be 7000—8000 feet high.

gneiss, come (in order of superposition) shivered and undulating strata of metamorphic quartz, hornstone, hornstone-porphyry, jaspers, &c. These are thrown up, by greenstone I believe, along the north and north-west boundary of the gneiss range, and are to be recognised as forming the rocks of Colgong, of Sultangunj, and of Monghyr, on the Ganges, as also various detached hills near Gyah, and along the upper course of the Soane. From these are derived the beautiful agates and cornelians, so famous under the name of Soane pebbles, and they are equally common on the Curruckpore range, as on the south bank of the Soane, so much so in the former position as to have been used in the decoration of the walls of the now ruined palaces near Bhagulpore.

In the route I had taken, I had crossed the eastern extremity alone of the range, commencing with a very gradual ascent, over the alluvial plains of the west bank of the Hoogly, then over laterite, succeeded by sandstone of the Indian coal era, which is succeeded by the granite table-land, properly so called. A little beyond the coal fields, the table-land reaches an average height of 1130 feet, which is continued for upwards of 100 miles, to the Dunwah pass. Here the descent is sudden to plains, which, continuous with those of the Ganges, run up the Soane till beyond Rotasghur. Except for the occasional ridges of metamorphic rocks mentioned above, and some hills of intruded greenstone, the lower plain is stoneless, its subjacent rocks being covered with a thicker stratum of the same alluvium which is thinly spread over the higher table-land above. This range is of great interest from its being the source of many important rivers,* and of all

* The chief rivers from this, the great water-shed of Western Bengal, flow north-west and south-east; a few comparatively insignificant streams running north to

those which water the country between the Soane, Hoogly, and Ganges, as well as from its deflecting the course of the latter river, which washes its base at Rajmahal, and forcing it to take a sinuous course to the sea. In its climate and botany it differs equally from the Gangetic plains to the north, and from the hot, damp, and exuberant forests of Orissa to the south. Nor are its geological features less different, or its concomitant and in part resultant characters of agriculture and native population. Still further west, the great rivers of the peninsula have their origin, the Nerbudda and Taptee flowing west to the gulf of Cambay, the Cane to the Jumna, the Soane to the Ganges, and the northern feeders of the Godavery to the Bay of Bengal.

On the 12th of February, we left Sheergotty (alt. 463 feet), crossing some small streams, which, like all else seen since leaving the Dunwah Pass, flow N. to the Ganges. Between Sheergotty and the Soane, occur many of the isolated hills of greenstone, mentioned above, better known to the traveller from having been telegraphic stations. Some are much impregnated with iron, and whether for their colour, the curious outlines of many, or their position, form quaint, and in some cases picturesque features in the otherwise tame landscape.

The road being highly cultivated, and the Date-palm becoming more abundant, we encamped in a grove of these trees. All were curiously distorted; the trunks growing zigzag, from the practice of yearly tapping the alternate sides for toddy. The incision is just below the

the Ganges. Amongst the former are the Rheru, the Kunner, and the Coyle, which contribute to the Soane; amongst the latter, the Dammooda, Adji, and Barakah, flow into the Hoogly, and the Subunrika, Brahminee, and Mahanuddee into the Bay of Bengal.

crown, and slopes upwards and inwards: a vessel is hung below the wound, and the juice conducted into it by a little piece of bamboo. This operation spoils the fruit, which, though eaten, is small, and much inferior to the African date.

At Mudunpore (alt. 440 feet) a thermometer, sunk 3 feet 4 inches in the soil, maintained a constant temperature of 71½°, that of the air varying from 77½°, at 3 P.M., to 62 at daylight the following morning; when we moved on to Nourunga (alt. 340 feet), where I bored to 3 feet 8 inches with a heavy iron jumper through an alluvium of such excessive tenacity, that eight natives were employed for four hours in the operation. In both this and another hole, 4 feet 8 inches, the temperature was 72° at 10 P.M.; and on the following morning 71½° in the deepest hole, and 70° in the shallower: that of the external air varied from 71° at 3 P.M., to 57° at daylight on the following morning. At the latter time I took the temperature of the earth near the surface, which showed,

Surface	53°	4 inches	62°
1 inch	57	7 "	64
2 "	58		

The following day we marched to Baroon (alt. 345 feet) on the alluvial banks of the Soane, crossing a deep stream by a pretty suspension bridge, of which the piers were visible two miles off, so level is the road. The Soane is here three miles wide, its nearly dry bed being a desert of sand, resembling a vast arm of the sea when the tide is out: the banks are very barren, with no trees near, and but very few in the distance. The houses were scarcely visible on the opposite side, behind which the Kymore mountains rise. The Soane is a classical river, being now satisfactorily identified with the Eranoboas of the ancients.*

The alluvium is here cut into a cliff, ten or twelve feet above the bed of the river, and against it the sand is blown in naked *dunes*. At 2 P.M., the surface-sand was heated to 110° where sheltered from the wind, and 104° in the open bed of the river. To compare the rapidity and depth to which the heat is communicated by pure sand, and by the tough alluvium, I took the temperature at some inches depth in both. That the alluvium absorbs the heat better, and retains it longer, would appear from the following, the only observations I could make, owing to the tenacity of the soil.

2 P.M.	Surface	104°	5 A.M. Surface	51°
	2½ inches,	93	28 inches,	68 5½
	5 ,,	88	Sand at this depth, 78°.	

Finding the fresh milky juice of *Calotropis* to be only 72°, I was curious to ascertain at what depth this temperature was to be obtained in the sand of the river-bed, where the plant grew.

Surface . . .	104½°	3½ inches .	85°	Compact.
1 inch. . . .	102	8 ,, . .	73	Wet.
2 ,, ...	94	15 ,, . .	72	Ditto.
2½ ,, . . .	90			

The power this plant exercises of maintaining a low temperature of 72°, though the main portion which is subterraneous is surrounded by a soil heated to between 90° and 104°, is very remarkable, and no doubt proximately due to the rapidity of evaporation from the foliage,

* The etymology of Eranoboas is undoubtedly *Hierrinia Vahu* (Sanskrit), the golden-armed. Sona is also the Sanskrit for *gold*. The stream is celebrated for its agates (Soane pebbles), which are common, but gold is not now obtained from it.

and consequent activity in the circulation. Its exposed leaves maintained a temperature of 80°, nearly 25° cooler than the similarly exposed sand and alluvium. On the same night the leaves were cooled down to 54°, when the sand had cooled to 51°. Before daylight the following morning the sand had cooled to 43°, and the leaves of the *Calotropis* to 45½°. I omitted to observe the temperature of the sap at the latter time; but the sand at the same depth (15 inches) as that at which its temperature and that of the plant agreed at mid-day, was 68°. And assuming this to be the heat of the plant, we find that the leaves are heated by solar radiation during the day 8°, and cooled by nocturnal radiation, 22½°.

Mr. Theobald (my companion in this and many other rambles) pulled a lizard from a hole in the bank. Its throat was mottled with scales of brown and yellow. Three ticks had fastened on it, each of a size covering three or four scales: the first was yellow, corresponding with the yellow colour of the animal's belly, where it lodged, the second brown, from the lizard's head; but the third, which was clinging to the parti-coloured scales of the neck, had its body parti-coloured, the hues corresponding with the individual scales which they covered. The adaptation of the two first specimens in colour to the parts to which they adhered, is sufficiently remarkable; but the third case was most extraordinary.

During the night of the 14th of February, I observed a beautiful display, apparently of the Aurora borealis, an account of which will be found in the Appendix.

February 15.—Our passage through the Soane sands was very tedious, though accomplished in excellent style, the elephants pushing forward the heavy waggons of mining tools with their foreheads. The wheels were sometimes buried to the axles in sand, and the draught bullocks were rather in the way than otherwise.

The body of water over which we ferried, was not above 80 yards wide. In the rains, when the whole space of three miles is one rapid flood, 10 or 12 feet deep, charged with yellow sand, this river must present an imposing spectacle. I walked across the dry portion, observing the sand-waves, all ranged in one direction, perpendicular to that of the prevailing wind, accurately representing the undulations of the ocean, as seen from a mast-head or high cliff. As the sand was finer or coarser, so did the surface resemble a gentle ripple, or an ocean-swell. The progressive motion of the waves was curious, and caused by the lighter particles being blown over the ridges, and filling up the hollows to leeward. There were a few islets in the sand, a kind of oases of mud and clay, in laminæ no thicker than paper, and these were at once denizened by various weeds. Some large spots were green with wheat and barley-crops, both suffering from smut.

We encamped close to the western shore, at the village of Dearee (alt. 330 feet); it marks the termination of the Kymore Hills, along whose S.E. bases our course now lay, as we here quitted the grand trunk road for a rarely visited country.

On the 16th we marched south up the river to Tilotho (alt. 395 feet), through a rich and highly cultivated country, covered with indigo, cotton, sugar-cane, safflower, castor-oil, poppy, and various grains. Dodders (*Cuscuta*) covered even tall trees with a golden web, and the *Capparis acuminata* was in full flower along the road side. Tilotho, a beautiful village, is situated in a superb grove of Mango, Banyan, Peepul, Tamarind, and *Bassia*. The Date or toddy-palm and fan-palm are very abundant and tall: each had a pot hung under the crown. The natives climb these trunks with a hoop or cord round the body and both ancles, and a bottle-gourd or other vessel hanging round the neck to receive the juice from the stock-bottle, in this aerial wine-cellar. These palms were so lofty that the climbers, as they paused in their ascent to gaze with wonder at our large retinue, resembled monkeys rather than men. Both trées yield a toddy, but in this district they stated that that from the *Phœnix* (Date) alone ferments, and is distilled; while in other parts of India, the *Borassus* (fan-palm) is chiefly employed. I walked to the hills, over a level cultivated country interspersed with occasional belts of low wood; in which the pensile nests of the weaver-bird were abundant, but generally hanging out of reach, in prickly *Acacias*.

The hills here present a straight precipitous wall of horizontally stratified sandstone, very like the rocks at the Cape of Good Hope, with occasionally a shallow valley, and a slope of débris at the base, densely clothed with dry jungle. The cliffs are about 1000 feet high, and the plants similar to those at the foot of Paras-nath, but stunted: I climbed to the top, the latter part by steps or ledges of sandstone. The summit was clothed with long grass, trees of *Diospyros* and *Terminalia*, and here and there the *Boswellia*. On the precipitous rocks the curious white-barked *Sterculia fœtida* "flung its arms abroad," leafless, and looking as if blasted by lightning.

A hole was sunk here again for the thermometers, and, as usual, with great labour; the temperatures obtained were—

	Air.	4 feet 6 inches, under good shade of trees.
9 P. M.	64½°	77°
11 P. M.	.	76°
5½ A. M.	58½°	76°

This is a very great rise (of 4°) above any of those

previously obtained, and certainly indicates a much higher mean temperature of the locality. I can only suppose it due to the radiation of heat from the long range of sandstone cliff, exposed to the south, which overlooks the flat whereon we were encamped, and which, though four or five miles off, forms a very important feature. The differences of temperature in the shade taken on this and the other side of the river are $2\frac{3}{4}^{\circ}$ higher on this side.

On the 17th we marched to Akbarpore (alt. 400 feet), a village overhung by the rocky precipice of Rotasghur, a spur of the Kymore, standing abruptly forward.

The range, in proceeding up the Soane valley, gradually approaches the river, and beds of non-fossiliferous limestone are seen protruding below the sandstone and occasionally rising into rounded hills, the paths upon which appear as white as do those through the chalk districts of England. The overlying beds of sandstone are nearly horizontal, or with a dip to the N. W ; the subjacent ones of limestone dip at a greater angle. Passing between the river and a detached conical hill of limestone, capped with a flat mass of sandstone, the spur of Rotas broke suddenly on the view, and very grand it was, quite realising my anticipations of the position of these eyrie-like hill-forts of India. To the left of the spur winds the valley of the Soane, with low-wooded hills on its opposite bank, and a higher range, connected with that of Behar, in the distance. To the right, the hills sweep round, forming an immense and beautifully wooded amphitheatre, about four miles deep, bounded with a continuation of the escarpment. At the foot of the crowned spur is the village of Akbarpore, where we encamped in a Mango tope ;* it occupies some

* On the 24th of June, 1848, the Soane rose to an unprecedented height, and laid this grove of Mangos three feet under water.

pretty undulating limestone hills, amongst which several streams flow from the amphitheatre to the Soane.

During our two days' stay here, I had the advantage of the society of Mr. C. E. Davis, who was our guide during some rambles in the neighbourhood, and to whose experience, founded on the best habits of observation, I am indebted for much information. At noon we started to ascend to the palace, on the top of the spur. On the way we passed a beautiful well, sixty feet deep, and with a fine flight of steps to the bottom. Now neglected and overgrown with flowering weeds and creepers, it afforded me many of the plants I had only previously obtained in a withered state; it was curious to observe there some of the species of the hill-tops, whose seeds doubtless are scattered abundantly over the surrounding plains, and only vegetate where they find a coolness and moisture resembling that of the altitude they elsewhere affect. A fine fig-tree growing out of the stone-work spread its leafy green branches over the well mouth, which was about twelve feet square; its roots assumed a singular form, enveloping two sides of the walls with a beautiful net-work, which at *high-water mark* (rainy season), abruptly divides into thousands of little brushes, dipping into the water which they fringe. It was a pretty cool place to descend to, from a temperature of 80° above, to 74° at the bottom, where the water was 60°; and most refreshing to look, either up the shaft to the green fig shadowing the deep profound, or along the sloping steps through a vista of flowering herbs and climbing plants, to the blue heaven of a burning sky.

The ascent to Rotas is over the dry hills of limestone, covered with a scrubby brushwood, to a crest where are the first rude and ruined defences. The limestone is succeeded by the sandstone cliff cut into steps, which led from ledge to ledge and gap to gap, well guarded with walls and an archway of solid masonry. Through this we passed on to the flat summit of the Kymore hills, covered with grass and forest, intersected by paths in all directions. The ascent is about 1200 feet—a long pull in the blazing sun of February. The turf consists chiefly of spear-grass and *Andropogon muricatus*, the kus-kus, which yields a favourite fragrant oil, used as a medicine in India. The trees are of the kinds mentioned before. A pretty octagonal summer-house, with its roof supported by pillars, occupies one of the highest points of the plateau, and commands a superb view of the scenery before described. From this a walk of three miles leads through the woods to the palace. The buildings are very extensive, and though now ruinous, bear evidence of great beauty in the architecture: light galleries, supported by slender columns, long cool arcades, screened squares and terraced walks, are the principal features. The rooms open out upon flat roofs, commanding views of the long endless table-land to the west, and a sheer precipice of 1000 feet on the other side, with the Soane, the amphitheatre of hills, and the village of Akbarpore below.

This and Beejaghur, higher up the Soane, were amongst the most recently reduced forts, and this was further the last of those wrested from Baber in 1542. Some of the rooms are still habitable, but the greater part are ruinous, and covered with climbers, both of wild flowers and of the naturalised garden plants of the adjoining shrubbery; the *Arbor-tristis*, with *Hibiscus*, *Abutilon*, &c., and above all, the little yellow-flowered *Linaria ramosissima*, crawling over every ruined wall, as we see the walls of our old English castles clothed with its congener *L. Cymbalaria*.

In the old dark stables I observed the soil to be covered with a copious evanescent efflorescence of nitrate of lime, like soap-suds scattered about.

I made Rotas Palace 1490 feet above the sea, so that this table-land is here only fifty feet higher than that I had crossed on the grand trunk road, before descending at the Dunwah pass. Its mean temperature is of course considerably (4°) below that of the valley, but though so cool, agues prevail after the rains. The extremes of temperature are less marked than in the valley, which becomes excessively heated, and where hot winds sometimes last for a week, blowing in furious gusts.

The climate of the whole neighbourhood has of late changed materially; and the fall of rain has much diminished, consequent on felling the forests; even within six years the hail-storms have been far less frequent and violent. The air on the hills is highly electrical, owing, no doubt, to the dryness of the atmosphere, and to this the frequent recurrence of hail-storms may be due.

The zoology of these regions is tolerably copious, but little is known of the natural history of a great part of the plateau; a native tribe, prone to human sacrifices, is talked of. Tigers are common, and bears are numerous; they have, besides, the leopard, panther, viverine cat, and civet; and of the dog tribe the pariah, jackal, fox, and wild dog, called Koa. Deer are very numerous, of six or seven kinds. A small alligator inhabits the hill streams, said to be a very different animal from either of the Soane species.

During our descent we examined several instances of ripple-mark (fossil waves' footsteps) in the sandstone; they resembled the fluting of the *Sigillaria* stems, in the coal-measures, and occurring as they did here, in sandstone, a little above great beds of limestone, had been taken for such, and as indications of coal.

On the following day we visited Rajghat, a steep ghat or pass leading up the cliff to Rotas Palace, a little higher up the river. We took the elephants to the mouth of the glen, where we dismounted, and whence we followed a stream abounding in small fish and aquatic insects (*Dytisci* and *Gyrini*), through a close jungle, to the foot of the cliffs, where there are indications of coal. The woods were full of monkeys, and amongst other plants I observed *Murraya exotica*, but it was scarce. Though the jungle was so dense, the woods were very dry, containing no Palm, *Aroideæ*, Peppers, *Orchideæ* or Ferns. Here, at the foot of the red cliffs, which towered imposingly above, as seen through the tree tops, are several small seams of coaly matter in the sandstone, with abundance of pyrites, sulphur, and copious efflorescences of salts of iron; but no coal. The springs from the cliffs above are charged with lime, of which enormous tuff beds are deposited on the sandstone, full of impressions of the leaves and stems of the surrounding trees, which, however, I found it very difficult to recognize, and could not help contrasting this circumstance with the fact that geologists, unskilled in botany, see no difficulty in referring equally imperfect remains of extinct vegetables to existing genera. In some parts of their course the streams take up quantities of the efflorescence, which they scatter over the sandstones in a singular manner.

At Akbarpore I had sunk two thermometers, one 4 feet 6 inches, the other 5 feet 6 inches; both invariably indicated 76°, the air varying from 56° to 79½°. Dew had formed every night since leaving Dunwah, the grass being here cooled 12° below the air.

On the 19th of February we marched up the Soane to

Tura, passing some low hills of limestone, between the cliffs of the Kymore and the river. On the shaded river-banks grew abundance of English genera—*Cynoglossum, Veronica, Potentilla, Ranunculus sceleratus, Rumex*, several herbaceous *Compositæ* and *Labiatæ; Tamarix* formed a small bush in rocky hillocks in the bed of the river, and in pools were several aquatic plants, *Zannichellia, Chara*, a pretty little *Vallisneria*, and *Potamogeton*. The Brahminee goose was common here, and we usually saw in the morning immense flocks of wild geese overhead, migrating northward.

Here I tried again the effect of solar and nocturnal radiation on the sand, at different depths, not being able to do so on the alluvium.

Noon, Temperature of air, 87°.	Daylight of following morning.	Noon.	Daylight.
Surface 110°	52°	4 inches 84°	67°
1 inch 102°	55°	8 ditto 77° Sand wet	73° Wet
2 ditto 93½°	58°	16 ditto 76° ditto.	74°

From Tura our little army again crossed the Soane, the scarped cliffs of the Kymore approaching close to the river on the west side. The bed is very sandy, and about one mile and a half across.

The elephants were employed again, as at Baroon, to push the cart: one of them had a bump in consequence, as large as a child's head, just above the trunk, and bleeding much; but the brave beast disregarded this, when the word of command was given by his driver.

The stream was very narrow, but deep and rapid, obstructed with beds of coarse agate, jasper, cornelian and chalcedony pebbles. A clumsy boat took us across to the village of Soanepore, a wretched collection of hovels. The crops were thin and poor, and I saw no palms or good trees.

Squirrels however abounded, and were busy laying up their stores; descending from the trees they scoured across a road to a field of tares, mounted the hedge, took an observation, foraged and returned up the tree with their booty, quickly descended, and repeated the operation of reconnoitering and plundering.

The bed of the river is here considerably above that at Dearee, where the mean of the observations with those of Baroon, made it about 300 feet. The mean of those taken here and on the opposite side, at Tura, gives about 400 feet, indicating a fall of 100 feet in only 40 miles.

Near this the sandy banks of the Soane were full of martins' nests, each one containing a pair of eggs. The deserted ones were literally crammed full of long-legged spiders (*Opilio*), which could be raked out with a stick, when they came pouring down the cliff like corn from a sack; the quantities are quite inconceivable. I did not observe the martin feed on them.

The entomology here resembled that of Europe, more than I had expected in a tropical country, where predaceous beetles, at least *Carabideæ* and *Staphylinideæ*, are generally considered rare. The latter tribes swarmed under the clods, of many species but all small, and so singularly active that I could not give the time to collect many. In the banks again, the round egg-like earthy chrysalis of the *Sphynx Atropos* (?) and the many-celled nidus of the leaf-cutter bee, were very common.

A large columnar *Euphorbia* (*E. ligulata*) is common all along the Soane, and I observed it to be used everywhere for fencing. I had not remarked the *E. neriifolia;* and the *E. tereticaulis* had been very rarely seen since leaving Calcutta. The *Cactus* is nowhere found; it is abundant in many parts of Bengal, but certainly not indigenous.

J. D. H. delt.

CROSSING THE SOANE, WITH THE KYMORE HILLS IN THE DISTANCE.

From this place onwards up the Soane, there was no road of any kind, and we were compelled to be our own road engineers. The sameness of the vegetation and lateness of the season made me regret this the less, for I was disappointed in my anticipations of finding luxuriance and novelty in these wilds. Before us the valley narrowed considerably, the forest became denser, the country on the south side was broken with rounded hills, and on the north the noble cliffs of the Kymore dipped down to the river. The villages were smaller, more scattered and poverty-stricken, with the Mahowa and Mango as the usual trees; the banyan, peepul, and tamarind being rare. The natives are of an aboriginal jungle race; and are tall, athletic, erect, much less indolent and more spirited than the listless natives of the plains.

February 21.—Started at daylight: but so slow and difficult was our progress through fields and woods, and across deep gorges from the hills, that we only advanced five miles in the day; the elephant's head too was aching too badly to let him push, and the cattle would not proceed when the draught was not equal. What was worse, it was impossible to get them to pull together up the inclined planes we cut, except by placing a man at the head of each of the six, eight, or ten in a team, and simultaneously screwing round their tails; when one tortured animal sometimes capsizes the vehicle. The small carts got on better, though it was most nervous to see them rushing down the steeps, especially those with our fragile instruments, &c.

Kosdera, where we halted, is a pretty place, elevated 440 feet, with a broad stream from the hills flowing past it. These hills are of limestone, and rounded, resting upon others of hornstone and jasper. Following up the stream

I came to some rapids, where the stream is crossed by large beds of hornstone and porphyry rocks, excessively hard, and pitched up at right angles, or with a bold dip to the north. The number of strata was very great, and only a few inches or even lines thick: they presented all varieties of jasper, hornstone, and quartz of numerous colours, with occasional seams of porphyry or breccia. The rocks were elegantly fringed with a fern I had not hitherto seen, *Polypodium proliferum*, which is the only species the Soane valley presents at this season.

Returning over the hills, I found *Hardwickia binata*, a most elegant leguminous tree, tall, erect, with an elongated coma, and the branches pendulous. These trees grew in a shallow bed of alluvium, enclosing abundance of agate pebbles and kunker, the former derived from the quartzy strata above noticed.

On the 23rd and 24th we continued to follow up the Soane, first to Panchadurma (alt. 490 feet), and thence to Pepura (alt. 587 feet), the country becoming densely wooded, very wild, and picturesque, the woods being full of monkeys, parrots, peacocks, hornbills, and wild animals. *Strychnos potatorum*, whose berries are used to purify water, forms a dense foliaged tree, 30 to 60 feet high, some individuals pale yellow, others deep green, both in apparent health. *Feronia Elephantum* and *Ægle marmelos** were very abundant, with *Sterculia*, and the dwarf date-palm.

One of my carts was here hopelessly broken down; advancing on the spokes instead of the tire of the wheels. By the banks of a deep gully here the rocks are well exposed: they consist of soft clay shales resting on the

* The Bhel fruit, lately introduced into English medical practice, as an astringent of great effect, in cases of diarrhœa and dysentery.

limestone, which is nearly horizontal; and this again, unconformably on the quartz and hornstone rocks, which are confused, and tilted up at all angles.

A spur of the Kymore, like that of Rotas, here projects to the bed of the river, and was blazing at night with the beacon-like fires of the natives, lighted to scare the tigers and bears from the spots where they cut wood and bamboo; they afforded a splendid spectacle, the flames in some places leaping zig-zag from hill to hill in front of us, and looking as if a gigantic letter W were written in fire.

The night was bright and clear, with much lightning, the latter attracted to the spur, and darting down as it were to mingle its fire with that of the forest; so many flashes appeared to strike on the flames, that it is probable the heated air in their neighbourhood attracted them. We were awakened between 3 and 4 A.M., by a violent dust-storm, which threatened to carry away the tents. Our position at the mouth of the gulley formed by the opposite hills, no doubt accounted for it. The gusts were so furious that it was impossible to observe the barometer, which I returned to its case on ascertaining that any indications of a rise or fall in the column must have been quite trifling. The night had been oppressively hot, with many insects flying about; amongst which I noticed earwigs, a genus erroneously supposed rarely to take to the wing in Britain.

At 8½ A.M. it suddenly fell calm, and we proceeded to Chanchee (alt. 500 feet), the native carts breaking down in their passage over the projecting beds of flinty rocks, or as they hurried down the inclined planes we cut through the precipitous clay banks of the streams. Near Chanchee we passed an alligator, just killed by two men, a foul beast, about nine feet long, of the mugger kind. More absorbing than its natural history was the circumstance of its having swallowed a child, that was playing in the water as its mother was washing her utensils in the river. The brute was hardly dead, much distended by the prey, and the mother was standing beside it. A very touching group was this: the parent with her hands clasped in agony, unable to withdraw her eyes from the cursed reptile, which still clung to life with that tenacity for which its tribe are so conspicuous; beside these the two athletes leaned on the bloody bamboo staffs, with which they had all but despatched the animal.

This poor woman earned a scanty maintenance by making catechu: inhabiting a little cottage, and having no property but two cattle to bring wood from the hills, and a very few household chattels; and how few of these they only know who have seen the meagre furniture of Danga hovels. Her husband cut the trees in the forest and dragged them to the hut, but at this time he was sick, and her only boy, her future stay, it was, whom the beast had devoured.

This province is famous for the quantity of catechu its dry forests yield. The plant (*Acacia*) is a little thorny tree, erect, and bearing a rounded head of well remembered prickly branches. Its wood is yellow, with a dark brick-red heart, most profitable in January and useless in June (for yielding the extract).

The *Butea frondosa* was abundantly in flower here, and a gorgeous sight. In mass the inflorescence resembles sheets of flame, and individually the flowers are eminently beautiful, the bright orange-red petals contrasting brilliantly against the jet-black velvety calyx. The nest of the *Megachile* (leaf-cutter bee) was in thousands in the cliffs, with Mayflies, Caddis-worms, spiders, and many predaceous

Pl. I.

Soane Valley and Kymor Hills
Cochlospermum gossypium & Butea frondosa
in flower

London John Murray Decbr 1853.

beetles. Lamellicorn beetles were very rare, even *Aphodius*, and of *Cetoniæ* I did not see one.

We marched on the 28th to Kota, at the junction of the river of that name with the Soane, over hills of flinty rock, which projected everywhere, to the utter ruin of the elephants' feet, and then over undulating hills of limestone; on the latter I found trees of *Cochlospermum*, whose curious thick branches spread out somewhat awkwardly, each tipped with a cluster of golden yellow flowers, as large as the palm of the hand, and very beautiful: it is a tropical Gum-Cistus in the appearance and texture of the petals, and their frail nature. The bark abounds in a transparent gum, of which the white ants seem fond, for they had killed many trees. Of the leaves the curious rude leaf-bellows are made, with which the natives of these hills smelt iron. Scorpions appeared very common here, of a small kind, 1½ inch long; several were captured, and one of our party was stung on the finger; the smart was burning for an hour or two, and then ceased.

At Kota we were nearly opposite the cliffs at Beejaghur, where coal is reported to exist; and here we again crossed the Soane, and for the last time. The ford is three miles up the river, and we marched to it through deep sand. The bed of the river is here 500 feet above the sea, and about three-quarters of a mile broad, the rapid stream being 50 or 60 yards wide, and breast deep. The sand is firm and siliceous, with no mica; nodules of coal are said to be washed down thus far from the coal-beds of Burdee, a good deal higher up, but we saw none.

The cliffs come close to the river on the opposite side, their bases clothed with woods which teemed with birds. The soil is richer, and individual trees, especially of *Bombax*, *Terminalia* and *Mahowa*, very fine; one tree of the *Hardwickia*, about 120 feet high, was as handsome a monarch of the forest as I ever saw, and it is not often that one sees trees in the tropics, which for a combination of beauty in outline, harmony of colour, and arrangement of branches and foliage, would form so striking an addition to an English park.

There is a large break in the Kymore hills here, beyond the village of Kunch, through which our route lay to Beejaghur, and the Ganges at Mirzapore; the cliffs leaving the river and trending to the north in a continuous escarpment flanked with low ranges of rounded hills, and terminating in an abrupt spur (Mungeesa Peak) whose summit was covered with a ragged forest. At Kunch we saw four alligators sleeping in the river, looking at a distance like logs of wood, all of the short-nosed or mugger kind, dreaded by man and beast; I saw none of the sharp-snouted (or garial), so common on the Ganges, where their long bills, with a garniture of teeth and prominent eyes peeping out of the water, remind one of geological lectures and visions of *Ichthyosauri.* Tortoises were frequent in the river, basking on the rocks, and popping into the water when approached.

On the 1st of March we left the Soane, and struck inland over a rough hilly country, covered with forest, fully 1000 feet below the top of the Kymore table-land, which here recedes from the river and surrounds an undulating plain, some ten miles either way, facing the south. The roads, or rather pathways, were very bad, and quite impassable for the carts without much engineering, cutting through forest, smoothing down the banks of the water-courses to be crossed, and clearing away the rocks as we best might. We traversed the empty bed of a mountain torrent, with perpendicular banks of alluvium 30 feet high, and thence plunged into a dense forest. Our course was directed towards Mungeesa Peak, the remarkable projecting spur, between which and a conical hill the path led. Whether on the elephants or on foot, the thorny jujubes, *Acacias*, &c. were most troublesome, and all our previous scratchings were nothing to this. Peacocks and jungle-fowl were very frequent, the squabbling of the former and the hooting of the monkeys constantly grating on the ear. There were innumerable pigeons and a few Floricans (a kind of bustard—considered the best eating game-bird in India). From the defile we emerged on an open flat, halting at Sulkun, a scattered village (alt. 684 feet), peopled by a bold-looking race (Coles)* who habitually carry the spear and shield. We had here the pleasure of meeting Mr. Felle, an English gentleman employed in the Revenue department; this being one of the roads along which the natives transport their salt, sugar, &c., from one province to another.

In the afternoon, I examined the conical hill, which, like that near Rotas, is of stratified beds of limestone, capped with sandstone. A stream runs round its base, cutting through the alluvium to the subjacent rock, which is exposed, and contains. flattened spheres of limestone. These spheres are from the size of a fist to a child's head, or even much larger; they are excessively hard, and neither laminated nor formed of concentric layers. At the top of the hill the sandstone cap was perpendicular on all sides, and its dry top covered with small trees, especially of *Cochlospermum*. A few larger trees of *Fici* clung to the edge of the rocks, and by forcing their roots into the interstices detached enormous masses, affording good dens

* The Coles, like the Danghas of the Rajmahal and Behar hills, and the natives of the mountains of the peninsula, form one of the aboriginal tribes of British India, and are widely different people from either the Hindoos or Mussulmen.

for bears and other wild animals. From the top, the view of rock, river, forest, and plain, was very fine, the eye ranging over a broad flat, girt by precipitous hills;—West, the Kymore or Vindhya range rose again in rugged elevations;—South, flowed the Soane, backed by ranges of wooded hills, smoking like volcanos with the fires of the natives;—below, lay the bed of the stream we had left at the foot of the hills, cutting its way through the alluvium, and following a deep gorge to the Soane, which was there hidden by the rugged heights we had crossed, on which the greater part of our camp might be seen still straggling onwards;—east, and close above us, the bold spur of Mungeesa shot up, terminating a continuous stretch of red precipices, clothed with forest along their bases, and over their horizontal tops.

From Sulkun the view of the famed fort and palace of Beejaghur is very singular, planted on the summit of an isolated hill of sandstone, about ten miles off. A large tree by the palace marks its site; for, at this distance, the buildings are themselves undistinguishable.

There are many tigers on these hills; and as one was close by, and had killed several cattle, Mr. Felle kindly offered us a chance of slaying him. Bullocks are tethered out, over-night, in the places likely to be visited by the brute; he kills one of them, and is from the spot tracked to his haunt by natives, who visit the stations early in the morning, and report the whereabouts of his lair. The sportsman then goes to the attack mounted on an elephant, or having a *roost* fixed in a tree, on the trail of the tiger, and he employs some hundred natives to drive the animal past the lurking-place.

On the present occasion, the locale of the tiger was doubtful; but it was thought that by beating over several miles of country he (or at any rate, some other game) might be driven past a certain spot. Thither, accordingly, the natives were sent, who built machans (stages) in the trees, high out of danger's reach; Mr. Theobald and myself occupied one of these perches in a *Hardwickia* tree, and Mr. Felle another, close by, both on the slope of a steep hill, surrounded by jungly valleys. We were also well thatched in with leafy boughs, to prevent the wary beast from espying the ambush, and had a whole stand of small arms ready for his reception.

When roosted aloft, and duly charged to keep profound silence (which I obeyed to the letter, by falling sound asleep), the word was passed to the beaters, who surrounded our post on the plain-side, extending some miles in line, and full two or three distant from us. They entered the jungle, beating tom-toms, singing and shouting as they advanced, and converging towards our position. In the noonday solitude of these vast forests, our situation was romantic enough: there was not a breath of wind, an insect or bird stirring; and the wild cries of the men, and the hollow sound of the drums broke upon the ear from a great distance, gradually swelling and falling, as the natives ascended the heights or crossed the valleys. After about an hour and a half, the beaters emerged from the jungle under our retreat; one by one, two by two, but preceded by no single living thing, either mouse, bird, deer, or bear, and much less tiger. The beaters received about a penny a-piece for the day's work; a rich guerdon for these poor wretches, whom necessity sometimes drives to feed on rats and offal.

We were detained three days at Sulkun, from inability to get on with the carts; and as the pass over the Kymore to the north (on the way to Mirzapore) was to be still worse, I took advantage of Mr. Felle's kind offer of camels and elephants to make the best of my way forward, accompanying that gentleman, *en route*, to his residence at Shahgunj, on the table-land.

Both the climate and natural history of this flat on which Sulkun stands, are similar to those of the banks of the Soane; the crops are wretched. At this season the dryness of the atmosphere is excessive: our nails cracked, and skins peeled, whilst all articles of wood, tortoiseshell, &c., broke on the slightest blow. The air, too, was always highly electrical, and the dew-point was frequently 40° below the temperature of the air.

The natives are far from honest: they robbed one of the tents placed between two others, wherein a light was burning. One gentleman in it was awake, and on turning saw five men at his bedside, who escaped with a bag of booty, in the shape of clothes, and a tempting strong brass-bound box, containing private letters. The clothes they dropped outside, but the box of letters was carried off. There were about a hundred people asleep outside the tents, between whose many fires the rogues must have passed, eluding also the guard, who were, or ought to have been, awake.

CHAPTER III.

Ek-powa Ghat—Sandstones—Shahgunj—Table-land, elevation, &c.—Gum-arabic—Mango—Fair—Aquatic plants—Rujubbund—Storm—False sunset and sunrise—Bind hills—Mirzapore—Manufactures, imports, &c.—Climate of—Thuggee — Chunar — Benares — Mosque — Observatory — Sar-nath — Ghazeepore—Rose-gardens—Manufactory of Attar—Lord Cornwallis' tomb—Ganges, scenery and natural history of—Pelicans—Vegetation—Insects—Dinapore—Patna—Opium godowns and manufacture—Mudar, white and purple—Monghyr islets—Hot Springs of Setakoond—Alluvium of Ganges—Rocks of Sultun-gunj—Bhaugulpore—Temples of Mt. Manden—Coles and native tribes—Bhaugulpore rangers—Horticultural gardens.

On the 3rd of March I bade farewell to Mr. Williams and his kind party, and rode over a plain to the village of Markunda, at the foot of the Ghat. There the country becomes very rocky and wooded, and a stream is crossed, which runs over a flat bed of limestone, cracked into the appearance of a tesselated pavement. For many miles there is no pass over the Kymore range, except this, significantly called "Ek-powa-Ghat" (one-foot Ghat). It is evidently a *fault*, or shifting of the rocks, producing so broken a cliff as to admit of a path winding over the shattered crags. On either side, the precipices are extremely steep, of horizontally stratified rocks, continued in an unbroken line, and the views across the plain and Soane valley, over which the sun was now setting, were superb. At the summit we entered on a dead flat plain or table-land, with no hills, except along the brim of the broad valley we had left, where are some curious broad pyramids, formed of slabs of sandstone arranged in steps. By dark we reached the village of Roump (alt. 1090 feet), beyond the top of the pass.

On the next day I proceeded on a small, fast, and wofully high-trotting elephant, to Shahgunj, where I enjoyed Mr. Felle's hospitality for a few days. The country here, though elevated, is, from the nature of the soil and formation, much more fertile than what I had left. Water is abundant, both in tanks and wells, and rice-fields, broad and productive, cover the ground; while groves of tamarinds and mangos, now loaded with blossoms, occur at every village.

It is very singular that the elevation of this table-land (1100 feet at Shahgunj) should coincide with that of the granite range of Upper Bengal, where crossed by the grand trunk road, though they have no feature but the presence of alluvium in common. Scarce a hillock varies the surface here, and the agricultural produce of the two is widely different. Here the flat ledges of sandstone retain the moisture, and give rise to none of those impetuous torrents which sweep it off the inclined beds of gneiss, or splintered quartz. Nor is there here any of the effloresced salts so forbidding to vegetation where they occur. Wherever the alluvium is deep on these hills, neither *Catechu*, *Olibanum*, *Butea*, *Terminalia*, *Diospyros*, dwarf-palm, or any of those plants are to be met with, which abound wherever the rock is superficial, and irrespectively of its mineral characters.

The gum-arabic *Acacia* is abundant here, though not seen below, and very rare to the eastward of this meridian, for I saw but little of it in Behar. It is a plant partial to a dry climate, and rather prefers a good soil. In its distribution it in some degree follows the range of the camel, which is its constant companion over thousands of leagues. In the valley of the Ganges I was told that neither the animal nor plant flourish east of the Soane, where I experienced a marked change in the humidity of the atmosphere on my passage down the Ganges. It was a circumstance I was interested in, having first met with the camel at Teneriffe and the Cape Verd Islands, the westernmost limit of its distribution; imported thither, however, as it now is into Australia, where, though there is no *Acacia Arabica*, four hundred other species of the genus are known.

The mango, which is certainly *the* fruit of India, (as the pine-apple is of the Eastern Islands, and the orange of the West,) was now blossoming, and a superb sight. The young leaves are purplish-green, and form a curious contrast to the deep lurid hue of the older foliage; especially when the tree is (which often occurs) dimidiate, one half the green, and the other the red shades of colours; when in full blossom, all forms a mass of yellow, diffusing a fragrance rather too strong and peculiar to be pleasant.

We passed a village where a large fair was being held, and singularly familiar its arrangements were to my early associations. The women and children are the prime customers; for the latter whirl-you-go-rounds, toys, and sweetmeats were destined; to tempt the former, little booths of gay ornaments, patches for the forehead, ear-rings of quaint shapes, bugles and beads. Here as at home, I remarked that the vendors of these superfluities occupy the approaches to this Vanity-Fair. As, throughout the East, the trades are congregated into particular quarters of the cities, so here the itinerants grouped themselves into little bazaars for each class of commodity. Whilst I was engaged in purchasing a few articles of native workmanship, my elephant made an attack on a sweetmeat stall, demolishing a magnificent erection of barley-sugar, before his proceedings could be put a stop to.

Mr. Felle's bungalow (whose garden smiled with roses in this wilderness) was surrounded by a moat (fed by a spring), which was full of aquatic plants, *Nymphæa, Damasonium, Villarsia cristata, Aponogeton,* three species of *Potamogeton,* two of *Naias, Chara* and *Zannichellia* (the two latter indifferently, and often together, used in the refinement of sugar). In a large tank hard by, wholly fed by rain water, I observed only the *Villarsia Indica,* no *Aponogeton, Nymphæa,* or *Damasonium,* nor did these occur in any of the other tanks I examined, which were otherwise well peopled with plants. This may not be owing to the quality of the water so much as to its varying quantity in the tank.

All around here, as at Roump, is a dead flat, except towards the crest of the ghats which overhang the valley of the Soane, and there the sandstone rock rises by steps into low hills. During a ride to a natural tank amongst these rocky elevations, I passed from the alluvium to the sandstone, and at once met with all the prevailing plants of the granite, gneiss, limestone and hornstone rocks previously examined, and which I have enumerated too often to require recapitulation; a convincing proof that the mechanical properties and not the chemical constitution of the rocks regulate the distribution of these plants.

Rujubbund (the pleasant spot), is a small tarn, or more properly the expanded bed of a stream, art having aided nature in its formation: it is edged by rocks and cliffs fringed with the usual trees of the neighbourhood; it is a wild and pretty spot, not unlike some birch-bordered pool in the mountains of Wales or Scotland, sequestered and picturesque. It was dark before I got back, with heavy clouds and vivid lightning approaching from the south-west. The day had been very hot (3 P.M., 90°), and the evening the same; but the barometer did not foretell the coming tempest, which broke with fury at 7 P.M., blowing open the doors, and accompanied with vivid lightning and heavy thunder, close by and all round, though no rain fell.

In the clear dry mornings of these regions, a curious optical phenomena may be observed, of a *sunrise* in the *west*, and *sunset* in the *east*. In either case, bright and well-defined beams rise to the zenith, often crossing to the opposite horizon. It is a beautiful feature in the firmament, and equally visible whether the horizon be cloudy or clear, the white beams being projected indifferently against a dark vapour or the blue serene. The zodiacal light shines from an hour or two after sunset till midnight, with singular brightness, almost equalling the milky way.

March 7.—Left Shahgunj for Mirzapore, following the road to Goorawal, over a dead alluvial flat without a feature to remark. Turning north from that village, the country undulates, exposing the rocky nucleus, and presenting the usual concomitant vegetation. Occasionally park-like views occurred, which, where diversified by the rocky valleys, resemble much the noble scenery of the Forest of Dean on the borders of Wales; the *Mahowa* especially representing the oak, with its spreading and often gnarled branches. Many of the exposed slabs of sandstone are beautifully waved on the surface with the *ripple-mark* impression.

Amowee, where I arrived at 9 P.M., is on an open grassy flat, about fifteen miles from the Ganges, which is seen from the neighbourhood, flowing among trees, with the white houses, domes, and temples of Mirzapore scattered around, and high above which the dust-clouds were coursing along the horizon.

Mr. Money, the magistrate of Mirzapore, kindly sent a mounted messenger to meet me here, who had vast trouble in getting bearers for my palkee. In it I proceeded the next day to Mirzapore, descending a steep ghat of the Bind hills by an excellent road, to the level plains of the Ganges. Unlike the Dunwah pass, this is wholly barren. At the foot the sun was intensely hot, the roads alternately rocky and dusty, the villages thronged with a widely different looking race from those of the hills, and the whole air of the outskirts, on a sultry afternoon, far from agreeable.

Mirzapore is a straggling town, said to contain 100,000 inhabitants. It flanks the river, and is built on an undulating alluvial bank, full of kunker, elevated 360 feet above the sea, and from 50 to 80 above the present level of the river. The vicinity of the Ganges and its green bank, and the numbers of fine trees around, render it a pleasing, though not a fine town. It presents the usual Asiatic contrast of squalor and gaudiness; consisting of large squares and broad streets, interspersed with acres of low huts and groves of trees. It is celebrated for its manufactory of carpets, which are admirable in appearance, and, save in durability, equal to the English. Indigo seed from Bundelkund is also a most extensive article of commerce, the best coming from the Doab. For cotton, lac, sugar, and saltpetre, it is one of the greatest marts in India. The articles of native manufacture are brass washing and cooking utensils, and stone deities worked out of the sandstone.

There is little native vegetation, the country being covered with cultivation and extensive groves of mango, and occasionally of guava. English vegetables are abundant and excellent, and the strawberries, which ripen in March, rival the European fruit in size, but hardly in flavour.

During the few days spent at Mirzapore with my kind friend, Mr. C. Hamilton, I was surprised to find the temperature of the day cooler by nearly 4° than that of the hills above, or of the upper part of the Soane valley; while on the other hand the nights were decidedly warmer. The dew-point again was even lower in proportion, (7½°) and the climate consequently drier. The atmosphere was extremely dry and electrical, the hair constantly crackling when combed. Further west, where the climate becomes still drier, the electricity of the air is even greater. Mr. Griffith mentions in his journal that in filling barometer tubes in Affghanistan, he constantly experienced a shock.

Here I had the pleasure of meeting Lieutenant Ward, one of the suppressors of Thuggee (*Thuggee*, in Hindostan, signifies a deceiver; fraud, not open force, being employed). This gentleman kindly showed me the approvers or king's evidence of his establishment, belonging to those three classes of human scourges, the Thug, Dakoit, and Poisoner. Of these the first was the Thug, a mild-looking man, who had been born and bred to the profession: he had committed many murders, saw no harm in them, and felt neither shame nor remorse. His organs of observation and destructiveness were large, and the cerebellum small. He explained to me how the gang waylay the unwary traveller, enter into conversation with him, and have him suddenly seized, when the superior throws his own linen girdle round the victim's neck and strangles him, pressing the knuckles against the spine.

Taking off his own, he passed it round my arm, and showed me the turn as coolly as a sailor once taught me the *hangman's knot*. The Thug is of any caste, and from any part of India. The profession have particular stations, which they generally select for murder, throwing the body of their victim into a well.

The Dakoit (*dakhee*, a robber) belongs to a class who rob in gangs, but never commit murder—arson and housebreaking also forming part of their profession. These are all high-class Rajpoots, originally from Guzerat; who, on being conquered, vowed vengeance on mankind. They speak both Hindostanee and the otherwise extinct Guzerat language; this is guttural in the extreme, and very singular in sound. They are a very remarkable people, found throughout India, and called by various names; their women dress peculiarly, and are utterly devoid of modesty. The man I examined was a short, square, but far from powerful Nepalese, with high arched eyebrows, and no organs of observation. These people are great cowards.

The Poisoners all belong to one caste, of Pasie, or dealers in toddy: they go singly or in gangs, haunting the travellers' resting-places, where they drop half a rupee weight of pounded or whole *Datura* seeds into his food, producing a twenty-hours' intoxication, during which he is robbed, and left to recover or sink under the stupifying effects of the narcotic. He told me that the *Datura* seed is gathered without ceremony, and at any time, place, or age of the plant. He was a dirty, ill-conditioned looking fellow, with no bumps behind his ears, or prominence of eyebrow region, but a remarkable cerebellum.

Though now all but extinct (except in Cuttack), through ten or fifteen years of unceasing vigilance on the part of

Government, and incredible activity and acuteness in the officers employed, the Thugs were formerly a wonderfully numerous body, who abstained from their vocation solely in the immediate neighbourhood of their own villages; which, however, were not exempt from the visits of other Thugs; so that, as Major Sleeman says,—"The annually returning tide of murder swept unsparingly over the whole face of India, from the Sutlej to the sea-coast, and from the Himalaya to Cape Comorin. One narrow district alone was free, the Concan, beyond the ghats, whither they never penetrated." In Bengal, river Thugs replace the travelling practitioner. Candeish and Rohilkund alone harboured no Thugs as residents, but they were nevertheless haunted by the gangs.

Their origin is uncertain, but supposed to be very ancient, soon after the Mahommedan conquest. They now claim a divine original, and are supposed to have supernatural powers, and to be the emissaries of the divinity, like the wolf, the tiger, and the bear. It is only lately that they have swarmed so prodigiously,—seven original gangs having migrated from Delhi to the Gangetic provinces about 200 years ago, and from these all the rest have sprung. Many belong to the most amiable, intelligent, and respectable classes of the lower and even middle ranks: they love their profession, regard murder as sport, and are never haunted with dreams, or troubled with pangs of conscience during hours of solitude, or in the last moments of life. The victim is an acceptable sacrifice to the goddess Davee, who by some classes is supposed to eat the lifeless body, and thus save her votaries the necessity of concealing it.

They are extremely superstitious, always consulting omens, such as the direction in which a hare or jackall crosses the road; and even far more trivial circumstances will determine the fate of a dozen of people, and perhaps of an immense treasure. All worship the pickaxe, which is symbolical of their profession, and an oath sworn on it binds closer than on the Koran. The consecration of this weapon is a most elaborate ceremony, and takes place only under certain trees. They rise through various grades: the lowest are scouts; the second, sextons; the third are holders of the victims' hands; the highest, stranglers.

Though all agree in never practising cruelty, or robbing previous to murder,—never allowing any but infants to escape (and these are trained to Thuggee), and never leaving a trace of such goods as may be identified,—there are several variations in their mode of conducting operations; some tribes spare certain castes, others none: murder of woman is against all rules; but the practice crept into certain gangs, and this it is which led to their discountenance by the goddess Davee, and the consequent downfall of the system. Davee, they say, allowed the British to punish them, because a certain gang had murdered the mothers to obtain their daughters to be sold to prostitution.

Major Sleeman has constructed a map demonstrating the number of "Bails," or regular stations for committing murder, in the kingdom of Oude alone, which is 170 miles long by 100 broad, and in which are 274, which are regarded by the Thug with as much satisfaction and interest as a game preserve is in England: nor are these "bails" less numerous in other parts of India. Of twenty assassins who were examined, one frankly confessed to having been engaged in 931 murders, and the least guilty of the number to 24. Sometimes 150 persons collected into one gang, and their profits have often been immense, the murder of six persons on one occasion yielding 82,000 rupees; upwards of 8000*l*.

Of the various facilities for keeping up the system, the most prominent are, the practice amongst the natives of travelling before dawn, of travellers mixing freely together, and taking their meals by the way-side instead of in villages; in the very Bails, in fact, to which they are inveigled by the Thug in the shape of a fellow-traveller; money remittances are also usually made by disguised travellers, whose treasure is exposed at the custom-houses, and, worst of all, the bankers will never own to the losses they sustain, which, as a visitation of God, would, if avenged, lead, they think, to future, and perhaps heavier punishment. Had the Thugs destroyed Englishmen, they would quickly have been put down; but the system being invariably practised on a class of people acknowledging the finger of the Deity in its execution, its glaring enormities were long in rousing the attention of the Indian Government.

A few examples of the activity exercised by the suppressors may be interesting. They act wholly through the information given by approvers, who are simply king's evidences. Of 600 Thugs engaged in the murder of 64 people, and the plunder of nearly 20,000*l*., all except seventy were captured in ten years, though separated into six gangs, and their operations continued from 1826 to 1830: the last party was taken in 1836. And again, between the years 1826 and 1835, 1562 Thugs were seized, of whom 382 were hanged, and 909 transported; so that now it is but seldom these wretches are ever heard of.

To show the extent of their operations I shall quote an anecdote from Sleeman's Reports (to which I am indebted for most of the above information). He states that he was for three years in charge of a district on the Nerbudda, and considered himself acquainted with every circumstance that occurred in the neighbourhood ; yet, during that time, 100 people were murdered and buried within less than a quarter of a mile of his own residence !

Two hundred and fifty boats full of river Thugs, in crews of fifteen, infested the Ganges between Benares and Calcutta, during five months of every year, under pretence of conveying pilgrims. Travellers along the banks were tracked, and offered a passage, which if refused in the first boat was probably accepted in some other. At a given signal the crews rushed in, doubled up the decoyed victim, broke his back, and threw him into the river, where floating corpses are too numerous to elicit even an exclamation.

At Mirzapore I engaged a boat to carry me down the river to Bhagulpore, whence I was to proceed to the Sikkim-Himalaya. The sketch at p. 88 will give some idea of this vessel, which, though slow and very shabby, had the advantage of being cooler and more commodious than the handsomer craft. Its appearance was not unlike that of a floating haystack, or thatched cottage: its length was forty feet, and breadth fifteen, and it drew a foot and a half of water : the deck, on which a kind of house, neatly framed of matting, was erected, was but a little above the water's edge. My portion of this floating residence was lined with a kind of reed-work formed of long culms of *Saccharum*. The crew and captain consisted of six naked Hindoos, one of whom steered by the huge rudder, sitting on a bamboo-stage astern ; the others pulled four oars in the very bows opposite my door, or tracked the boat along the river-bank.

In my room (for cabin I cannot call it) stood my palkee,

fitted as a bed, with mosquito curtains; a chair and table. On one side were placed all my papers and plants, under arrangement to go home; on the other, my provisions, rice, sugar, curry-powder, a preserved ham, and cheese, &c. Around hung telescope, botanical box, dark lantern, barometer, and thermometer, &c., &c. Our position was often *ashore*, and, Hindoo-like, on the lee-shore, going bump, bump, bump, so that I could hardly write. I considered myself fortunate in having to take this slow conveyance down, it enabling me to write and arrange all day long.

I left on the 15th of March, and in the afternoon of the same day passed Chunar.* This is a tabular mass of sandstone, projecting into the river, and the eastern termination of the Kymore range. There is not a rock between this and the Himalaya, and barely a stone all the way down the Ganges, till the granite and gneiss rocks of the Behar range are again met with. The current of the Ganges is here very strong, and its breadth much lessened: the river runs between high banks of alluvium, containing much kunker. At Benares it expands into a broad stream, with a current which during the rains is said to flow eight miles an hour, when the waters rise 43 feet. The fall hence is 300 feet to its junction with the Hooghly, viz., one foot to every mile. My observations made that from Mirzapore to Benares considerably greater.

Benares is the Athens of India. The variety of buildings along the bank is incredible. There are temples of every shape in all stages of completion and dilapidation, and at all angles of inclination; for the banks give way so much that many of these edifices are fearfully out of the perpendicular.

* The first station at which Henry Martyn laboured in India.

The famed mosque, built by Aurungzebe on the site of a Hindoo temple, is remarkable for its two octagonal minarets, 232 feet above the Ganges. The view from it over the town, especially of the European Resident's quarter, is fine; but the building itself is deficient in beauty or ornament: it commands the muddy river with its thousands of boats, its waters peopled with swimmers and bathers, who spring in from the many temples, water-terraces, and ghats on the city side: opposite is a great sandy plain. The town below looks a mass of poor, square, flat-roofed houses, of which 12,000 are brick, and 16,000 mud and thatch, through the crowd of which, and of small temples, the eye wanders in vain for some attractive feature or evidence of the wealth, the devotion, the science, or the grandeur of a city celebrated throughout the East for all these attributes. Green parrots and pigeons people the air

The general appearance of an oriental town is always more or less ruinous; and here the eye is fatigued with bricks and crumbling edifices, and the ear with prayer-bells. The bright meadows and green trees which adorn the European Resident's dwelling, some four miles back from the river, alone relieve the monotony of the scene. The streets are so narrow that it is difficult to ride a horse through them; and the houses are often six stories high, with galleries crossing above from house to house. These tall, gaunt edifices sometimes give place to clumps of cottages, and a mass of dusty ruins, the unsavoury retreats of vermin and filth, where the *Calotropis arborea* generally spreads its white branches and glaucous leaves—a dusty plant. Here, too, enormous spiders' webs hang from the crumbling walls, choked also with dust, and resembling curtains of coarse muslin, being often some yards across, and not arranged in radii and arcs, but spun like weaver's woofs. Paintings, remarkable only for their hideous proportions and want of perspective, are daubed in vermilion, ochre, and indigo. The elephant, camel, and porpoise of the Ganges, dog, shepherd, peacock, and horse, are especially frequent, and so is a running pattern of a hand spread open, with a blood-red spot on the palm. A still less elegant but frequent object is the fuel, which is composed of the manure collected on the roads of the city, moulded into flat cakes, and stuck by the women on the walls to dry, retaining the sign-manual of the artist in the impressed form of her outspread hand. The cognizance of the Rajah, two fish chained together, appears over the gates of public buildings.

The hundreds of temples and shrines throughout the city are its most remarkable feature: sacred bulls, and lingams of all sizes, strewed with flowers and grains of rice, meet the eye at every turn; and the city's boast is the possession of one million idols, which, of one kind and another, I can well believe. The great Hindoo festival of the *Holi* was now celebrating, and the city more than ordinarily crowded; throwing red powder (lac and flour), with rose-water, is the great diversion at a festival more childish by far than a carnival.

Through the kindness of Mr. Reade (the Commissioner), I obtained admission to the Bishishar-Kumardil, the "holiest of holies." It was a small, low, stone building, daubed with red inside, and swarming with stone images of Brahminee bulls, and various disgusting emblems. A fat old Brahmin, naked to the waist, took me in, but allowed no followers; and what with my ignorance of his phraseology, the clang of bells and din of voices, I gained but little information. Some fine bells from Nepal were evidently the lion of the temple. I emerged, adorned with a chaplet of magnolia flowers, and with my hands full of *Calotropis* and *Nyctanthes* blossoms. It was a horrid place for noise, smell, and sights. Thence I went to a holy well, rendered sacred because Siva, when stepping from the Himalaya to Ceylon, accidentally let a medicine chest fall into it. The natives frequent it with little basins or baskets of rice, sugar, &c., dropping in a little of each while they mutter prayers.

1. EQUATORIAL SUN-DIAL.
(DIAMETER OF FACE OF DIAL, 2 FEET 2 INCHES.)

The observatory at Benares, and those at Delhi, Matra on the Jumna, and Oujein, were built by Jey-Sing, Rajah of Jayanagar, upwards of 200 years ago; his skill in mathematical science was so well known, that the Emperor Mahommed Shah employed him to reform the calendar. Mr. Hunter, in the "Asiatic Researches," gives a translation of the lucubrations of this really enlightened man, as contained in the introduction to his own almanac.

2. EQUINOCTIAL SUN-DIAL.
(LENGTH OF GNOMON, 39 FEET; OF EACH QUADRANT, 9 FEET.)

Of the more important instruments I took sketches; No. 1, is the Naree-wila, or Equatorial dial; No. 2, the Semrat-yunta, or Equinoctial dial; No. 3, an Equatorial, probably a Kranti-urit, or Azimuth circle.* Jey-Sing's genius and love of science seem, according to Hunter, to have descended to some of his family, who died early in this century, when "Urania fled before the brazen-fronted

3. BRASS AZIMUTH CIRCLE.
(DIAMETER 2 FEET.)

Mars, and the best of the observatories, that of Oujein, was turned into an arsenal and cannon foundry."

The observatory is still the most interesting object in

* Hunter, in As Soc. Researches, 177 (Calcutta); Sir R. Barker in Phil. Trans., lxvii. 608 (1777); J. L. Williams, Phil. Trans., lxxxiii. 45 (1793).

Benares, though it is now dirty and ruinous, and the great stone instruments are rapidly crumbling away. The building is square, with a central court and flat roof, round which the astrolabes, &c. are arranged. A half naked Astronomer-Royal, with a large sore on his stomach, took me round—he was a pitiful object, and told me he was very hungry. The observatory is nominally supported by the Rajah of Jeypore, who doles out a too scanty pittance to his scientific corps.

In the afternoon Mr. Reade drove me to the Sar-nath, a singular Boodhist temple, a cylindrical mass of brickwork, faced with stone, the scrolls on which were very beautiful, and as sharp as if freshly cut: it is surmounted by a tall dome, and is altogether about seventy or a hundred feet high. Of the Boodh figures only one remains, the others having been used by a recent magistrate of Benares in repairing a bridge over the Goomtee! From this place the Boodhist monuments, Hindoo temple, Mussulman mosque, and English church, were all embraced in one *coup d'œil.* On our return, we drove past many enormous mounds of earth and brick-work, the vestiges of Old Benares, but whether once continued to the present city or not is unknown. Remains are abundant, eighteen feet below the site of the present city.

Benares is the Mecca of the Hindoos, and the number of pilgrims who visit it is incalculable. Casi (its ancient name, signifying splendid), is alleged to be no part of this world, which rests on eternity, whereas Benares is perched on a prong of Siva's trident, and is hence beyond the reach of earthquakes.* Originally built of gold, the

* Probably an allusion to the infrequency of these phenomena in this meridian; they being common both in Eastern Bengal, and in Western India beyond the Ganges.

sins of the inhabitants were punished by its transmutation into stone, and latterly into mud and thatch: whoever enters it, and especially visits its principal idol (Siva fossilised) is secure of heaven.

On the 18th I left Benares for Ghazepore, a pretty town situated on the north bank of the river, celebrated for its manufacture of rose-water, the tomb of Lord Cornwallis, and a site of the Company's stud. The Rose gardens surround the town: they are fields, with low bushes of the plant grown in rows, red with blossoms in the morning, all of which are, however, plucked long before midday. The petals are put into clay stills, with twice their weight of water, and the produce exposed to the fresh air, for a night, in open vessels. The unskimmed water affords the best, and it is often twice and even oftener distilled; but the fluid deteriorates by too much distillation. The Attar is skimmed from the exposed pans, and sells at 10*l.* the rupee weight, to make which 20,000 flowers are required. It is frequently adulterated with sandal-wood oil.

Lord Cornwallis' mausoleum is a handsome building, modelled by Flaxman after the Sybil's Temple. The allegorical designs of Hindoos and sorrowing soldiers with reversed arms, which decorate two sides of the enclosed tomb, though perhaps as good as can be, are under any treatment unclassical and uncouth. The simple laurel and oak-leaf chaplets on the alternating faces are far more suitable and suggestive.

March 21.—I left Ghazepore and dropped down the Ganges; the general features of which are soon described. A strong current four or five miles broad, of muddy water, flows between a precipitous bank of alluvium or sand on one side, and a flat shelving one of sand or more rarely mud, on the other. Sand-banks are frequent in the river, especially where the great affluents débouche; and there generally are formed vast expanses of sand, small "Saharas," studded with stalking pillars of sand, raised seventy or eighty feet high by gusts of wind, erect, stately, grave-looking columns, all shaft, with neither basement nor capital, the genii of the "Arabian Nights." The river is always dotted with boats of all shapes, mine being perhaps of the most common description; the great square, Yankee-like steamers, towing their accommodation-boats (as the passengers' floating hotels are called), are the rarest. Trees are few on the banks, except near villages, and there is hardly a palm to be seen above Patna. Towns are unfrequent, such as there are being mere collections of huts, with the ghat and boats at the bottom of the bank; and at a respectful distance from the bazaar, stand the neat bungalows of the European residents, with their smiling gardens, hedgings and fencings, and loitering servants at the door. A rotting charpoy (or bedstead) on the banks is a common sight,—the "*sola reliquia*" of some poor Hindoo, who departs this life by the side of the stream, to which his body is afterwards committed.

Shoals of small goggled-eyed fish are seen, that spring clear out of the water, and are preyed upon by terns and other birds; a few insects skim the surface; turtle and porpoises tumble along, all forming a very busy contrast to the lazy alligator, sunning his green and scaly back near the shore, with his ichthyosaurian snout raised high above the water. Birds are numerous, especially early and late in the day. Along the silent shore the hungry Pariah dog may be seen tearing his meal from some stranded corpse, whilst the adjutant-bird, with his head sunk on his body and one leg tucked up, patiently awaits his turn. At night the beautiful Brahminee geese alight, one by one, and seek total solitude; ever since having disturbed a god in his slumbers, these birds are fated to pass the night in single blessedness. The gulls and terns, again, roost in flocks, as do the wild geese and pelicans,—the latter, however, not till after making a hearty and very noisy supper. These birds congregate by the sides of pools, and beat the water with violence, so as to scare the fish, which thus become an easy prey; a fact which was, I believe, first indicated by Pallas, during his residence on the banks of the Caspian Sea. Shells are scarce, and consist of a few small bivalves; their comparative absence is probably due to the paucity of limestone in the mountains whence the many feeders flow. The sand is pure white and small-grained, with fragments of hornblende and mica, the latter varying in abundance as a feeder is near or far away. Pink sand* of garnets is very common, and deposited in layers interstratified with the white quartz sand. Worm-marks, ripple-marks, and the footsteps of alligators, birds and beasts, abound in the wet sand. The vegetation of the banks consists of annuals which find no permanent resting-place. Along the sandy shores the ever-present plants are mostly English, as Dock, a *Nasturtium*, *Ranunculus sceleratus*, *Fumitory*, *Juncus bufonius*, Common Vervain, *Gnaphalium luteo-album*, and very frequently *Veronica Anagallis*. On the alluvium grow the same, mixed with Tamarisk, *Acacia Arabica*, and a few other bushes.

Withered grass abounds; and wheat, dhal (*Cajanus*) and gram (*Cicer arietinum*), *Carthamus*, vetches, and rice are the staple products of the country. Bushes are few, except the universally prevalent Adhatoda and *Calotropis*.

* I have seen the same garnet sand covering the bottom of the Himalayan torrents, where it is the produce of disintegrated gneiss, and whence it is transported to the Ganges.

Trees, also, are rare, and of stunted growth; Figs, the *Artocarpus* and some *Leguminosæ* prevail most. I saw but two kinds of palm, the fan-palm, and *Phœnix:* the latter is characteristic of the driest locality. Then, for the animal creation, men, women, and children abound, both on the banks, and plying up and down the Ganges. The humped cow (of which the ox is used for draught) is common. Camels I occasionally observed, and more rarely the elephant; poneys, goats, and dogs muster strong. Porpoises and alligators infest the river, even above Benares. Flies and mosquitos are terrible pests; and so are the odious flying-bugs,* which insinuate themselves between one's skin and clothes, diffusing a dreadful odour, which is increased by any attempt to touch or remove them. In the evening it was impossible to keep insects out of the boat, or to hinder their putting the lights out; and of these the most intolerable was the above-mentioned flying-bug. Saucy crickets, too, swarm, and spring up at one's face, whilst mosquitos maintain a constant guerilla warfare, trying to the patience no less than to the nerves. Thick webs of the gossamer spider float across the river during the heat of the day, as coarse as fine thread, and being inhaled keep tickling the nose and lips.

On the 18th, the morning commenced with a dust-storm, the horizon was about 20 yards off, and ashy white with clouds of sand; the trees were scarcely visible, and everything in my boat was covered with a fine coat of impalpable powder, collected from the boundless alluvial plains through which the Ganges flows. Trees were scarcely discernible, and so dry was the wind that drops of water vanished like magic. Neither ferns, mosses, nor lichens grow along the banks of the Ganges, they cannot survive the

* Large Hemipterous insects, of the genus *Derecteryx.*

transition from parching like this to the three months' floods at midsummer, when the country is for miles under water.

March 23.—Passed the mouth of the Soane, a vast expanse of sand dotted with droves of camels; and soon after, the wide-spread spits of sand along the north bank announced the mouth of the Gogra, one of the vastest of the many Himalayan affluents of the Ganges.

On the 25th of March I reached Dinapore, a large military station, sufficiently insalubrious, particularly for European troops, the barracks being so misplaced that the inmates are suffocated: the buildings run east and west instead of north and south, and therefore lose all the breeze in the hottest weather. From this place I sent the boat down to Patna, and proceeded thither by land to the house of Dr. Irvine, an old acquaintance and botanist, from whom I received a most kind welcome. On the road, Bengal forms of vegetation, to which I had been for three months a stranger, reappeared; likewise groves of fan and toddy palms, which are both very rare higher up the river; clumps of large bamboo, orange, *Acacia Sissoo, Melia, Guatteria longifolia, Spondias mangifera, Odina, Euphorbia pentagona, neriifolia* and *trigona*, were common road-side plants. In the gardens, Papaw, *Croton, Jatropha, Buddleia, Cookia*, Loquat, Litchi, Longan, all kinds of the orange tribe, and the cocoa-nut, some from their presence, and many from their profusion, indicated a decided change of climate, a receding from the desert north-west of India, and its dry winds, and an approach to the damper regions of the many-mouthed Ganges.

My main object at Patna being to see the opium Godowns (stores), I waited on Dr. Corbett, the Assistant-Agent, who kindly explained everything to me, and to whose obliging attentions I am much indebted.

The E. I. Company grant licences for the cultivation of the poppy, and contract for all the produce at certain rates, varying with the quality. No opium can be grown without this licence, and an advance equal to about two-thirds of the value of the produce is made to the grower. This produce is made over to district collectors, who approximately fix the worth of the contents of each jar, and forward it to Patna, where rewards are given for the best samples, and the worst are condemned without payment; but all is turned to some account in the reduction of the drug to a state fit for market.

The poppy flowers in the end of January and beginning of February, and the capsules are sliced in February and March with a little instrument like a saw, made of three iron plates with jagged edges, tied together. The cultivation is very carefully conducted, nor are there any very apparent means of improving this branch of commerce and revenue. During the N. W., or dry winds, the best opium is procured, the worst during the moist, or E. and N.E., when the drug imbibes moisture, and a watery bad solution of opium collects in cavities of its substance, and is called Passewa, according to the absence of which the opium is generally prized.

At the end of March the opium jars arrive at the stores by water and by land, and continue accumulating for some weeks. Every jar is labelled and stowed in a proper place, separately tested with extreme accuracy, and valued. When the whole quantity has been received, the contents of all the jars are thrown into great vats, occupying a very large building, whence the mass is distributed, to be made up into balls for the markets. This operation is carried on in a long paved room, where every man is ticketed, and many overseers are stationed to see that the work is properly conducted. Each workman sits on a stool, with a double stage and a tray before him. On the top stage is a tin basin, containing opium sufficient for three balls; in the lower another basin, holding water: in the tray stands a brass hemispherical cup, in which the ball is worked. To the man's right hand is another tray, with two compartments, one containing thin pancakes of poppy petals pressed together, the other a cupful of sticky opium-water, made from refuse opium. The man takes the brass cup, and places a pancake at the bottom, smears it with opium-water, and with many plies of the pancakes makes a coat for the opium. Of this he takes about one-third of the mass before him, puts it inside the petals, and agglutinates many other coats over it the balls are then again weighed, and reduced or increased to a certain weight if necessary. At the day's end, each man takes his work to a rack with numbered compartments, and deposits it in that which answers to his own number, thence the balls (each being put in a clay cup) are carried to an enormous drying-room, where they are exposed in tiers, and constantly examined and turned, to prevent their being attacked by weevils, which are very prevalent during moist winds, little boys creeping along the racks all day long for this purpose. When dry, the balls are packed in two layers of six each in chests, with the stalks, dried leaves, and capsules of the plant, and sent down to Calcutta. A little opium is prepared of very fine quality for the Government Hospitals, and some for general sale in India; but the proportion is trifling, and such is made up into square cakes. A good workman will prepare from thirty to fifty balls a day, the total produce being 10,000 to 12,000 a day; during one working season 1,353,000 balls are manufactured for the Chinese market alone.

The poppy-petal *pancakes*, each about a foot radius, are made in the fields by women, by the simple operation of pressing the fresh petals together. They are brought in large baskets, and purchased at the commencement of the season. The liquor with which the pancakes are agglutinated together by the ball-maker, and worked into the ball, is merely inspissated opium-water, the opium for which is derived from the condemned opium, (Passewa,) the washing of the utensils, and of the workmen, every one of whom is nightly laved before he leaves the establishment, and the water is inspissated. Thus not a particle of opium is lost. To encourage the farmers, the refuse stalks, leaves, and heads are bought up, to pack the balls with; but this is far from an economical plan, for it is difficult to keep the refuse from damp and insects.

A powerful smell of opium pervaded these vast buildings, which Dr. Corbett* assured me did not affect himself or the assistants. The men work ten hours a day, becoming sleepy in the afternoon; but this is only natural in the hot season: they are rather liable to eruptive diseases, possibly engendered by the nature of their occupation.

Even the best East Indian opium is inferior to the Turkish, and owing to peculiarities of climate, will probably always be so. It never yields more than five per cent. of morphia, whence its inferiority, but is as good in other respects, and even richer in narcotine.

The care and attention devoted to every department of collecting, testing, manipulating, and packing, is quite extraordinary; and the result has been an impulse to the trade, beyond what was anticipated. The natives have

* I am greatly indebted to Mr. Oldfield, the Opium Agent, and to Dr. Corbett, for a complete set of specimens, implements, and drawings, illustrating the cultivation and manufacture of Opium. They are exhibited in the Kew Museum of Economic Botany.

been quick at apprehending and supplying the wants of the market, and now there are more demands for licences to grow opium than can be granted. All the opium eaten in India is given out with a permit to licensed dealers, and the drug is so adulterated before it reaches the retailers in the bazaars, that it does not contain one-thirtieth part of the intoxicating power that it did when pure.

Patna is the stronghold of Mahommedanism, and from its central position, its command of the Ganges, and its proximity to Nepal (which latter has been aptly compared to a drawn dagger, pointed at the heart of India), it is an important place. For this reason there are always a European and several Native Regiments stationed there. In the neighbourhood there is little to be seen, and the highly cultivated flat country is unfavourable to native vegetation.

The *mudar* plant (*Calotropis*) was abundant here, but I found that its properties and nomenclature were far from settled points. On the banks of the Ganges, the larger, white-flowered, sub-arboreous species prevailed; in the interior, and along my whole previous route, the smaller purple-flowered kind only was seen. Mr. Davis, of Rotas, was in the habit of using the medicine copiously, and vouched for the cure of eighty cases, chiefly of leprosy, by the *white mudar*, gathered on the Ganges, whilst the purple of Rotas and the neighbourhood was quite inert: Dr. Irvine, again, used the purple only, and found the white inert. The European and native doctors, who knew the two plants, all gave the preference to the *white;* except Dr. Irvine, whose experience over various parts of India is entitled to great weight.

March 29.—Dropped down the river, experiencing a succession of east and north-east winds during the whole remainder of the voyage. These winds are very prevalent throughout the month of March, and they rendered the passage in my sluggish boat sufficiently tedious. In other respects I had but little bad weather to complain of: only one shower of rain occurred, and but few storms of thunder and lightning. The stream is very strong, and its action on the sand-banks conspicuous. All night I used to hear the falling cliffs precipitated with a dull heavy splash into the water,—a pretty spectacle in the day-time, when the whirling current is seen to carry a cloud of white dust, like smoke, along its course.

The Curruckpore hills, the northern boundary of the gneiss and granite range of Paras-nath, are seen first in the distance, and then throwing out low loosely timbered spurs towards the river; but no rock or hill comes close to the banks till near Monghyr, where two islets of rock rise out of the bed of the river. They are of stratified quartz, dipping, at a high angle, to the south-east; and, as far as I could observe, quite barren, each crowned with a little temple. The swarm of boats from below Patna to this place was quite incredible.

April 1.—Arrived at Monghyr, by far the prettiest town I had seen on the river, backed by a long range of wooded hills,—detached outliers of which rise in the very town. The banks are steep, and they appear more so owing to the fortifications, which are extensive. A number of large, white, two-storied houses, some very imposing, and perched on rounded or conical hills, give a European aspect to the place.

Monghyr is celebrated for its iron manufactures, especially of muskets, in which respect it is the Birmingham of Bengal. Generally speaking, these weapons are poor, though stamped with the first English names. A native workman will, however, if time and sufficient reward be given, turn out a first rate fowling-piece. The inhabitants are reported to be sad drunkards, and the abundance of toddy-palms was quite remarkable. The latter, (here the *Phœnix sylvestris,*) I never saw wild, but it is considered to be so in N.W. India; it is still a doubtful point whether it is the same as the African species. In the morning of the following day I went to the hot springs of Seeta-koond (wells of Seeta), a few miles south of the town.

MONGHYR ON THE GANGES, WITH THE CURRUCKPORE HILLS IN THE DISTANCE.

The hills are hornstone and quartz, stratified and dipping southerly with a very high angle; they are very barren, and evidently identical with those on the south bank of the Soane; skirting, in both cases, the granite and gneiss range of Paras-nath. The alluvium on the banks of the Ganges is obviously an aqueous deposit subsequent to the elevation of these hills, and is perfectly plane up to their bases. The river has its course through the alluvium, like the Soane. The depth of the former is in many places upwards of 100 feet, and the kunker pebbles it contains are often disposed in parallel undulating bands. It nowhere contains sand pebbles or fossils; concretions of lime (kunker) alone interrupting its uniform consistence. It attains its greatest thickness in the valleys of the Ganges and the Soane, gradually sloping up to the Himalaya and Curruckpore hills on either flank. It is, however, well developed on the Kymore and Paras-nath hills, 1200 to 1500 feet above the Ganges valley, and I have no doubt was deposited in very deep water, when the relative positions of these mountains to the Ganges and Soane valleys were the same that they are now. Like every other part of the surface of India, it has suffered much from denudation, especially on the above-named mountains, and around their bases, where various rocks protrude through it. Along the Ganges again, its surface is an unbroken level between Chunar and the rocks of Monghyr. The origin of its component mineral matter must be sought in the denudation of the Himalayas within a very recent geological period. The contrast between the fertility of the alluvium and the sterility of the protruded quartzy rocks is very striking, cultivation running up to these fields of stones, and suddenly stopping.

Unlike the Soorujkoond hot-springs, those of Seeta-koond rise in a plain, and were once covered by a handsome temple. All the water is collected in a tank, some yards square, with steps leading down to it. The water, which is clear and tasteless (temp. 104°), is so pure as to be exported copiously, and the Monghyr manufactory of soda-water presents the anomaly of owing its purity to Seeta's ablutions.

On my passage down the river I passed the picturesque rocks of Sultangunj; they are similar to those of Monghyr, but very much larger and loftier. One, a round-headed mass, stands on the bank, capped with a triple-domed Mahommedan tomb, palms, and figs. The other, which is far more striking, rises isolated in the bed of the river, and is crowned with a Hindoo temple, its pyramidal cone surmounted with a curious pile of weathercocks, and two little banners. The current of the Ganges is here very strong, and runs in deep black eddies between the rocks.

Though now perhaps eighty or a hundred yards from the shore, the islet must have been recently a peninsula, for it retains a portion of the once connecting bank of alluvium, in the form of a short flat-topped cliff, about thirty feet above the water. Some curious looking sculptures on the rocks are said to represent Naragur (or Vishnu), Surce and Sirooj; but to me they were quite unintelligible. The temple is dedicated to Naragur, and inhabited by Fakirs; it is the most holy on the Ganges.

April 5.—I arrived at Bhagulpore, and took up my quarters with my friend Dr. Grant, till he should arrange my dawk for Sikkim.

The town has been supposed to be the much-sought Palibothra, and a dirty stream hard by (the Chundum), the Eranoboas; but Mr. Ravenshaw has now brought all existing proofs to bear on Patna and the Soane. It is, like most hilly places in India, S. of the Himalaya, the seat of much Jain worship; and the temples on Mount Manden,* a few miles off, are said to have been 540 in number. At the assumed summer-palaces of the kings of Palibothra the ground is covered with agates, brought from the

* For the following information about Bhagulpore and its neighbourhood, I am indebted chiefly to Col. Francklin's essay in the Asiatic Researches; and the late Major Napleton and Mr. Pontet.

neighbouring hills, which were, in a rough state, let into the walls of the buildings. These agates perfectly resemble the Soane pebbles, and they assist in the identification of these flanking hills with those of the latter river.

Again, near the hills, the features of interest are very numerous. The neighbouring mountains of Curruckpore, which are a portion of the Rajmahal and Paras-nath range, are peopled by tribes representing the earliest races of India, prior to the invasion of young Rama, prince of Oude, who, according to the legend, spread Brahminism with his conquests, and won the hand of King Jannuk's daughter, Seėta, by bending her father's bow. These people are called Coles, a middle-sized, strong, very dark, and black-haired race, with thick lips: they have no vocation but collecting iron from the soil, which occurs abundantly in nodules. They eat flesh, whether that of animals killed by themselves, or of those which have died a natural death, and mix with Hindoos, but not with Mussulmen. There are other tribes, vestiges of the Tamulian race, differing somewhat in their rites from these, and approaching, in their habits, more to Hindoos; but all are timorous and retiring.

The hill-rangers, or Bhagulpore-rangers, are all natives of the Rajmahal hills, and form a local corps maintained by the Company for the protection of the district. For many years these people were engaged in predatory excursions, which, owing to the nature of the country, were checked with great difficulty. The plan was therefore conceived, by an active magistrate in the district, of embodying a portion into a military force, for the protection of the country from invasions of their own tribes; and this scheme has answered perfectly.

To me the most interesting object in Bhagulpore was the Horticultural Gardens, whose origin and flourishing condition are due to the activity and enterprise of the late Major Napleton, commander of the hill-rangers. The site is good, consisting of fifteen acres, that were, four years ago, an indigo field, but form now a smiling garden. About fifty men are employed; and the number of seeds and vegetables annually distributed is very great. Of trees the most conspicuous are the tamarind, *Tecoma jasminoides*, *Erythrina*, *Adansonia*, *Bombax*, teak, banyan, peepul, *Sissoo*, *Casuarina*, *Terminalia*, *Melia*, *Bauhinia*. Of introduced species English and Chinese flat peaches (pruned to the centre to let the sun in), Mangos of various sorts, *Eugenia Jambos*, various Anonas, Litchi, Loquat and Longan, oranges, *Sapodilla*; apple, pear, both succeeding tolerably; various Cabool and Persian varieties of fruit-trees; figs, grapes, guava, apricots, and jujube. The grapes looked extremely well, but they require great skill and care in the management. They form a long covered walk, with a row of plantains on the W. side, to diminish the effects of the hot winds, but even with this screen, the fruit on that side are inferior to that on the opposite trellis. Easterly winds, again, being moist, blight these and other plants, by favouring the abundant increase of insects, and causing the leaves to curl and fall off; and against this evil there is no remedy. With a clear sky the mischief is not great; under a cloudy one the prevalence of such winds is fatal to the crop. The white ant sometimes attacks the stems, and is best checked by washing the roots with lime-water, yellow arsenic, or tobacco-water. Numerous Cerealia, and the varieties of cotton, sugar-cane, &c. all thrive extremely well; so do many of our English vegetables. Cabbages, peas, and beans are much injured by the caterpillars of a *Pontia*, like our English "White;" raspberries, currants, and gooseberries will not grow at all.

The seeds were all deposited in bottles, and hung round the walls of a large airy apartment; and for cleanliness and excellence of kind they would bear comparison with the best seedsman's collection in London. Of English garden vegetables, and varieties of the Indian Cerealia, and leguminous plants, Indian corn, millets, rice, &c., the collections for distribution were extensive.

The manufacture of economic products is not neglected. Excellent coffee is grown; and arrow-root, equal to the best West Indian, is prepared, at 1*s.* 6*d.* per bottle of twenty-four ounces,—about a fourth of the price of that article in Calcutta.

In most respects the establishment is a model of what such institutions ought to be in India; not only of real practical value, in affording a good and cheap supply of the best culinary and other vegetables that the climate can produce, but as showing to what departments efforts are best directed. Such gardens diffuse a taste for the most healthy employments, and offer an elegant resource for the many unoccupied hours which the Englishman in India finds upon his hands. They are also schools of gardening; and a simple inspection of what has been done at Bhagulpore is a valuable lesson to any person about to establish a private garden of his own.

I often heard complaints made of the seeds distributed from these gardens not vegetating freely in other parts of India, and it is not to be expected that they should retain their vitality unimpaired through an Indian rainy season; but on the other hand I almost invariably found that the planting and tending had been left to the uncontrolled management of native gardeners, who with a certain amount of skill in handicraft are, from habits and prejudices, singularly unfit for the superintendence of a garden.

CHAPTER IV.

Leave Bhagulpore—Kunker—Colgong—Himalaya, distant view of—Cosi, mouth of—Difficult navigation—Sand storms—Caragola-Ghat—Purnea—Ortolans—Mahanuddee, transport of pebbles, &c.—Betel-pepper, cultivation of—Titalya—Siligoree—View of outer Himalaya—Terai—Mechis—Punkabaree—Foot of mountains—Ascent to Dorjiling—Cicadas—Leeches—Animals—Kursiong, spring vegetation of—Pacheem—Arrive at Dorjiling—Dorjiling, origin and settlement of—Grant of land from Rajah—Dr. Campbell appointed superintendent—Dewan, late and present—Aggressive conduct of the latter—Increase of the station—Trade—Titalya fair—Healthy climate for Europeans and children—Invalids, diseases prejudicial to.

I TOOK as it were, a new departure, on Saturday, April the 8th, my dawk being laid on that day from Caragola-Ghat, about thirty miles down the river, for the foot of the Himalaya range and Dorjiling.

Passing the pretty villa-like houses of the English residents, the river-banks re-assumed their wonted features: the hills receded from the shore; and steep clay cliffs, twenty to fifty feet high, on one side, opposed long sandy shelves on the other. Kunker was still most abundant, especially in the lower bed of the banks, close to the (now very low) water. The strata containing it were much undulated, but not uniformly so; horizontal layers over or under-lying the disturbed ones. At Colgong, conical hills appear, and two remarkable sister-rocks start out of the river, the same in structure with those of Sultangunj. A boisterous current swirls round them, strong even at this season, and very dangerous in the rains, when the swollen river is from twenty-eight to forty feet deeper than now. We landed opposite the rocks, and proceeded to the residence of Mr. G. Barnes, prettily situated on one of the conical elevations characteristic of the geology of the district. The village we passed through had been recently destroyed by fire; and nothing but the clay outer walls and curious-looking partition walls remained, often white-washed and daubed with figures in red of the palm of the hand, elephant, peacock, and tiger,—a sort of rude fresco-painting. We did not arrive till past mid-day, and the boat, with my palkee and servant, not having been able to face the gale, I was detained till the middle of the following day. Mr. Barnes and his brother proved most agreeable companions,—very luckily for me, for it requires no ordinary philosophy to bear being storm-stayed on a voyage, with the prospect of paying a heavy demurrage for detaining the dawk, and the worse one of finding the bearers given to another traveller when you arrive at the rendezvous. The view from Mr. Barnes' house is very fine: it commands the river and its rocks; the Rajmahal hills to the east and south; broad acres of indigo and other crops below; long lines of palm-trees, and groves of mango, banana, tamarind, and other tropical trees, scattered close around and in the distance. In the rainy season, and immediately after, the snowy Himalaya are distinctly seen on the horizon, fully 170 miles off. Nearly opposite, the Cosi river enters the Ganges, bearing (considering its short course) an enormous volume of water, comprising the drainage of the whole Himalaya between the two giant peaks of Kinchinjunga in Sikkim, and Gossain-Than in Nepal. Even at this season, looking from Mr. Barnes' eyrie over the bed of the Ganges, the enormous expanses of sand, the numerous shifting islets, and the long spits of mud betray the proximity of some very restless and resistless power. During the rains, the scene must indeed be extraordinary, when the Cosi lays many miles of land under water, and pours so vast a quantity of detritus into the bed of the Ganges that long islets are heaped up and swept away in a few hours; and the latter river becomes all but unnavigable. Boats are caught in whirlpools, formed without a moment's warning, and sunk ere they have spun round thrice in the eddies; and no part of the inland navigation of India is so dreaded or dangerous, as the Ganges at its junction with the Cosi.

Rain generally falls in partial showers at this season, and they are essential to the well-being of the spring crops of indigo. The stormy appearance of the sky, though it proved fallacious, was hailed by my hosts as predicting a fall, which was much wanted. The wind however seemed but to aggravate the drought, by the great body of sand it lifted and swept up the valleys, obscuring the near horizon, and especially concealing the whole delta of the Cosi, where the clouds were so vast and dense, and ascended so high as to resemble another element.

All night the gale blew on, accompanied with much thunder and lightning, and it was not till noon of the 9th that I descried my palkee-boat toiling down the stream. Then I again embarked, taking the lagging boat in tow of my own. Passing the mouths of the Cosi, the gale and currents were so adverse that we had to bring up on the sand, when the quantity which drifted into the boat rendered the delay as disagreeable as it was tedious. The particles penetrated everywhere, up my nose and down my back, drying my eyelids, and gritting between my teeth. The craft kept bumping on the banks, and being both crazy and leaky, the little comfortless cabin became the refuge of scared rats and cockroaches. In the evening I shared a meal with these creatures, on some provisions my kind friends had put into the boat, but the food was so sandy that I had to bolt my supper!

At night the storm lulled a little, and I proceeded to Caragola Ghat and took up my dawk, which had been twenty-eight hours expecting me, and was waiting, in despair of my arrival, for another traveller on the opposite bank, who however could not cross the river.

Having accomplished thirty miles, I halted at 9 A.M. on the following morning at Purnea, quitting it at noon for Kishengunj. The whole country wore a greener garb than I had seen anywhere south of the Ganges: the climate was evidently more humid, and had been gradually becoming so from Mirzapore. The first decided change was a few miles below the Soane mouth, at Dinapore and Patna; and the few hygrometrical observations I took at Bhagulpore confirmed the increase of moisture. The proximity to the sea and great Delta of the Ganges sufficiently accounts for this; as does the approach to the hills for the still greater dampness and brighter verdure of Purnea. I was glad to feel myself within the influence of the long-looked-for Himalaya; and I narrowly watched every change in the character of the vegetation. A fern, growing by the roadside, was the first and most tangible evidence of this; together with the rarity or total absence of *Butea*, *Boswellia*, *Catechu*, *Grislea*, *Carissa*, and all the companions of my former excursion.

Purnea is a large station, and considered very unhealthy during and after the rains. From it the road passed through some pretty lanes, with groves of planted Guava and a rattan palm (*Calamus*), the first I had seen. Though no hills are nearer than the Himalaya, from the constant alteration of the river-beds, the road undulates remarkably for this part of India, and a jungly vegetation ensues, consisting of the above plants, with the yellow-flowered Cactus replacing the Euphorbias, which were previously much more common. Though still 100 miles distant from the hills, mosses appeared on the banks, and more ferns were just sprouting above ground.

The Bamboo was a very different species from any I had hitherto met with, forming groves of straight trees fifteen to twenty feet high, thin of foliage, and not unlike poplars.

Thirty-six miles from Purnea brought me to Kishengunj, when I found that no arrangements whatever had been made for my dawk, and I was fairly stranded. Luckily a thoughtful friend had provided me with letters to the scattered residents along the road, and I proceeded with one to Mr. Perry, the assistant magistrate of the district,—a gentleman well known for his urbanity, and the many aids he affords to travellers on this neglected line of road. Owing to this being some festival or holiday, it was impossible to get palkee-bearers; the natives were busy catching fish in all the muddy pools around. Some of Mr. Perry's own family also were about to proceed to Dorjiling, so that I had only to take patience, and be thankful for having to exercise it in such pleasant quarters. The Mahanuddee, a large stream from the hills, flows near this place, strewing the surrounding neighbourhood with sand, and from the frequent alterations in its course, causing endless disputes amongst the landholders. A kind of lark called an Ortolan was abundant: this is not, however, the European delicacy of that name, though a migratory bird; the flocks are large, and the birds so fat, that they make excellent table game. At this time they were rapidly disappearing; to return from the north in September.

I had just got into bed at night, when the bearers arrived; so bidding a hurried adieu to my kind host, I proceeded onwards.

April 12.—I awoke at 4 A.M., and found my palkee on the ground, and the bearers coolly smoking their hookahs under a tree (it was raining hard): they had carried me the length of their stage, twelve miles, and there were no others to take me on. I had paid twenty-four pounds for my dawk, from Caragola to the hills, to which I had been obliged to add a handsome douceur; so I lost all patience. After waiting and entreating during several hours, I found the head-man of a neighbouring village, and by a further disbursement induced six out of the twelve bearers to carry the empty palkee, whilst I should walk to the next stage; or till we should meet some others. They agreed, and cutting the thick and spongy sheaths of the banana, used them for shoulder-pads: they also wrapped them round the palkee-poles, to ease their aching clavicles. Walking along I picked up a few plants, and fourteen miles further on came again to the banks of the Mahanuddee, whose bed was strewn with pebbles and small boulders, brought thus far from the mountains (about thirty miles distant). Here, again, I had to apply to the head-man of a village, and pay for bearers to take me to Titalya, the next stage (fourteen miles). Some curious long low sheds puzzled me very much, and on examining them they proved to be for the growth of Pawn or Betel-pepper, another indication of the moisture of the climate. These sheds are twenty to fifty yards long, eight or twelve or so broad, and scarcely five high; they are made of bamboo, wattled all round and over the top. Slender rods are placed a few feet apart, inside, up which the Pepper Vines climb, and quickly fill the place with their deep green glossy foliage. The native enters every morning by a little door, and carefully cleans the plants. Constant heat, damp, and moisture, shelter from solar beams, from scorching heat, and from nocturnal radiation, are thus all procured for the plant, which would certainly not live twenty-four hours, if exposed to the climate of this treeless district. Great attention is paid to the cultivation, which is very profitable. Snakes frequently take up their quarters in these hot-houses, and cause fatal accidents.

Titalya was once a military station of some importance, and from its proximity to the hills has been selected by Dr. Campbell (the Superintendent of Dorjiling) as the site for an annual fair, to which the mountain tribes resort, as well as the people of the plains. The Calcutta road to Dorjiling by Dinajpore meets, near here, that by which I had come; and I found no difficulty in procuring bearers to proceed to Siligoree, where I arrived at 6 A.M. on the 13th. Hitherto I had not seen the mountains, so uniformly had they been shrouded by dense wreaths of vapour: here, however, when within eight miles of their base, I caught a first glimpse of the outer range—sombre masses, of far from picturesque outline, clothed everywhere with a dusky forest.

Siligoree stands on the verge of the Terai, that low malarious belt which skirts the base of the Himalaya, from the Sutlej to Brahma-koond in Upper Assam. Every feature, botanical, geological, and zoological, is new on entering this district. The change is sudden and immediate: sea and shore are hardly more conspicuously different; nor from the edge of the Terai to the limit of perpetual snow is any botanical region more clearly marked than this, which is the commencement of Himalayan vegetation. A sudden descent leads to the Mahanuddee river, flowing in a shallow valley, over a pebbly bottom: it is a rapid river, even at this season; its banks are fringed with bushes, and it is as clear and sparkling as a trout stream in Scotland. Beyond it the road winds through a thick brushwood, choked with long grasses, and with but few trees, chiefly of *Acacia*, *Dalbergia Sissoo*, and a scarlet-fruited *Sterculia*. The soil is a red, friable clay and gravel. At this season only a few spring plants were in flower, amongst which a very sweet-scented *Crinum*, Asphodel, and a small *Curcuma*, were in the greatest profusion. Leaves of terrestrial Orchids appeared, with ferns and weeds of hot damp regions. I crossed the beds of many small streams: some were dry, and all very tortuous; their banks were richly clothed with brushwood and climbers of Convolvulus, Vines, *Hiræa*, *Leea*, *Menispermeæ*, *Cucurbitaceæ*, and *Bignoniaceæ*. Their pent-up waters, percolating the gravel beds, and partly carried off by evaporation through the stratum of ever-increasing vegetable mould, must be one main agent in the production of the malarious vapours of this pestilential region. Add to this, the detention of the same amongst the jungly herbage, the amount of vapour in the humid atmosphere above, checking the upward passage of that from the soil, the sheltered nature of the locality at the immediate base of lofty mountains; and there appear to me to be here all necessary elements, which, combined, will produce stagnation and deterioration in an atmosphere loaded with vapour. Fatal as this district is, and especially to Europeans, a race inhabit it with impunity, who, if not numerous, do not owe their paucity to any climatic causes. These are the Mechis, often described as a squalid, unhealthy people, typical of the region they frequent; but who are, in reality, more robust than the Europeans in India, and whose disagreeably sallow complexion is deceptive as indicating a sickly constitution. They are a mild, inoffensive people, industrious for Orientals, living by annually burning the Terai jungle and cultivating the cleared spots; and, though so sequestered and isolated, they rather court than avoid intercourse with those whites whom they know to be kindly disposed.

After proceeding some six miles along the gradually ascending path, I came to a considerable stream, cutting its way through stratified gravel, with cliffs on each side fifteen to twenty feet high, here and there covered with ferns, the little *Oxalis sensitiva*, and other herbs. The road here suddenly ascends a steep gravelly hill, and opens out on a short flat, or spur, from which the Himalaya rise abruptly, clothed with forest from the base: the little bungalow of Punkabaree, my immediate destination, nestled in the woods, crowning a lateral knoll, above which, to east and west, as far as the eye could reach, were range after range of wooded mountains, 6000 to 8000 feet high. I here met with the India-rubber tree (*Ficus elastica*); it abounds in Assam, but this is its western limit.

From this steppe, the ascent to Punkabaree is sudden and steep, and accompanied with a change in soil and vegetation. The mica slate and clay slate protrude everywhere, the former full of garnets. A giant forest replaces the stunted and bushy timber of the Terai Proper; of which the *Duabanga* and *Terminalias* form the prevailing trees, with *Cedrela* and the *Gordonia Wallichii*. Smaller timber and shrubs are innumerable; a succulent character pervades the bushes and herbs, occasioned by the prevalence of *Urticeæ*. Large bamboos rather crest the hills than court the deeper shade, and of the latter there is abundance, for the torrents cut a straight, deep, and steep course down the hill flanks: the gulleys they traverse are choked with vegetation and bridged by fallen trees, whose trunks are richly clothed with *Dendrobium Pierardi* and other epiphytical Orchids, with pendulous *Lycopodia* and many ferns, *Hoya*, *Scitamineæ*, and similar types of the hottest and dampest climates.

The bungalow at Punkabaree was good—which was well, as my luggage-bearers were not come up, and there were no signs of them along the Terai road, which I saw winding below me. My scanty stock of paper being full of plants, I was reduced to the strait of botanising, and throwing away my specimens. The forest was truly magnificent along the steep mountain sides. The apparently large proportion of deciduous trees was far more considerable than I had expected; partly, probably, due to the abundance of the *Dillenia*, *Cassia*, and *Sterculia*, whose copious fruit was all the more conspicuous from the leafless condition of the plant. The white or lilac blossoms of the convolvulus-like *Thunbergia*, and other *Acanthaceæ*, were the predominant features of the shrubby vegetation, and very handsome.

All around, the hills rise steeply five or six thousand feet, clothed in a dense deep-green dripping forest. Torrents rush down the slopes, their position indicated by the dipping of the forest into their beds, or the occasional cloud of spray rising above some more boisterous part of their course. From the road, at and a little above Punkabaree, the view is really superb, and very instructive. Behind (or north) the Himalaya rise in steep confused masses. Below, the hill on which I stood, and the ranges as far as the eye can reach east and west, throw spurs on to the plains of India. These are very thickly wooded, and enclose broad, dead-flat, hot and damp valleys, apparently covered with a dense forest. Secondary spurs of clay and gravel, like that immediately below Punkabaree, rest on the bases of the mountains, and seem to form an intermediate neutral ground between flat and mountainous India. The Terai district forms a very irregular belt, scantily clothed, and intersected by innumerable rivulets from the hills, which unite and divide again on the flat, till, emerging from the region of many trees, they enter the plains, following devious courses, which glisten like silver threads. The whole horizon is bounded by the sea-like expanse of the plains, which stretch away into the region of sunshine and fine weather, in one boundless flat.

In the distance, the courses of the Teesta and Cosi, the great drainers of the snowy Himalayas, and the recipients of innumerable smaller rills, are with difficulty traced at this, the dry season. The ocean-like appearance of this southern view is even more conspicuous in the heavens than on the land, the clouds arranging themselves after a singularly sea-scape fashion. Endless strata run in parallel ribbons over the extreme horizon; above these, scattered cumuli, also in horizontal lines, are dotted against a clear grey sky, which gradually, as the eye is lifted, passes into a deep cloudless blue vault, continuously clear to the zenith; there the cumuli, in white fleecy masses, again appear; till, in the northern celestial hemisphere, they thicken and assume the leaden hue of nimbi, discharging their moisture on the dark forest-clad hills around. The breezes are south-easterly, bringing that vapour from the Indian Ocean, which is rarefied and suspended aloft over the heated plains, but condensed into a drizzle when it strikes the cooler flanks of the hills, and into heavy rain when it meets their still colder summits. Upon what a gigantic scale does nature here operate! Vapours, raised

J. D. H. delt.

PUNKABAREE BUNGALOW AND BASE OF THE HIMALAYA.

from an ocean whose nearest shore is more than 400 miles distant, are safely transported without the loss of one drop of water, to support the rank luxuriance of this far distant region. This and other offices fulfilled, the waste waters are returned, by the Cosi and Teesta, to the ocean, and again exhaled, exported, expended, re-collected, and returned.

The soil and bushes everywhere swarmed with large and troublesome ants, and enormous earthworms. In the evening, the noise of the great *Cicadæ* in the trees was almost deafening. They burst suddenly into full chorus, with a voice so harshly croaking, so dissonant, and so unearthly, that in these solitary forests I could not help being startled. In general character the note was very similar to that of other *Cicadæ*. They ceased as suddenly as they commenced. On the following morning my baggage arrived, and, leaving my palkee, I mounted a pony kindly sent for me by Mr. Hodgson, and commenced a very steep ascent of about 3000 feet, winding along the face of a steep, richly-wooded valley. The road zigzags extraordinarily in and out of the innumerable lateral ravines, each with its water course, dense jungle, and legion of leeches; the bite of these blood-suckers gives no pain, but is followed by considerable effusion of blood. They puncture through thick worsted stockings, and even trousers, and, when full, roll in the form of a little soft ball into the bottom of the shoe, where their presence is hardly felt in walking.

Not only are the roadsides rich in plants, but native paths, cutting off all the zigzags, run in straight lines up the steepest hill-faces, and thus double the available means for botanising; and it is all but impossible to leave the paths of one kind or other, except for a yard or two up the rocky ravines. Elephants, tigers, and occasionally the rhinoceros, inhabit the foot of these hills, with wild boars, leopards, &c.; but none are numerous. The elephant's path is an excellent specimen of engineering—the opposite of the native track, for it winds judiciously.

At about 1000 feet above Punkabaree, the vegetation is very rich, and appears all the more so from the many turnings of the road, affording glorious prospects of the foreshortened tropical forests. The prevalent timber is gigantic, and scaled by climbing *Leguminosæ*, as *Bauhinias* and *Robinias*, which sometimes sheath the trunks, or span the forest with huge cables, joining tree to tree. Their trunks are also clothed with parasitical Orchids, and still more beautifully with Pothos (*Scindapsus*), Peppers, *Gnetum*, Vines, Convolvulus, and *Bignoniæ*. The beauty of the drapery of the Pothos-leaves is pre-eminent, whether for the graceful folds the foliage assumes, or for the liveliness of its colour. Of the more conspicuous smaller trees, the wild banana is the most abundant, its crown of very beautiful foliage contrasting with the smaller-leaved plants amongst which it nestles; next comes a screw-pine (*Pandanus*) with a straight stem and a tuft of leaves, each eight or ten feet long, waving on all sides. *Araliaceæ*, with smooth or armed slender trunks, and *Mappa*-like *Euphorbiaceæ*, spread their long petioles horizontally forth, each terminated with an ample leaf some feet in diameter. Bamboo abounds everywhere: its dense tufts of culms, 100 feet and upwards high, are as thick as a man's thigh at the base. Twenty or thirty species of ferns (including a tree-fern) were luxuriant and handsome. Foliaceous lichens and a few mosses appeared at 2000 feet. Such is the vegetation of the roads through the tropical forests of the Outer-Himalaya.

At about 4000 feet the road crossed a saddle, and ran along the narrow crest of a hill, the top of that facing the plains of India, and over which is the way to the interior ranges, amongst which Dorjiling is placed, still twenty-five miles off. A little below this a great change had taken place in the vegetation,—marked, first, by the appearance of a very English-looking bramble, which, however, by way of proving its foreign origin, bore a very good yellow fruit, called here the "yellow raspberry." Scattered oaks, of a noble species, with large lamellated cups and magnificent foliage, succeeded; and along the ridge of the mountain to Kursiong (a dawk bungalow at about 4800 feet), the change in the flora was complete.

The spring of this region and elevation most vividly recalled that of England. The oak flowering, the birch bursting into leaf, the violet, *Chrysosplenium*, *Stellaria* and *Arum*, *Vaccinium*, wild strawberry, maple, geranium, bramble. A colder wind blew here: mosses and lichens carpeted the banks and roadsides: the birds and insects were very different from those below; and everything proclaimed the marked change in elevation, and not only in this, but in season, for I had left the winter of the tropics and here encountered the spring of the temperate zone.

The flowers I have mentioned are so notoriously the harbingers of a European spring that their presence carries one home at once; but, as species, they differ from their European prototypes, and are accompanied at this elevation (and for 2000 feet higher up) with tree-fern, Pothos, bananas, palms, figs, pepper, numbers of epiphytal Orchids, and similar genuine tropical genera. The uniform temperature and humidity of the region here favour the extension of tropical plants into a temperate region; exactly as the same conditions cause similar forms to reach higher latitudes in the southern hemisphere (as in New Zealand, Tasmania, South Chili, &c.) than they do in the northern.

Along this ridge I met with the first tree-fern. This species seldom reaches the height of forty feet; the black trunk is but three or four in girth, and the feathery crown is ragged in comparison with the species of many other countries: it is the *Alsophila gigantea*, and ascends nearly to 7000 feet elevation.

Kursiong bungalow, where I stopped for a few hours, is superbly placed, on a narrow mountain ridge. The west window looks down the valley of the Balasun river, the east into that of the Mahanuddee: both of these rise from the outer range, and flow in broad, deep, and steep valleys (about 4000 feet deep) which give them their respective names, and are richly wooded from the Terai to their tops. Till reaching this spur, I had wound upwards along the western slope of the Mahanuddee valley. The ascent from the spur at Kursiong, to the top of the mountain (on the northern face of which Dorjiling is situated), is along the eastern slope of the Balasun.

From Kursiong a very steep zigzag leads up the mountain, through a magnificent forest of chesnut, walnut, oaks, and laurels. It is difficult to conceive a grander mass of vegetation:—the straight shafts of the timber-trees shooting aloft, some naked and clean, with grey, pale, or brown bark; others literally clothed for yards with a continuous garment of epiphytes, one mass of blossoms, especially the white Orchids *Cœlogynes*, which bloom in a profuse manner, whitening their trunks like snow. More bulky trunks were masses of interlacing climbers, *Araliaceæ*, *Leguminosæ*, Vines, and *Menispermeæ*, Hydrangea, and Peppers, enclosing a hollow, once filled by the now strangled supporting tree, which had long ago decayed away. From the sides and summit of these, supple branches hung forth, either leafy or naked; the latter resembling cables flung from one tree to another, swinging in the breeze, their rocking motion increased by the weight of great bunches of ferns or Orchids, which were perched aloft in the loops. Perpetual moisture nourishes this dripping forest: and pendulous mosses and lichens are met with in profusion.

Two thousand feet higher up, near Mahaldiram (whence the last view of the plains is gained), European plants appear,—Berberry, *Paris*, &c.; but here, night gathered round, and I had still ten miles to go to the nearest bungalow, that of Pacheem. The road still led along the eastern slope of the Balasun valley, which was exceedingly steep, and so cut up by ravines, that it winds in and out of gulleys almost narrow enough to be jumped across.

It was very late before I arrived at Pacheem bungalow, the most sinister-looking rest-house I ever saw, stuck on a little cleared spur of the mountain, surrounded by dark forests, overhanging a profound valley, and enveloped in mists and rain, and hideous in architecture, being a miserable attempt to unite the Swiss cottage with the suburban gothic;—it combined a maximum of discomfort with a minimum of good looks or good cheer. I was some time in finding the dirty housekeeper, in an outhouse hard by, and then in waking him. As he led me up the crazy verandah, and into a broad ghostly room, without glass in the windows, or fire, or any one comfort, my mind recurred to the stories told of the horrors of the Hartz forest, and of the benighted traveller's situation therein. Cold sluggish beetles hung to the damp walls,—and these I immediately secured. After due exertions and perseverance with the damp wood, a fire smoked lustily, and, by cajoling the gnome of a housekeeper, I procured the usual roast fowl and potatos, with the accustomed sauce of a strong smoky and singed flavour.*

Pacheem stands at an elevation of nearly 7300 feet, and as I walked out on the following morning I met with English looking plants in abundance, but was too early in the season to get aught but the foliage of most. *Chrysosplenium*, violet, *Lobelia*, a small geranium, strawberry, five or six kinds of bramble, *Arum*, *Paris*, *Convallaria*, *Stellaria*, *Rubia*, *Vaccinium*, and various *Gnaphalia*. Of small bushes, cornels, honeysuckles, and the ivy tribe predominated, with *Symplocos* and *Skimmia*, *Eurya*, bushy brambles, having simple or compound green or beautifully silky foliage; *Hypericum*, Berberry, Hydrangea, Wormwood, *Adamia cyanea*, *Viburnum*, Elder, dwarf bamboo, &c.

The climbing plants were still *Panax* or *Aralia*, *Kadsura*, *Sauranja*, *Hydrangea*, Vines, *Smilax*, *Ampelopsis*, *Polygona*, and, most beautiful of all, *Stauntonia*, with pendulous racemes of lilac blossoms. Epiphytes were rarer, still I found white and purple *Cœlogynes*, and other Orchids, and a most noble white Rhododendron, whose truly enormous and delicious lemon-scented blossoms strewed the ground. The trees were one half oaks, one quarter Magnolias, and nearly another quarter laurels, amongst which grew Himalayan kinds of birch, alder, maple, holly, bird-cherry, common cherry, and apple. The absence of *Leguminosæ* was most remarkable, and the most prominent botanical feature in the vegetation of this region: it is too high for the tropical tribes of the warmer elevations, too low for the Alpines, and probably too moist for those of temperate regions; cool, equable, humid climates being generally

* Since writing the above a comfortable house has been erected at Senadah, the name now given to what was called Pacheem Bungalow.

unfavourable to that order. Clematis was rare, and other *Ranunculaceæ* still more so. *Cruciferæ* were absent, and, what was still more remarkable, I found very few native species of grasses. Both *Poa annua* and white Dutch clover flourished where accidentally disseminated, but only in artificially cleared spots. Of ferns I collected about sixty species, chiefly of temperate genera. The supremacy of this temperate region consists in the infinite number of forest trees, in the absence (in the usual proportion, at any rate) of such common orders as *Compositæ, Leguminosæ, Cruciferæ*, and *Ranunculaceæ*, and of Grasses amongst Monocotyledons, and in the predominance of the rarer and more local families, as those of Rhododendron, Camellia, Magnolia, Ivy, Cornel, Honeysuckle, Hydrangea, Begonia, and Epiphytic orchids.

From Pacheem, the road runs in a northerly direction to Dorjiling, still along the Balasun valley, till the saddle of the great mountain Sinchul is crossed. This is narrow, stretching east and west, and from it a spur projects northwards for five or six miles, amongst the many mountains still intervening between it and the snows. This saddle (alt. 7,400 feet) crossed, one is fairly amongst the mountains: the plains behind are cut off by it; and in front, the snows may be seen when the weather is propitious. The valleys on this side of the mountain run northwards, and discharge their streams into great rivers, which, coming from the snow, wind amongst the hills, and debouche into the Teesta, to the east, where it divides Sikkim from Bhotan.

Dorjiling station occupies a narrow ridge, which divides into two spurs, descending steeply to the bed of the Great Rungeet river, up whose course the eye is carried to the base of the great snowy mountains. The ridge itself is very narrow at the top, along which most of the houses are perched, while others occupy positions on its flanks, where narrow *locations* on the east, and broader ones on the west, are cleared from wood. The valleys on either side are at least 6000 feet deep, forest-clad to the bottom, with very few and small level spots, and no absolute precipice; from their flanks project innumerable little spurs, occupied by native clearings.

My route lay along the east flank, overhanging the valley of the Rungmo river. Looking east, the amphitheatre of hills from the ridge I had crossed was very fine; enclosing an area some four miles across and 4000 feet deep, clothed throughout with an impenetrable, dark forest: there was not one clear patch except near the very bottom, where were some scattered hamlets of two or three huts each The rock is everywhere near the surface, and the road has been formed by blasting at very many places. A wooded slope descends suddenly from the edge of the road, while, on the other hand, a bank rises abruptly to the top of the ridge, alternately mossy, rocky, and clayey, and presenting a good geological section, all the way along, of the nucleus of Dorjiling spur, exposing broken masses of gneiss. As I descended, I came upon the upper limit of the chesnut, a tree second in abundance to the oak; gigantic, tall, and straight in the trunk.

I arrived at Dorjiling on the 16th of April; a showery, cold month at this elevation. I was so fortunate as to find Mr. Charles Barnes (brother of my friend at Colgong), the sole tenant of a long, cottage-like building, divided off into pairs of apartments, which are hired by visitors. It is usual for Europeans to bring a full establishment of servants (with bedding, &c.) to such stations, but I had not done so, having been told that there was a furnished hotel in Dorjiling; and I was, therefore, not a little indebted to Mr. Barnes for his kind invitation to join his mess. As he was an active mountaineer, we enjoyed many excursions together, in the two months and a half during which we were companions.

Dr. Campbell procured me several active native (Lepcha) lads as collectors, at wages varying from eight to twenty shillings a month; these either accompanied me on my excursions, or went by themselves into the jungles to collect plants, which I occupied myself in drawing, dissecting, and ticketing: while the preserving of them fell to the Lepchas, who, after a little training, became, with constant superintendence, good plant-driers. Even at this season (four weeks before the setting in of the rains) the weather was very uncertain, so that the papers had generally to be dried by the fire.

The hill-station or Sanatarium of Dorjiling owes its origin (like Simla, Mussooree, &c.) to the necessity that exists in India, of providing places where the health of Europeans may be recruited by a more temperate climate. Sikkim proved an eligible position for such an establishment, owing to its proximity to Calcutta, which lies but 370 miles to the southward; whereas the north-west stations mentioned above are upwards of a thousand miles from that city. Dorjiling ridge varies in height from 6500 to 7500 feet above the level of the sea; 8000 feet being the elevation at which the mean temperature most nearly coincides with that of London, viz., 50°.

Sikkim was, further, the only available spot for a Sanatarium throughout the whole range of the Himalaya, east of the extreme western frontier of Nepal; being a protected state, and owing no allegiance, except to the British government; which, after the Rajah had been driven from the country by the Ghorkas, in 1817, replaced him on his throne, and guaranteed him the sovereignty. Our main object in doing this was to retain Sikkim as a fender between Nepal and Bhotan: and but for this policy, the aggressive Nepalese would, long ere this, have possessed themselves of Sikkim, Bhotan, and the whole Himalaya, eastwards to the borders of Burmah.*

From 1817 to 1828 no notice was taken of Sikkim, till a frontier dispute occurred between the Lepchas and Nepalese, which was referred (according to the terms of the treaty) to the British Government. During the arrangement of this, Dorjiling was visited by a gentleman of high scientific attainments, Mr. J. W. Grant, who pointed out its eligibility as a site for a Sanatarium to Lord William Bentinck, then Governor-General; dwelling especially upon its climate, proximity to Calcutta, and accessibility; on its central position between Tibet, Bhotan, Nepal, and British India; and on the good example a peaceably-conducted and well-governed station would be to our turbulent neighbours in that quarter. The suggestion was cordially received, and Major Herbert (the late eminent Surveyor-General of India) and Mr. Grant were employed to report further on the subject.

The next step taken was that of requesting the Rajah to cede a tract of country which should include Dorjiling, for an equivalent in money or land. His first demand was unreasonable; but on further consideration he surrendered Dorjiling unconditionally, and a sum of 300*l*. per

* Of such being their wish the Nepalese have never made any secret, and they are said to have asked permission from the British to march an army across Sikkim for the purpose of conquering Bhotan, offering to become more peaceable neighbours to us than the Bhotanese are. Such they would doubtless have proved, but the Nepal frontier is considered broad enough already.

annum was granted to him as an equivalent for what was then a worthless uninhabited mountain. In 1840 Dr. Campbell was removed from Nepal as superintendent of the new station, and was entrusted with the charge of the political relations between the British and Sikkim government.

Once established, Dorjiling rapidly increased. Allotments of land were purchased by Europeans for building dwelling-houses; barracks and a bazaar were formed, with accommodation for invalid European soldiers; a few official residents, civil and military, formed the nucleus of a community, which was increased by retired officers and their families, and by temporary visitors in search of health, or the luxury of a cool climate and active exercise.

For the first few years matters went on smoothly with the Rajah, whose minister (or Dewan) was upright and intelligent: but the latter, on his death, was succeeded by the present Dewan, a Tibetan, and a relative of the Ranee (or Rajah's wife); a man unsurpassed for insolence and avarice, whose aim was to monopolise the trade of the country, and to enrich himself at its expense. Every obstacle was thrown by him in the way of a good understanding between Sikkim and the British government. British subjects were rigorously excluded from Sikkim; every liberal offer for free trade and intercourse was rejected,. generally with insolence; merchandise was taxed, and notorious offenders, refugees from the British territories, were harboured; despatches were detained; and the Vakeels, or Rajah's representatives, were chosen for their insolence and incapacity. The conduct of the Dewan throughout was Indo-Chinese; assuming, insolent, aggressive, never perpetrating open violence, but by petty insults effectually preventing all good understanding. He was met by neglect or forbearance on the part of the Calcutta government; and by patience and passive resistance at Dorjiling. Our inaction and long-suffering were taken for weakness, and our concessions for timidity. Such has been our policy in China, Siam, and Burmah, and in each instance the result has been the same. Had it been insisted that the terms of the treaty should be strictly kept, and had the first act of insolence been noticed, we should have maintained the best relations with Sikkim, whose people and rulers (with the exception of the Dewan and his faction) have proved themselves friendly throughout, and most anxious for unrestricted communication.

These political matters have not, however, prevented the rapid increase of Dorjiling; the progress of which, during the two years I spent in Sikkim, resembled that of an Australian colony, not only in amount of building, but in the accession of native families from the surrounding countries. There were not a hundred inhabitants under British protection when the ground was transferred; there are now four thousand. At the former period there was no trade whatever; there is now a very considerable one, in musk, salt, gold-dust, borax, soda, woollen cloths, and especially in poneys, of which the Dewan in one year brought on his own account upwards of 50 into Dorjiling.* The trade has been greatly increased by the annual fair which Dr. Campbell has established at the foot of the hills, to which many thousands of natives flock from all quarters, and which exercises a most beneficial influence throughout the neighbouring territories. At this, prizes (in medals, money, and kind) are given for agricultural implements

* The Tibetan pony, though born and bred 10,000 to 14,000 feet above the sea, is one of the most active and useful animals in the plains of Bengal, powerful and hardy, and when well trained early, docile, although by nature vicious and obstinate.

and produce, stock, &c., by the originator and a few friends; a measure attended with eminent success.

In estimating in a sanitory point of view the value of any health-station, little reliance can be placed on the general impressions of invalids, or even of residents; the opinion of each varies with the nature and state of his complaint, if ill, or with his idiosyncracy and disposition, if well. I have seen prejudiced invalids rapidly recovering in spite of themselves, and all the while complaining in unmeasured terms of the climate of Dorjiling, and abusing it as killing them. Others are known who languish under the heat of the plains at one season, and the damp at another; and who, though sickening and dying under its influence, yet consistently praise a tropical climate to the last. The opinions of those who resort to Dorjiling in health, differ equally; those of active minds invariably thoroughly enjoy it, while the mere lounger or sportsman mopes. The statistical tables afford conclusive proofs of the value of the climate to Europeans suffering from acute diseases, and they are corroborated by the returns of the medical officer in charge of the station. With respect to its suitability to the European constitution I feel satisfied, and that much saving of life, health, and money would be effected were European troops drafted thither on their arrival in Bengal, instead of being stationed in Calcutta, exposed to disease, and temptation to those vices which prove fatal to so many hundreds. This, I have been given to understand, was the view originally taken by the Court of Directors, but it has never been carried out.

I believe that children's faces afford as good an index as any to the healthfulness of a climate, and in no part of the world is there a more active, rosy, and bright young community, than at Dorjiling It is incredible what a few weeks of that mountain air does for the India-born children of European parents: they are taken there sickly, pallid or yellow, soft and flabby, to become transformed into models of rude health and activity.

There are, however, disorders to which the climate (in common with all damp ones) is not at all suited; such are especially dysentery, bowel complaints, and liver complaints of long standing; which are not benefited by a residence on these hills, though how much worse they might have become in the plains is not shown. I cannot hear that the climate aggravates. but it certainly does not remove them. Whoever is suffering from the debilitating effects of any of the multifarious acute maladies of the plains, finds instant relief, and acquires a stock of health that enables him to resist fresh attacks, under circumstances similar to those which before engendered them.

Natives of the low country, and especially Bengalees, are far from enjoying the climate as Europeans do, being liable to sharp attacks of fever and ague, from which the poorly clad natives are not exempt. It is, however, difficult to estimate the effects of exposure upon the Bengalees, who sleep on the bare and often damp ground, and adhere, with characteristic prejudice, to the attire of a torrid climate, and to a vegetable diet, under skies to which these are least of all adapted.

It must not be supposed that Europeans who have resided in the plains can, on their first arrival, expose themselves with impunity to the cold of these elevations; this was shown in the winter of 1848 and 1849, when troops brought up to Dorjiling were cantoned in newly-built dwellings, on a high exposed ridge 8000 feet above the sea, and lay, insufficiently protected, on a floor of loosely laid planks, exposed to the cold wind, when the ground without was covered with snow. Rheumatisms, sharp febrile attacks, and dysenteries ensued, which were attributed in the public prints to the unhealthy nature of the climate of Dorjiling.

The following summary of hospital admissions affords the best test of the healthiness of the climate, embracing, as the period does, the three most fatal months to European troops in India. Out of a detachment (105 strong) of H. M. 80th Regiment stationed at Dorjiling, in the seven months from January to July inclusive, there were sixty-four admissions to the hospital, or, on the average, $4\frac{1}{3}$ per cent. per month; and only two deaths, both of dysentery. Many of these men had suffered frequently in the plains from acute dysentery and hepatic affections, and many others had aggravated these complaints by excessive drinking, and two were cases of delirium tremens. During the same period, the number of entries at Calcutta or Dinapore would probably have more than trebled this.

CHAPTER V.

View from Mr. Hodgson's of range of snowy mountains—Their extent and elevation—Delusive appearance of elevation—Sinchul, view from and vegetation of—Chumulari—Magnolias, white and purple—Rhododendron Dalhousiæ, arboreum and argenteum—Natives of Dorjiling—Lepchas, origin, tradition of flood, morals, dress, arms, ornaments, diet—cups, origin and value—Marriages—Diseases—Burial—Worship and religion—Bijooas—Kampa Rong, or Arratt—Limboos, origin, habits, language, &c.—Moormis—Magras—Mechis—Comparison of customs with those of the natives of Assam, Khasia, &c.

THE summer, or rainy season of 1848, was passed at or near Dorjiling, during which period I chiefly occupied myself in forming collections, and in taking meteorological observations. I resided at Mr Hodgson's for the greater part of the time, in consequence of his having given me a hospitable invitation to consider his house my home. The view from his windows is one quite unparalleled for the scenery it embraces, commanding confessedly the grandest known landscape of snowy mountains in the Himalaya, and hence in the world.* Kinchinjunga (forty-five miles distant) is the prominent object, rising 21,000 feet above the level of the observer out of a sea of intervening wooded hills; whilst, on a line with its snows, the eye descends below the horizon, to a narrow gulf 7000 feet deep in the mountains, where the Great Rungeet, white with foam, threads a tropical forest with a silver line.

* For an account of the geography of these regions, and the relation of the Sikkim Himalaya to Tibet, &c., see Appendix.

To the north-west towards Nepal, the snowy peaks of Kubra and Junnoo (respectively 24,005 feet and 25,312 feet) rise over the shoulder of Singalelah; whilst eastward the snowy mountains appear to form an unbroken range, trending north-east to the great mass of Donkia (23,176 feet) and thence south-east by the fingered peaks of Tunkola and the silver cone of Chola, (17,320 feet) gradually sinking into the Bhotan mountains at Gipmoochi (14,509 feet).

The most eloquent descriptions I have read fail to convey to my mind's eye the forms and colours of snowy mountains, or to my imagination the sensations and impressions that rivet my attention to these sublime phenomena when they are present in reality; and I shall not therefore obtrude any attempt of the kind upon my reader. The latter has probably seen the Swiss Alps, which, though barely possessing half the sublimity, extent, or height of the Himalaya, are yet far more beautiful. In either case he is struck with the precision and sharpness of their outlines, and still more with the wonderful play of colours on their snowy flanks, from the glowing hues reflected in orange, gold and ruby, from clouds illumined by the sinking or rising sun, to the ghastly pallor that succeeds with twilight, when the red seems to give place to its complementary colour green. Such dissolving-views elude all attempts at description, they are far too aërial to be chained to the memory, and fade from it so fast as to be gazed upon day after day, with undiminished admiration and pleasure, long after the mountains themselves have lost their sublimity and apparent height.

The actual extent of the snowy range seen from Mr. Hodgson's windows is comprised within an arc of 80° (from north 30° west to north 50° east), or nearly a quarter of the horizon, along which the perpetual snow forms an unbroken girdle or crest of frosted silver; and in winter, when the mountains are covered down to 8000 feet, this white ridge stretches uninterruptedly for more than 160°. No known view is to be compared with this in extent, when the proximity and height of the mountains are considered; for within the 80° above mentioned more than twelve peaks rise above 20,000 feet, and there are none below 15,000 feet, while Kinchin is 28,178, and seven others above 22,000. The nearest perpetual snow is on Nursing, a beautifully sharp conical peak 19,139 feet high and thirty-two miles distant; the most remote mountain seen is Donkia, 23,176 feet high, and seventy-three miles distant; whilst Kinchin, which forms the principal mass both for height and bulk, is exactly forty-five miles distant.

On first viewing this glorious panorama, the impression produced on the imagination by their prodigious elevation is, that the peaks tower in the air and pierce the clouds, and such are the terms generally used in descriptions of similar alpine scenery; but the observer, if he look again, will find that even the most stupendous occupy a very low position on the horizon, the top of Kinchin itself measuring only 4° 31 above the level of the observer! Donkia again, which is 23,176 feet above the sea, or about 15,700 above Mr. Hodgson's, rises only 1° 55′ above the horizon; an angle which is quite inappreciable to the eye, when unaided by instruments.*

This view may be extended a little by ascending Sinchul, which rises a thousand feet above the elevation of Mr. Hodgson's house, and is a few miles south-east of

* These are the apparent angles which I took from Mr. Hodgson's house (alt. 7300 feet) with an excellent theodolite, no deduction being made for refraction.

Dorjiling: from its summit Chumulari (23,929 feet) is seen to the north-east, at eighty-four miles distance, rearing its head as a great rounded mass over the snowy Chola range, out of which it appears to rise, although in reality lying forty miles beyond;—so deceptive is the perspective of snowy mountains. To the north-west again, at upwards of 100 miles distance, a beautiful group of snowy mountains rises above the black Singalelah range, the chief being, perhaps, as high as Kinchinjunga, from which it is fully eighty miles distant to the westward; and between them no mountain of considerable altitude intervenes; the Nepalese Himalaya in that direction sinking remarkably towards the Arun river, which there enters Nepal from Tibet.

The top of Sinchul is a favourite excursion from Dorjiling, being very easy of access, and the path abounding in rare and beautiful plants, and passing through magnificent forests of oak, magnolia, and rhododendron; while the summit, besides embracing this splendid view of the snowy range over the Dorjiling spur in the foreground, commands also the plains of India, with the courses of the Teesta, Mahanuddee, Balasun and Mechi rivers. In the months of April and May, when the magnolias and rhododendrons are in blossom, the gorgeous vegetation is, in some respects, not to be surpassed by anything in the tropics; but the effect is much marred by the prevailing gloom of the weather. The white-flowered magnolia (*M. excelsa*, Wall,) forms a predominant tree at 7000 to 8000 feet; and in 1848 it blossomed so profusely, that the forests on the broad flanks of Sinchul, and other mountains of that elevation, appeared as if sprinkled with snow. The purple-flowered kind again (*M. Campbellii*) hardly occurs below 8000 feet, and forms an immense, but very ugly, black-barked, sparingly branched tree, leafless in winter and also during the flowering season, when it puts forth from the ends of its branches great rose-purple cup-shaped flowers, whose fleshy petals strew the ground. On its branches, and on those of oaks and laurels, *Rhododendron Dalhousiæ* grows epiphytically, a slender shrub, bearing from three to six white lemon-scented bells, four and a half inches long and as many broad, at the end of each branch. In the same woods the scarlet rhododendron (*R. arboreum*) is very scarce, and is outvied by the great *R. argenteum*, which grows as a tree forty feet high, with magnificent leaves twelve to fifteen inches long, deep green, wrinkled above and silvery below, while the flowers are as large as those of *R. Dalhousiæ*, and grow more in a cluster. I know nothing of the kind that exceeds in beauty the flowering branch of *R. argenteum*, with its wide spreading foliage and glorious mass of flowers.

Oaks, laurels, maples, birch, chesnut, hydrangea, a species of fig (which is found on the very summit), and three Chinese and Japanese genera, are the principal features of the forest; the common bushes being *Aucuba*, *Skimmia*, and the curious *Helwingia*, which bears little clusters of flowers on the centre of the leaf, like butcher's-broom. In spring immense broad-leaved arums spring up, with green or purple-striped hoods, that end in tail-like threads, eighteen inches long, which lie along the ground; and there are various kinds of *Convallaria*, *Paris*, *Begonia*, and other beautiful flowering herbs. Nearly thirty ferns may be gathered on this excursion, including many of great beauty and rarity, but the tree-fern does not ascend so high. Grasses are very rare in these woods, excepting the dwarf bamboo, now cultivated in the open air in England.

Before proceeding to narrate my different expeditions into

Sikkim and Nepal from Dorjiling, I shall give a sketch of the different peoples and races composing the heterogeneous population of Sikkim and the neighbouring mountains.

The Lepcha is the aboriginal inhabitant of Sikkim, and the prominent character in Dorjiling, where he undertakes all sorts of out-door employment. The race to which he belongs is a very singular one; markedly Mongolian in features, and a good deal too, by imitation, in habit; still he differs from his Tibetan prototype, though not so decidedly as from the Nepalese and Bhotanese, between whom he is hemmed into a narrow tract of mountain country, barely 60 miles in breadth. The Lepchas possess a tradition of the flood, during which a couple escaped to the top of a mountain (Tendong) near Dorjiling. The earliest traditions which they have of their history date no further back than some three hundred years, when they describe themselves as having been long-haired, half-clad savages. At about that period they were visited by Tibetans, who introduced Boodh worship, the platting of their hair into pig-tails, and very many of their own customs. Their physiognomy is however so Tibetan in its character, that it cannot be supposed that this was their earliest intercourse with the trans-nivean races: whether they may have wandered from beyond the snows before the spread of Boodhism and its civilisation, or whether they are a cross between the Tamulian of India and the Tibetan, has not been decided. Their language, though radically identical with Tibetan, differs from it in many important particulars. They, or at least some of their tribes, call themselves Rong, and Arratt, and their country Dijong: they once possessed a great part of East Nepal, as far west as the Tambur river, and at a still earlier period they penetrated as far west as the Arun river.

An attentive examination of the Lepcha in one respect entirely contradicts our preconceived notions of a mountaineer, as he is timid, peaceful, and no brawler; qualities which are all the more remarkable from contrasting so strongly with those of his neighbours to the east and west: of whom the Ghorkas are brave and warlike to a proverb, and the Bhotanese quarrelsome, cowardly, and cruel. A group of Lepchas is exceedingly picturesque. They are of short stature—four feet eight inches to five feet—rather broad in the chest, and with muscular arms, but small hands and slender wrists.* The face is broad, flat, and of eminently Tartar character, flat-nosed and oblique-eyed, with no beard, and little moustache; the complexion is sallow, or often a clear olive; the hair is collected into an immense tail, plaited flat or round. The lower limbs are powerfully developed, befitting genuine mountaineers: the feet are small. Though never really handsome, and very womanish in the cast of countenance, they have invariably a mild, frank, and even engaging expression, which I have in vain sought to analyse, and which is perhaps due more to the absence of anything unpleasing, than to the presence of direct grace or beauty. In like manner, the girls are often very engaging to look upon, though without one good feature: they are all smiles and good-nature; and the children are frank, lively, laughing urchins. The old women are thorough hags. Indolence, when left to themselves, is their besetting sin; they detest any fixed employment, and their foulness of person and garments renders them disagreeable inmates: in this rainy climate they are supportable out of doors. Though fond of bathing when

* I have seldom been able to insert my own wrist (which is smaller than the average) into the wooden guard which the Lepcha wears on his left, as a protection against the bow-string: it is a curved ring of wood with an opening at one side, through which, by a little stretching, the wrist is inserted.

they come to a stream in hot weather, and expert, even admirable swimmers, these people never take to the water for the purpose of ablution. In disposition they are

LEPCHA GIRL AND BHOODIST LAMA.

amiable and obliging, frank, humorous, and polite, without the servility of the Hindoos; and their address is free and unrestrained. Their intercourse with one another and with Europeans is scrupulously honest; a present is divided equally amongst many, without a syllable of discontent or grudging look or word: each, on receiving his share, coming up and giving the donor a brusque bow and thanks. They have learnt to overcharge already, and use extortion in dealing, as is the custom with the people of the plains; but it is clumsily done, and never accompanied with the grasping air and insufferable whine of the latter. They are constantly armed with a long, heavy, straight knife,* but never draw it on one another: family and political feuds are alike unheard of amongst them.

The Lepcha is in morals far superior to his Tibet and Bhotan neighbours, polyandry being unknown, and polygamy rare. This is no doubt greatly due to the conventual system not being carried to such an excess as in Bhotan, where the ties of relationship even are disregarded.

Like the New Zealander, Tasmanian, Fuegian, and natives of other climates, which, though cold, are moist and equable, the Lepcha's dress is very scanty, and when we are wearing woollen under-garments and hose, he is content with one cotton vesture, which is loosely thrown round the body, leaving one or both arms free; it reaches to the knee, and is gathered round the waist: its fabric is close, the ground colour white, ornamented with longitudinal blue stripes, two or three fingers broad, prettily worked with red and white. When new and clean, this garb is remarkably handsome and gay, but not showy. In cold weather an upper garment with loose sleeves is added. A long knife, with a common wooden handle, hangs by the side, stuck in a sheath; he has often also a quiver of poisoned arrows and a bamboo† bow across his back. On his right wrist is a curious wooden guard for the

* It is called "Ban," and serves equally for plough, toothpick, table-knife, hatchet, hammer, and sword.

† The bamboo, of which the quiver is made, is thin and light: it is brought from Assam, and called Tulda, or Dulwa, by the Bengalees.

bowstring; and a little pouch, containing aconite poison and a few common implements, is suspended to his girdle. A hat he seldom wears, and when he does, it is often extravagantly broad and flat-brimmed, with a small hemispherical crown. It is made of leaves of *Scitamineæ*, between two thin plates of bamboo-work, clumsy and heavy; this is generally used in the rainy weather, while in the dry a conical one is worn, also of platted slips of bamboo, with broad flakes of talc between the layers, and a peacock's feather at the side. The umbrella consists of a large hood, much like the ancient boat called a coracle, which being placed over the head reaches to the thighs behind. It is made of platted bamboo, enclosing broad leaves of *Phrynium*. A group of Lepchas with these on, running along in the pelting rain, are very droll figures; they look like snails with their shells on their backs. All the Lepchas are fond of ornaments, wearing silver hoops in their ears, necklaces made of cornelian, amber, and turquoise, brought from Tibet, and pearls and corals from the south, with curious silver and golden charm-boxes or amulets attached to their necks or arms. These are of Tibetan workmanship, and often of great value: they contain little idols, charms and written prayers, or the bones, hair, or nail-parings of a Lama: some are of great beauty, and highly ornamented. In these decorations, and in their hair, they take some pride, the ladies frequently dressing the latter for the gentlemen: thus one may often see, the last thing at night, a damsel of discreet port, demurely go behind a young man, unplait his pig-tail, teaze the hair, thin it of some of its lively inmates, braid it up for him, and retire. The women always wear two braided pig-tails, and it is by this they are most readily distinguished from their effeminate-looking partners, who wear only one.* When in full dress, the woman's costume is extremely ornamental and picturesque; besides the shirt and petticoat she wears a small sleeveless woollen cloak, of gay pattern, usually covered with crosses, and fastened in front by a girdle of silver chains. Her neck is loaded with silver chains, amber necklaces, &c., and her head adorned with a coronet of scarlet cloth, studded with seed-pearls, jewels, glass beads, &c. The common dress is a long robe of indi, a cloth of coarse silk, spun from the cocoon of a large caterpillar that is found wild at the foot of the hills, and is also cultivated: it feeds on many different leaves, Sal (*Shorea*), castor-oil, &c.

In diet, they are gross feeders; † rice, however, forming their chief sustenance; it is grown without irrigation, and produces a large, flat, coarse grain, which becomes gelatinous, and often pink, when cooked. Pork is a staple dish: and they also eat elephant, and all kinds of animal food. When travelling, they live on whatever they can find, whether animal or vegetable. Fern-tops, roots of *Scitamineæ*, and their flower-buds, various leaves (it is difficult to say what not), and fungi, are chopped up, fried with a little oil, and eaten. Their cooking is coarse and dirty. Salt is costly, but prized; pawn (Betel pepper) is never eaten. Tobacco they are too poor to buy, and too indolent to grow and cure. Spices, oil, &c. are relished.

They drink out of little wooden cups, turned from knots of maple, or other woods; these are very curious on several accounts; they are very pretty, often polished, and mounted with silver. Some are supposed to be antidotes against

* Ermann (Travels in Siberia, ii. p. 204) mentions the Buraet women as wearing two tails, and fillets with jewels, and the men as having one queue only.

† Dr. Campbell's definition of the Lepcha's *Flora cibaria*, is, that he eats, or must have eaten, everything soft enough to chew; for, as he knows whatever is poisonous, he must have tried all; his knowledge being wholly empirical.

poison, and hence fetch an enormous price; they are of a peculiar wood, rarer and paler-coloured. I have paid a guinea for one such, hardly different from the common sort, which cost but 4*d.* or 6*d.* MM. Huc and Gabet graphically allude to this circumstance, when wishing to purchase cups at Lhassa, where their price is higher, as they are all imported from the Himalaya. The knots from which they are formed, are produced on the roots of oaks, maples, and other mountain forest trees, by a parasitical plant, known to botanists as *Balanophora.*

Their intoxicating drink, which seems more to excite than to debauch the mind, is partially fermented Murwa grain (*Eleusine Coracana*). Spirits are rather too strong to be relished raw, and when a glass of wine is given to one of a party, he sips it, and hands it round to all the rest. A long bamboo flute, with four or six burnt holes far below the mouth-hole, is the only musical instrument I have seen in use among them. When travelling, and the fatigues of the day are over, the Lepchas will sit for hours chatting, telling stories, singing in a monotonous tone, or blowing this flute. I have often listened with real pleasure to the simple music of this rude instrument; its low and sweet tones are singularly Æolian, as are the airs usually played, which fall by octaves: it seems to harmonize with the solitude of their primæval forests, and he must have a dull car who cannot draw from it the indication of a contented mind, whether he may relish its soft musical notes or not. Though always equipped for the chase, I fancy the Lepcha is no great sportsman; there is little to be pursued in this region, and he is not driven by necessity to follow what there is.

Their marriages are contracted in childhood, and the wife purchased by money, or by service rendered to the future father-in-law, the parties being often united before the woman leaves her parents' roof, in cases where the payment is not forthcoming, and the bridegroom prefers giving his and his wife's labour to the father for a stated period in lieu. On the time of service expiring, or the money being paid up, the marriage is publicly celebrated by feasting and riot. The females are generally chaste, and the marriage-tie is strictly kept, its violation being heavily punished by divorce, beating, slavery, &c. In cases of intermarriage with foreigners, the children belong to the father's country. All the labours of the house, the field, and march, devolve on the women and children, or slaves if they have them.

Small-pox is dreaded, and infected persons often cruelly shunned: a suspicion of this or of cholera frequently emptying a village or town in a night. Vaccination has been introduced by Dr. Pearson, and it is much practised by Dr. Campbell; it being eagerly sought. Cholera is scarcely known at Dorjiling, and when it has been imported thither has never spread. Disease is very rare amongst the Lepchas; and ophthalmia, elephantiasis, and leprosy, the scourges of hot climates, are rarely known. Goitre prevails,* though not so conspicuously as amongst Bhoteeas,

* May not the use of the head instead of the shoulder-strap in carrying loads be a predisposing cause of goître, by inducing congestion of the laryngeal vessels? The Lepcha is certainly far more free from this disease than any of the tribes of E. Nepal I have mixed with, and he is both more idle and less addicted to the head-strap as a porter. I have seen it to be almost universal in some villages of Bhoteeas, where the head-strap alone is used in carrying in both summer and winter crops; as also amongst the salt-traders, or rather those families who carry the salt from the passes to the Nepalese villages, and who very frequently have no shoulder-straps, but invariably head-bands. I am far from attributing all goitre, even in the mountains, to this practice, but I think it is proved, that the disease is most prevalent in the mountainous regions of both the old and new world, and that in these the practice of supporting enormous loads by the cervical muscles is frequent. It is also found in the Himalayan sheep and goats which accompany the salt-traders, and whose loads are supported in ascending, by a band passing under the throat.

Bhotanese, and others. Rheumatism is frequent, and intermittent fevers, with ague; also violent and often fatal remittents, almost invariably induced by sleeping in the hot valleys, especially at the beginning and end of the rains. The European complaints of liver and bowel disease are all but unknown. Death is regarded with horror. The dead are burnt or buried, sometimes both; much depending on custom and position. Omens are sought in the entrails of fowls, &c., and other vestiges of their savage origin are still preserved, though now gradually disappearing.

The Lepchas profess no religion, though acknowledging the existence of good and bad spirits. To the good they pay no heed; "Why should we?" they say, "the good spirits do us no harm; the evil spirits, who dwell in every rock, grove, and mountain, are constantly at mischief, and to them we must pray, for they hurt us." Every tribe has a priest-doctor; he neither knows nor attempts to practise the healing art, but is a pure exorcist; all bodily ailments being deemed the operations of devils, who are cast out by prayers and invocations. Still they acknowledge the Lamas to be very holy men, and were the latter only moderately active, they would soon convert all the Lepchas. Their priests are called "Bijooas:" they profess mendicancy, and seem intermediate between the begging friars of Tibet, whose dress and attributes they assume, and the exorcists of the aboriginal Lepchas: they sing, dance (masked and draped like harlequins), beg, bless, curse, and are merry mountebanks; those that affect more of the Lama Boodhist carry the "Mani," or revolving praying machine, and wear rosaries and amulets; others again are all tatters and rags. They are often employed to carry messages, and to transact little knaveries. The natives stand in some awe of them, and being besides of a generous disposition, keep the wallet of the Bijooa always full.

Such are some of the prominent features of this people, who inhabit the sub-Himalayas, between the Nepalese and Bhotan frontiers, at elevations of 3000 to 6000 feet. In their relations with us, they are conspicuous for their honesty, their power as carriers and mountaineers, and their skill as woodsmen; for they build a waterproof house with a thatch of banana leaves in the lower, or of bamboo in the elevated regions, and equip it with a table and bedsteads for three persons, in an hour, using no implement but their heavy knife. Kindness and good humour soon attach them to your person and service. A gloomy-tempered or morose master they avoid, an unkind one they flee. If they serve a good hills-man like themselves, they will follow him with alacrity, sleep on the cold, bleak mountain exposed to the pitiless rain, without a murmur, lay down the heavy burden to carry their master over a stream, or give him a helping hand up a rock or precipice—do anything, in short, but encounter a foe, for I believe the Lepcha to be a veritable coward.* It is well, perhaps, he is so: for if a race, numerically so weak, were to embroil itself by resenting the injuries of the warlike Ghorkas, or dark Bhotanese, the folly would soon lead to destruction.

Before leaving the Lepchas, it may be worth mentioning that the northern parts of the country, towards the Tibet frontier, are inhabited by Sikkim Bhoteeas † (or Kumpas),

* Yet, during the Ghorka war, they displayed many instances of courage: when so hard pressed, however, that there was little choice of evils.

† Bhote is the general name for Tibet (not Bhotan), and Kumpa is a large province, or district, in that country. The Bhotanese, natives of Bhotan, or of the Dhurma country, are called Dhurma people, in allusion to their spiritual chief, the Dhurma Rajah. They are a darker and more powerful race, rude, turbulent, and Tibetan in language and religion, with the worst features of those people exaggerated. The various races of Nepal are too numerous to be alluded

a mixed race calling themselves Kumpa Rong, or Kumpa Lepchas; but they are emigrants from Tibet, having come with the first rajah of Sikkim. These people are more turbulent and bolder than the Lepchas, and retain much of their Tibetan character, and even of that of the very province from which they came; which is north-east of Lhassa, and inhabited by robbers. All the accounts I have received of it agree with those given by MM. Huc and Gabet.

Next to the Lepchas, the most numerous tribe in Sikkim is that of the Limboos (called "Chung" by the Lepchas); they abound also in East Nepal, which they once ruled, inhabiting elevations from 2000 feet to 5000 feet. They are Boodhists, and though not divided into castes, belong to several tribes. All consider themselves as the earliest inhabitants of the Tambur Valley, though they have a tradition of having originally emigrated from Tibet, which their Tartar countenance confirms. They are more slender and sinewy than the Lepchas, and neither plait their hair nor wear ornaments; instead of the ban they use the Nepal curved knife, called "cookree," while for the striped kirtle of the Lepcha are substituted loose cotton trousers and a tight jacket; a sash is worn round the middle, and on the head a small cotton cap. When they ruled over East Nepal, their system was feudal; and on their uniting against the Nepalese, they were with difficulty dislodged from their strongholds. They are said to be equally brave and cruel in battle, putting the old and weak to the sword,

to here: they are all described in various papers by Mr. Hodgson, in the "Journal of the Asiatic Society of Bengal." The Dhurma people are numerous at Dorjiling; they are often runaways, but invariably prove more industrious settlers than the Lepchas. In the Himalaya the name Bhotan is unknown amongst the Tibetans; it signifies literally (according to Mr. Hodgson) the end of Bhote, or Tibet, being the eastern extreme of that country. The Lepchas designate Bhotan as Ayeu, or Aieu, as do often the Bhotanese themselves. Sikkim, again, is called Lhop, or Lho', by the Lepchas and Bhotanese.

carrying the younger to slavery, and killing on the march such captives as are unable to proceed. Many enlist at Dorjiling, which the Lepchas never do; and the rajah of Nepal employs them in his army, where, however, they seldom obtain promotion, this being reserved for soldiers of Hindoo tribes. Latterly Jung Bahadur levied a force of 6000 of them, who were cantoned at Katmandoo, where the cholera breaking out, carried off some hundreds, causing many families who dreaded conscription to flock to Dorjiling. Their habits are so similar to those of the Lepchas, that they constantly intermarry. They mourn, burn, and bury their dead, raising a mound over the corpse, erecting a headstone, and surrounding the grave with a little paling of sticks; they then scatter eggs and pebbles over the ground. In these offices the Bijooa of the Lepchas is employed, but the Limboo has also priests of his own, called "Phedangbos," who belong to rather a higher order than the Bijooas. They officiate at marriages, when a cock is put into the bridegroom's hands, and a hen into those of the bride; the Phedangbo then cuts off the birds' heads, when the blood is caught on a plantain leaf, and runs into pools from which omens are drawn. At death, guns are fired, to announce to the gods the departure of the spirit; of these there are many, having one supreme head, and to them offerings and sacrifices are made. They do not believe in metempsychosis.

The Limboo language is totally different from the Lepcha, with less of the *z* in it, and more labials and palatals, hence more pleasing. Its affinities I do not know; it has no peculiar written character, the Lepcha or Nagri being used. Dr. Campbell, from whom I have derived most of my information respecting these people, was informed,*

* See "Dorjiling Guide," p. 89. Calcutta, 1845.

on good authority, that they had once a written language, now lost; and that it was compounded from many others by a sage of antiquity. The same authority stated that their Lepcha name "Chung" is a corruption of that of their place of residence; possibly the "Tsang" province of Tibet.

The Moormis are the only other native tribe remaining in any numbers in Sikkim, except the Tibetans of the loftier mountains (whom I shall mention at a future period), and the Mechis of the pestilential Terai, the forests of which they never leave. The Moormis are a scattered people, respecting whom I have no information, except from the authority quoted above. They are of Tibetan origin, and called "Nishung," from being composed of two branches, respectively from the districts of Nimo and Shung, both on the road between Sikkim and Lhassa. They are now most frequent in central and eastern Nepal, and are a pastoral and agricultural people, inhabiting elevations of 4000 to 6000 feet, and living in stone houses, thatched with grass. They are a large, powerful, and active race, grave, very plain in features, with little hair on the face. Both their language and religion are purely Tibetan.

The Magras, a tribe now confined to Nepal west of the Arun, are aborigines of Sikkim, whence they were driven by the Lepchas westward into the country of the Limboos, and by these latter further west still. They are said to have been savages, and not of Tibetan origin, and are now converted to Hindooism. A somewhat mythical account of a wild people still inhabiting the Sikkim mountains, will be alluded to elsewhere.

It is curious to observe that these mountains do not appear to have afforded refuge to the Tamulian * aborigines

* The Tamulians are the Coles, Dangas, &c., of the mountains of Central India and the peninsula, who retired to mountain fastnesses, on the invasion of their

of India proper; all the Himalayan tribes of Sikkim being markedly Mongolian in origin. It does not, however, follow that they are all of Tibetan extraction; perhaps, indeed, none but the Moormis are so. The Mechi of the Terai is decidedly Indo-Chinese, and of the same stock as the savage races of Assam, the north-east and east frontier of Bengal, Arracan, Burmah, &c. Both Lepchas and Limboos had, before the introduction of Lama Boodhism from Tibet, many features in common with the natives of Arracan, especially in their creed, sacrifices, faith in omens, worship of many spirits, absence of idols, and of the doctrine of metempsychosis. Some of their customs, too, are the same; the form of their houses and of some of their implements, their striped garments, their constant and dexterous use of the bamboo for all utensils, their practice of night-attacks in war, of using poisoned arrows only in the chase, and that of planting "crow-feet" of sharp bamboo stakes along the paths an enemy is expected to follow. Such are but a few out of many points of resemblance, most of which struck me when reading Lieutenant Phayre's account of Arracan,* and when travelling in the districts of Khasia and Cachar.

The laws affecting the distribution of plants, and the lower animals, materially influence the migrations of man also; and as the botany, zoology, and climate of the Malayan and Siamese peninsula advance far westwards into India, along the foot of the Himalaya, so do also the varieties of the human race. These features are most conspicuously displayed in the natives of Assam, on both sides of the Burrampooter, as far as the great bend of that river, beyond which they gradually disappear; and none of the

country by the Indo-Germanic conquerors, who are now represented by the Hindoos.

* "Journal of the Asiatic Society of Bengal."

Himalayan tribes east of that point practise the bloody and brutal rites in war that prevail amongst the Looties, Khasias, Garrows, and other Indo-Chinese tribes of the mountain forests of Assam, Eastern Bengal, and the Malay peninsula.

I have not alluded to that evidence of the extraction of the Sikkim races, which is to be derived from their languages, and from which we may hope for a clue to their origin; the subject is at present under discussion, and involved in much obscurity.

That six or seven different tribes, without any feudal system or coercive head, with different languages and customs, should dwell in close proximity and in peace and unity, within the confined territory of Sikkim, even for a limited period, is an anomaly; the more especially when it is considered that except for a tincture of the Boodhist religion among some few of the people, they are all but savages, as low in the scale of intellect as the New Zealander or the Tahitian, and beneath those races in ingenuity and skill as craftsmen. Wars have been waged amongst them, but they were neither sanguinary nor destructive, and the fact remains no less remarkable, that at the period of our occupying Dorjiling, friendship and unanimity existed amongst all these tribes; from the Tibetan at 14,000 feet, to the Mechi of the plains; under a sovereign whose temporal power was wholly unsupported by even the semblance of arms, and whose spiritual supremacy was acknowledged by very few.

CHAPTER VI.

Excursion from Dorjiling to Great Rungeet—Zones of vegetation—Tree-ferns—Palms, upper limit of—Leebong, tea plantations—Ging—Boodhist remains—Tropical vegetation—Pines—Lepcha clearances—Forest fires—Boodhist monuments—Fig—Cane bridge and raft over Rungeet—Sago-palm—India-rubber—Yel Pote—Butterflies and other insects—Snakes—Camp—Temperature and humidity of atmosphere—Junction of Teesta and Rungeet—Return to Dorjiling—Tonglo, excursion to—Bamboo flowering—Oaks—Gordonia—Maize, hermaphrodite flowered—Figs—Nettles—Peepsa—Simonbong, cultivation at—European fruits at Dorjiling—Plains of India.

A VERY favourite and interesting excursion from Dorjiling is to the cane bridge over the Great Rungeet river, 6000 feet below the station. To this an excellent road has been cut, by which the whole descent of six miles, as the crow flies, is easily performed on pony-back; the road distance being only eleven miles. The scenery is, of course, of a totally different description from that of Sinchul, or even of the foot of the hills, being that of a deep mountain-valley. I several times made this trip; on the excursion about to be described, and in which I was accompanied by Mr. Barnes, I followed the Great Rungeet to the Teesta, into which it flows.

In descending from Dorjiling, the zones of vegetation are well marked between 6000 and 7000 feet by—1. The oak, chesnut, and Magnolias, the main features from 7000 to 10,000 feet.—2. Immediately below 6,500 feet, the tree-fern appears (*Alsophila gigantea*, Wall.), a widely-distributed plant, common to the Himalaya, from Nepal eastward to the Malayan peninsula, Java, and Ceylon.—3. Of palms, a species of *Calamus*, and *Plectocomia*, the "Rhenoul" of the Lepchas. The latter, though not a very large plant, climbs lofty trees, and extends about 40 yards through the forest; 6,500 feet is the upper limit of palms in the Sikkim Himalaya, the Rhenoul alone attaining this elevation.*—4. The fourth striking feature is a wild plantain, which ascends to nearly the same elevation ("Lukhlo," Lepcha). This is replaced by another, and rather larger species, at lower elevations; both ripen austere and small fruits, which are full of seeds, and quite uneatable; that commonly grown in Sikkim is an introduced stock (nor have the wild species ever been cultivated); it is very large, but poor in flavour, and does not bear seeds. The zones of these conspicuous plants are very clearly defined, and especially if the traveller, standing on one of the innumerable spurs which project from the Dorjiling ridge, cast his eyes up the gorges of green on either hand.

At 1000 feet below Dorjiling a fine wooded spur projects, called Leebong. This beautiful spot is fully ten degrees warmer than Mr. Hodgson's house, and enjoys considerably more sunshine; peaches and English fruit-trees flourish extremely well, but do not ripen fruit. The tea-plant

Four other *Calami* range between 1000 and 6000 feet on the outer hills, some of them being found forty miles distant from the plains of India. The other palms of Sikkim are, "Simong" (*Caryota urens*); it is rare, and ascends to nearly 5000 feet. *Phœnix* (probably *P acaulis*, Buch.), a small, stemless species, which grows on the driest soil in the deep valleys; it is the "Schaap" of the Lepchas, who eat the young seeds, and use the feathery fronds as screens in hunting. *Wallichia oblongifolia*, the "Ooh" of the Lepchas, who make no use of it; Dr. Campbell and myself, however, found that it is an admirable fodder for horses, who prefer it to any other green food to be had in these mountains. *Areca gracilis* and *Licuala peltata* are the only other palms in Sikkim; but *Cycas pectinata*, with the India-rubber fig, occurs in the deepest and hottest valleys—the western limit of both these interesting plants. Of *Pandanus* there is a graceful species at elevations of 1000 to 4000 feet ("Borr," Lepcha).

succeeds here admirably, and might be cultivated to great profit, and be of advantage in furthering a trade with Tibet. It has been tried on a large scale by Dr. Campbell at his residence (alt. 7000 feet), but the frosts and snow of that height injure it, as do the hailstorms in spring.

Below Leebong is the village of Ging, surrounded by steeps, cultivated with maize, rice, and millet. It is rendered very picturesque by a long row of tall poles, each bearing a narrow, vertically elongated banner, covered with Boodhist inscriptions, and surmounted by coronet-like ornaments, or spear-heads, rudely cut out of wood, or formed of basket-work, and adorned with cotton fringe. Ging is peopled by Bhotan emigrants, and when one dies, if his relations can afford to pay for them, two additional poles and flags are set up by the Lamas in honour of his memory, and that of Sunga, the third member of the Boodhist Trinity.

Below this the *Gordonia* commences, with *Cedrela toona*, and various tropical genera, such as abound near Punkabarec. The heat and hardness of the rocks cause the streams to dry up on these abrupt hills, especially on the eastern slope, and the water is therefore conveyed along the sides of the path, in conduits ingeniously made of bamboo, either split in half, or, what is better, whole, except at the septum, which is removed through a lateral hole. The oak and chesnut of this level (3000 feet), are both different from those which grow above, as are the brambles. The *Arums* are replaced by *Caladiums*. Tree-ferns cease below 4000 feet, and the large bamboo abounds.

At about 2000 feet, and ten miles distant from Dorjiling, we arrived at a low, long spur, dipping down to the bed of the Rungeet, at its junction with the Rungmo. This is close to the boundary of the British ground, and there is a guard-house, and a sepoy or two at it; here we halted. It took the Lepchas about twenty minutes to construct a table and two bedsteads within our tent; each was made of four forked sticks, stuck in the ground, supporting as many side-pieces, across which were laid flat split pieces of bamboo, bound tightly together by strips of rattan palm-stem. The beds were afterwards softened by many layers of bamboo-leaf, and if not very downy, they were dry, and as firm as if put together with screws and joints.

This spur rises out of a deep valley, quite surrounded by lofty mountains; it is narrow, and covered with red clay, which the natives chew as a cure for goître. North, it looks down into a gully, at the bottom of which the Rungeet's foamy stream winds through a dense forest. In the opposite direction, the Rungmo comes tearing down from the top of Sinchul, 7000 feet above; and though its roar is heard, and its course is visible throughout its length, the stream itself is nowhere seen, so deep does it cut its channel. Except on this, and a few similarly hard rocky hills around, the vegetation is a mass of wood and jungle. At this spot it is rather scanty and dry, with abundance of the *Pinus longifolia* and Sal. The dwarf date-palm (*Phœnix acaulis*) also, was very abundant.

The descent to the river was exceedingly steep, the banks presenting an impenetrable jungle. The pines on the arid crests of the hills around formed a remarkable feature: they grow like the Scotch fir, the tall, red trunks springing from the steep and dry slopes. But little resin exudes from the stem, which, like that of most pines, is singularly free from lichens and mosses; its wood is excellent, and the charcoal of the burnt leaves is used as a pigment. Being confined to dry soil, this pine is local in Sikkim, and the elevation it attains here is not above 3000 feet. In Bhotan, where there is more dry country, its range is about the same, and in the north-west Himalaya, from 2,500 to 7000 feet.

The Lepcha never inhabits one spot for more than three successive years, after which an increased rent is demanded by the Rajah. He therefore *squats* in any place which he can render profitable for that period, and then moves to another. His first operation, after selecting a site, is to burn the jungle; then he clears away the trees, and cultivates between the stumps. At this season, firing the jungle is a frequent practice, and the effect by night is exceedingly fine; a forest, so dry and full of bamboo, and extending over such steep hills, affording grand blazing spectacles. Heavy clouds canopy the mountains above, and, stretching across the valleys, shut out the firmament; the air is a dead calm, as usual in these deep gorges, and the fires, invisible by day, are seen raging all around, appearing to an inexperienced eye in all but dangerous proximity. The voices of birds and insects being hushed, nothing is audible but the harsh roar of the rivers, and occasionally, rising far above it, that of the forest fires. At night we were literally surrounded by them; some smouldering, like the shale-heaps at a colliery, others fitfully bursting forth, whilst others again stalked along with a steadily increasing and enlarging flame, shooting out great tongues of fire, which spared nothing as they advanced with irresistible might. Their triumph is in reaching a great bamboo clump, when the noise of the flames drowns that of the torrents, and as the great stem-joints burst, from the expansion of the confined air, the report is as that of a salvo from a park of artillery. At Dorjiling the blaze is visible, and the deadened reports of the bamboos bursting is heard throughout the night; but in the valley, and within a mile of the scene of destruction, the effect is the most grand, being heightened by the glare reflected from the masses of mist which hover above.

On the following morning we pursued a path to the bed of the river; passing a rude Booddhist monument, a pile of slate-rocks, with an attempt at the mystical hemisphere at top. A few flags or banners, and slabs of slate, were inscribed with "Om Mani Padmi om." Placed on a jutting angle of the spur, backed with the pine-clad hills, and flanked by a torrent on either hand, the spot was wild and picturesque; and I could not but gaze with a feeling of deep interest on these emblems of a religion which perhaps numbers more votaries than any other on the face of the globe. Booddhism in some form is the predominating creed, from Siberia and Kamschatka to Ceylon, from the Caspian steppes to Japan, throughout China, Burmah, Ava, and a part of the Malayan Archipelago. Its associations enter into every book of travels over these vast regions, with Booddha, Dhurma, Sunga, Jos, Fo, and praying-wheels. The mind is arrested by the names, the imagination captivated by the symbols; and though I could not worship in the grove, it was impossible to deny to the inscribed stones such a tribute as is commanded by the first glimpse of objects which have long been familiar to our minds, but not previously offered to our senses. My head Lepcha went further: to a due observance of demon-worship he united a deep reverence for the Lamas, and he venerated their symbols rather as theirs than as those of their religion. He walked round the pile of stones three times from left to right repeating his "Om Mani," &c., then stood before it with his head hung down and his long queue streaming behind, and concluded by a votive offering of three pine-cones. When done, he looked round at me, nodded, smirked, elevated the angles of his little turned-up eyes, and seemed to think we were safe from all perils in the valleys yet to be explored.

In the gorge of the Rungeet the heat was intolerable, though the thermometer did not rise above 95°. The

PINES (PINUS LONGIFOLIA), RUNGEET VALLEY.

mountains leave but a narrow gorge between them, here and there bordered by a belt of strong soil, supporting a towering crop of long cane-like grasses and tall trees. The troubled river, about eighty yards across, rages along over a gravelly bed. Crossing the Rungmo, where it falls into the Rungeet, we came upon a group of natives drinking fermented Murwa liquor, under a rock; I had a good deal of difficulty in getting my people past, and more in inducing one of the topers to take the place of a Ghorka (Nepalese) of our party who was ill with fever. Soon afterwards, at a most wild and beautiful spot, I saw, for the first time, one of the most characteristic of Himalayan objects of art, *a cane bridge*. All the spurs, round the bases of which the river flowed, were steep and rocky, their flanks clothed with the richest tropical forest, their crests tipped with pines. On the river's edge, the Banana, *Pandanus*, and *Bauhinia*, were frequent, and Figs prevailed. One of the latter (of an exceedingly beautiful species) projected over the stream, growing out of a mass of rock, its roots interlaced and grasping at every available support, while its branches, loaded with deep glossy foliage, hung over the water. This tree formed one pier for the canes; that on the opposite bank was constructed of strong piles, propped with large stones; and between them swung the

bridge,* about eighty yards long, ever rocking over the torrent (forty feet below). The lightness and extreme simplicity of its structure were very remarkable. Two parallel canes, on the same horizontal plane, were stretched across the stream; from them others hung in loops, and along

* A sketch of one of these bridges will be found in Vol. ii.

the loops were laid one or two bamboo stems for flooring; cross pieces below this flooring, hung from the two upper canes, which they thus served to keep apart. The traveller grasps one of the canes in either hand, and walks along the loose bamboos laid on the swinging loops: the motion is great, and the rattling of the loose dry bamboos is neither a musical sound, nor one calculated to inspire confidence; the whole structure seeming as if about to break down. With shoes it is not easy to walk; and even with bare feet it is often difficult, there being frequently but one bamboo, which, if the fastening is loose, tilts up, leaving the pedestrian suspended over the torrent by the slender canes. When properly and strongly made, with good fastenings, and a floor of bamboos laid *transversely*, these bridges are easy to cross. The canes are procured from a species of *Calamus;* they are as thick as the finger, and twenty or thirty yards long, knotted together; and the other pieces are fastened to them by strips of the same plant. A Lepcha, carrying one hundred and forty pounds on his back, crosses without hesitation, slowly but steadily, and with perfect confidence.

A deep broad pool below the bridge was made available for a ferry: the boat was a triangular raft of bamboo stems, with a stage on the top, and it was secured on the opposite side of the stream, having a cane reaching across to that on which we were. A stout Lepcha leapt into the boiling flood, and boldly swam across, holding on by the cane, without which he would have been carried away. He unfastened the raft, and we drew it over by the cane, and, seated on the stage, up to our knees in water, we were pulled across; the raft bobbing up and down over the rippling stream.

We were beyond British ground, on the opposite bank,

where any one guiding Europeans is threatened with punishment: we had expected a guide to follow us, but his non-appearance caused us to delay for some hours; four roads, or rather forest paths, meeting here, all of which were difficult to find. After a while, part of a marriage-procession came up, headed by the bridegroom, a handsome young Lepcha, leading a cow for the marriage feast; and after talking to him a little, he volunteered to show us the path. On the flats by the stream grew the Sago palm (*Cycas pectinata*), with a stem ten feet high, and a beautiful crown of foliage; the contrast between this and the Scotch-looking pine (both growing with oaks and palms) was curious. Much of the forest had been burnt, and we traversed large blackened patches, where the heat was intense, and increased by the burning trunks of prostrate trees, which smoulder for months, and leave a heap of white ashes. The larger timber being hollow in the centre, a current of air is produced, which causes the interior to burn rapidly, till the sides fall in, and all is consumed. I was often startled, when walking in the forest, by the hot blast proceeding from such, which I had approached without a suspicion of their being other than cold dead trunks.

Leaving the forest, the path led along the river bank, and over the great masses of rock which strewed its course. The beautiful India-rubber fig was common, as was *Bassia butyracea*, the "Yel Pote" of the Lepchas, from the seeds of which they express a concrete oil, which is received and hardens in bamboo vessels. On the forest-skirts, *Hoya*, parasitical *Orchideæ*, and Ferns, abounded; the Chaulmoogra, whose fruit is used to intoxicate fish, was very common; as was an immense mulberry tree, that yields a milky juice and produces a long green sweet fruit. Large fish, chiefly

Cyprinoid, were abundant in the beautifully clear water of the river. But by far the most striking feature consisted in the amazing quantity of superb butterflies, large tropical swallow-tails, black, with scarlet or yellow eyes on their wings. They were seen everywhere, sailing majestically through the still hot air, or fluttering from one scorching rock to another, and especially loving to settle on the damp sand of the river-edge; where they sat by thousands, with erect wings, balancing themselves with a rocking motion, as their heavy sails inclined them to one side or the other; resembling a crowded fleet of yachts on a calm day. Such an entomological display cannot be surpassed. *Cicindelæ* were very numerous, and incredibly active, as were *Grylli;* and the great *Cicadeæ* were everywhere lighting on the ground, when they uttered a short sharp creaking sound, and anon disappeared, as if by magic. Beautiful whip-snakes were gleaming in the sun: they hold on by a few coils of the tail round a twig, the greater part of their body stretched out horizontally, occasionally retracting, and darting an unerring aim at some insect. The narrowness of the gorge, and the excessive steepness of the bounding hills, prevented any view, except of the opposite mountain face, which was one dense forest, in which the wild Banana was conspicuous.

Towards evening we arrived at another cane-bridge, still more dilapidated than the former, but similar in structure. For a few hundred yards before reaching it, we lost the path, and followed the precipitous face of slate-rocks overhanging the stream, which dashed with great violence below. Though we could not walk comfortably, even with our shoes off, the Lepchas, bearing their enormous loads, proceeded with perfect indifference.

Anxious to avoid sleeping at the bottom of the valley,

we crawled, very much fatigued, through burnt dry forest, up a very sharp ridge, so narrow that the tent sate astride on it, the ropes being fastened to the tops of small trees on either slope. The ground swarmed with black ants, which got into our tea, sugar, &c., while it was so covered with charcoal, that we were soon begrimed. Our Lepchas preferred remaining on the river-bank, whence they had to bring up water to us, in great bamboo "chungis," as they are called. The great dryness of this face is owing to its southern exposure: the opposite mountains, equally high and steep, being clothed in a rich green forest.

At nine the next morning, the temperature was 78°, but a fine cool easterly wind blew. Descending to the bed of the river, the temperature was 84°. The difference in humidity of the two stations (with about 300 feet difference in height) was more remarkable; at the upper, the wet bulb thermometer was 67½°, and consequently the saturation point, 0·713; at the lower, the wet bulb was 68°, and saturation, 0·599. The temperature of the river was, at all hours of the preceding day, and this morning, 67½°.*

Our course down the river was by so rugged a path, that, giddy and footsore with leaping from rock to rock, we at last attempted the jungle, but it proved utterly impervious. On turning a bend of the stream, the mountains of Bhotan suddenly presented themselves, with the Teesta flowing at their base; and we emerged at the angle formed by the junction of the Rungeet, which we

* At this hour, the probable temperature at Dorjiling (6000 feet above this) would be 56°, with a temperature of wet bulb 55°, and the atmosphere loaded with vapour. At Calcutta, again, the temperature was at the observatory 91·3°, wet bulb, 81·8°, and saturation = 0·737. The dryness of the air, in the damper-looking and luxuriant river-bed, was owing to the heated rocks of its channel; while the humidity of the atmosphere over the drier-looking hill where we encamped, was due to the moisture of the wind then blowing.

had followed from the west, of the Teesta, coming from the north, and of their united streams flowing south.

We were not long before enjoying the water, when I was surprised to find that of the Teesta singularly cold; its temperature being 7° below that of the Rungeet.* At the salient angle (a rocky peninsula) of their junction, we could almost place one foot in the cold stream and the other in the warmer. There is a no less marked difference in the colour of the two rivers; the Teesta being sea-green and muddy, the Great Rungeet dark green and very clear; and the waters, like those of the Arve and Rhone at Geneva, preserve their colours for some hundred yards; the line separating the two being most distinctly drawn. The Teesta, or main stream, is much the broadest (about 80 or 100 yards wide at this season), the most rapid and deep. The rocks which skirt its bank were covered with a silt or mud deposit, which I nowhere observed along the Great Rungeet, and which, as well as its colour and coldness, was owing to the vast number of then melting glaciers drained by this river. The Rungeet, on the other hand, though it rises amongst the glaciers of Kinchinjunga and its sister peaks, is chiefly supplied by the rainfall of the outer ranges of Sinchul and Singalelah, and hence its waters are clear, except during the height of the rains.

From this place we returned to Dorjiling, arriving on the afternoon of the following day.

The most interesting trip to be made from Dorjiling, is that to the summit of Tonglo, a mountain on the Singalelah

* This is, no doubt, due partly to the Teesta flowing south, and thus having less of the sun, and partly to its draining snowy mountains throughout a much longer portion of its course. The temperature of the one was 67½°, and that of the other 60½°.

range, 10,079 feet high, due west of the station, and twelve miles in a straight line, but fully thirty by the path.*

Leaving the station by a native path, the latter plunges at once into a forest, and descends very rapidly, occasionally emerging on cleared spurs, where are fine crops of various millets, with much maize and rice. Of the latter grain as many as eight or ten varieties are cultivated, but seldom irrigated, which, owing to the dampness of the climate, is not necessary: the produce is often eighty-fold, but the grain is large, coarse, reddish, and rather gelatinous when boiled. After burning the timber, the top soil is very fertile for several seasons, abounding in humus, below which is a stratum of stiff clay, often of great thickness, produced by the disintegration of the rocks; † the clay makes excellent bricks, and often contains nearly 30 per cent. of alumina.

At about 4000 feet the great bamboo ("Pao" Lepcha) abounds; it flowers every year, which is not the case with all others of this genus, most of which flower profusely over large tracts of country, once in a great many years, and then die away; their place being supplied by seedlings, which grow with immense rapidity. This well-known fact is not due, as some suppose, to the life of the species being of such a duration, but to favourable circumstances in the season. The Pao attains a height of 40 to 60 feet, and the culms average in thickness the human thigh; it is used for large water-vessels, and its leaves form admirable thatch, in universal use for European houses at Dorjiling. Besides this, the Lepchas are acquainted with nearly a dozen kinds of bamboo; these occur at various elevations below 12,000

* A full account of the botanical features noticed on this excursion (which I made in May, 1848, with Mr. Barnes) has appeared in the "London Journal of Botany," and the "Horticultural Society's Journal," and I shall, therefore, recapitulate its leading incidents only.

† An analysis of the soil will be found in the Appendix.

feet, forming, even in the pine-woods, and above their zone, in the skirts of the *Rhododendron* scrub, a small and sometimes almost impervious jungle. In an economical point of view they may be classed as those which split readily, and

LEPCHA WATER-CARRIER WITH A BAMBOO CHUNGI.

those which do not. The young shoots of several are eaten, and the seeds of one are made into a fermented drink, and into bread in times of scarcity; but it would take many pages to describe the numerous purposes to which the various species are put.

Gordonia is their most common tree (*G. Wallichii*), much prized for ploughshares and other purposes requiring a hard wood: it is the "Sing-brang-kun" of the Lepchas, and ascends to 4000 feet. Oaks at this elevation occur as solitary trees, of species different from those of Dorjiling. There are three or four with a cup-shaped involucre, and three with spinous involucres enclosing an eatable sweet nut; these generally grow on a dry clayey soil.

Some low steep spurs were well cultivated, though the angle of the field was upwards of 25°; the crops, chiefly maize, were just sprouting. This plant is occasionally hermaphrodite in Sikkim, the flowers forming a large drooping panicle and ripening small grains; it is, however, a rare occurrence, and the specimens are highly valued by the people.

The general prevalence of figs,* and their allies, the nettles,† is a remarkable feature in the botany of the Sikkim Himalaya, up to nearly 10,000 feet. Of the former there were here five species, some bearing eatable and very palatable fruit of enormous size, others with the fruit small and borne on prostrate, leafless branches, which spring from the root and creep along the ground.

A troublesome, dipterous insect (the "Peepsa," a species of *Siamulium*) swarms on the banks of the streams; it is very small and black, floating like a speck before the eye; its bite leaves a spot of extravasated blood under the cuticle, very irritating if not opened.

Crossing the Little Rungeet river, we camped on the base of Tonglo The night was calm and clear, with faint

* One species of this very tropical genus ascends almost to 9000 feet on the outer ranges of Sikkim.

† Of two of these cloth is made, and of a third, cordage. The tops of two are eaten, as are several species of *Procris*. The "Poa" belongs to this order, yielding that kind of grass cloth fibre, now abundantly imported into England from the Malay Islands, and used extensively for shirting.

cirrus, but no dew. A thermometer sunk two feet in rich vegetable mould stood at 78° two hours after it was lowered, and the same on the following morning. This probably indicates the mean temperature of the month at that spot, where, however, the dark colour of the exposed loose soil must raise the temperature considerably.

May 20th.—The temperature at sunrise was 67°; the morning bright, and clear over head, but the mountains looked threatening. Dorjiling, perched on a ridge 5000 feet above us, had a singular appearance. We ascended the Simonbong spur of Tonglo, so called from a small village and Lama temple of that name on its summit; where we arrived at noon, and passing some chaits* gained the Lama's residence.

Two species of bamboo, the "Payong" and "Praong" of the Lepchas, here replace the Pao of the lower regions. The former was flowering abundantly, the whole of the culms (which were 20 feet high) being a diffuse panicle of inflorescence. The "Praong" bears a round head of flowers at the ends of the leafy branches. Wild strawberry, violet, geranium, &c., announced our approach to the temperate zone. Around the temple were potato crops and peach-trees, rice, millet, yam, brinjal (egg-apple), fennel, hemp (for smoking its narcotic leaves), and cummin, &c. The potato thrives extremely well as a summer crop, at 7000 feet, in Sikkim, though I think the root (from the Dorjiling stock) cultivated as a winter crop in the plains, is superior both in size and flavour. Peaches never ripen in this part of Sikkim, apparently from the want of sun; the tree

* The chait of Sikkim, borrowed from Tibet, is a square pedestal, surmounted with a hemisphere, the convex end downwards, and on it is placed a cone, with a crescent on the top. These are erected as tombs to Lamas, and as monuments to illustrious persons, and are venerated accordingly, the people always passing them from left to right, often repeating the invocation, "Om Mani Padmi om."

grows well at from 3000 to 7000 feet elevation, and flowers abundantly; the fruit making the nearest approach to maturity (according to the elevation) from July to October. At Dorjiling it follows the English seasons, flowering in March and fruiting in September, when the scarce reddened and still hard fruit falls from the tree. In the plains of India, both this and the plum ripen in May, but the fruits are very acid.

It is curious that throughout this temperate region, there is hardly an eatable fruit except the native walnut, and some brambles, of which the "yellow" and "ground raspberry" are the best, some insipid figs, and a very austere crab-apple. The European apple will scarcely ripen,* and the pear not at all. Currants and gooseberries show no disposition to thrive, and strawberries are the only fruits that ripen at all, which they do in the greatest abundance. Vines, figs, pomegranates, plums, apricots, &c., will not succeed even as trees. European vegetables again grow, and thrive remarkably well throughout the summer of Dorjiling, and the produce is very fair, sweet and good, but inferior in flavour to the English.

Of tropical fruits cultivated below 4000 feet, oranges and indifferent bananas alone are frequent, with lemons of various kinds. The season for these is, however, very short; though that of the plantain might with care be prolonged; oranges abound in winter, and are excellent, but neither so large nor free of white pulp as those of the Khasia hills, the West Indies, or the west coast of Africa. Mangos are brought from the plains, for though wild in Sikkim, the cultivated kinds do not thrive; I have

* This fruit, and several others, ripen at Katmandoo, in Nepal (alt. 4000 feet), which place enjoys more sunshine than Sikkim. I have, however, received very different accounts of the produce, which, on the whole, appears to be inferior.

seen the pine-apple plant, but I never met with good fruit on it.

A singular and almost total absence of the light, and of the direct rays of the sun in the ripening season, is the cause of this dearth of fruit. Both the farmer and orchard gardener in England know full well the value of a bright sky as well as of a warm autumnal atmosphere. Without this corn does not ripen, and fruit-trees are blighted. The winter of the plains of India being more analogous in its distribution of moisture and heat to a European summer, such fruits as the peach, vine, and even plum, fig, strawberry, &c., may be brought to bear well in March, April, and May, if they are only carefully tended through the previous hot and damp season, which is, in respect to the functions of flowering and fruiting, their winter.

Hence it appears that, though some English fruits will turn the winter solstice of Bengal (November to May) into summer, and then flower and fruit, neither these nor others will thrive in the summer of 7000 feet on the Sikkim Himalaya, (though its temperature so nearly approaches that of England,) on account of its rain and fogs. Further, they are often exposed to a winter's cold equal to the average of that of London, the snow lying for a week on the ground, and the thermometer descending to 25°. It is true that in no case is the extreme of cold so great here as in England, but it is sufficient to check vegetation, and to prevent fruit-trees from flowering till they are fruiting in the plains. There is in this respect a great difference between the climate of the central and eastern and western Himalaya, at equal elevations. In the western (Kumaon, &c.) the winters are colder than in Sikkim—the summers warmer and less humid. The rainy season is shorter, and the sun shines so much more frequently between the heavy showers, that the apple and other fruits are brought to a much better state. It is true that the rain-gauge may show as great a fall there, but this is no measure of the humidity of the atmosphere, and still less so of the amount of the sun's direct light and heat intercepted by aqueous vapour, for it takes no account of the quantity of moisture suspended in the air, nor of the depositions from fogs, which are far more fatal to the perfecting of fruits than the heaviest brief showers.

The Indian climate, which is marked by one season of excessive humidity and the other of excessive drought, can never be favourable to the production either of good European or tropical fruits. Hence there is not one of the latter peculiar to the country, and perhaps but one which arrives at full perfection; namely, the mango. The plantains, oranges, and pine-apples are less abundant, of inferior kinds, and remain a shorter season in perfection than they do in South America, the West Indies, or Western Africa.

LEPCHA AMULET.

CHAPTER VII.

Continue the ascent of Tonglo—Trees—Lepcha construction of hut—Simsibong—Climbing-trees—Frogs—Magnolias, &c.—Ticks—Leeches—Cattle, murrain amongst—Summit of Tonglo—Rhododendrons—Skimmia—Yew—Rose—Aconite—Bikh poison—English genera of plants—Ascent of tropical orders—Comparison with south temperate zone—Heavy rain—Temperature, &c.—Descent—Simonbong temple—Furniture therein—Praying-cylinder—Thigh-bone trumpet—Morning orisons—Present of Murwa beer, &c.

CONTINUING the ascent of Tonglo, we left cultivation and the poor groves of peaches at 4000 to 5000 feet (and this on the eastern exposure, which is by far the sunniest), the average height which agriculture reaches in Sikkim.

Above Simonbong, the path up Tonglo is little frequented: it is one of the many routes between Nepal and Sikkim, which cross the Singalelah spur of Kinchinjunga at various elevations between 7000 and 15,000 feet. As usual, the track runs along ridges, wherever these are to be found, very steep, and narrow at the top, through deep humid forests of oaks and Magnolias, many laurels, both *Tetranthera* and *Cinnamomum*, one species of the latter ascending to 8,500 feet, and one of *Tetranthera* to 9000. Chesnut and walnut here appeared, with some leguminous trees, which however did not ascend to 6000 feet. Scarlet flowers of *Vaccinium serpens*, an epiphytical species, were strewed about, and the great blossoms of *Rhododendron Dalhousiæ* and of a Magnolia (*Talauma Hodgsoni*) lay together on the ground. The latter forms a large tree, with very dense foliage, and deep shining green leaves, a foot to eighteen inches long. Most of its flowers drop unexpanded from the tree, and diffuse a very aromatic smell; they are nearly as large as the fist, the outer petals purple, the inner pure white.

Heavy rain came on at 3 P.M., obliging us to take insufficient shelter under the trees, and finally to seek the nearest camping-ground. For this purpose we ascended to a spring, called Simsibong, at an elevation of 6000 feet. The narrowness of the ridge prevented our pitching the tent, small as it was; but the Lepchas rapidly constructed a house, and thatched it with bamboo and the broad leaves of the wild plantain. A table was then raised in the middle, of four posts and as many cross pieces of wood, lashed with strips of bamboo. Across these, pieces of bamboo were laid, ingeniously flattened, by selecting cylinders, crimping them all round, and then slitting each down one side, so that it opens into a flat slab. Similar but longer and lower erections, one on each side the table, formed bed or chair; and in one hour, half a dozen men, with only long knives and active hands, had provided us with a tolerably water-tight furnished house. A thick flooring of bamboo leaves kept the feet dry, and a screen of that and other foliage all round rendered the habitation tolerably warm.

At this elevation we found great scandent trees twisting around the trunks of others, and strangling them: the latter gradually decay, leaving the sheath of climbers as one of the most remarkable vegetable phenomena of these mountains. These climbers belong to several orders, and may be roughly classified in two groups.—(1.) Those whose stems merely twine, and by constricting certain parts of their support, induce death.—(2.) Those which form a net-work round the trunk, by the coalescence of their lateral branches and aerial roots, &c.: these wholly envelop and often conceal the tree they enclose, whose branches

CLASPING ROOTS OF WIGHTIA.

appear rising far above those of its destroyer. To the first of these groups belong many natural orders, of which the most prominent are—*Leguminosæ*, ivies, hydrangea, vines, *Pothos*, &c. The inosculating ones are almost all figs and

Wightia: the latter is the most remarkable, and I add a cut of its grasping roots, sketched at our encampment.

Except for the occasional hooting of an owl, the night was profoundly still during several hours after dark—the cicadas at this season not ascending so high on the mountain. A dense mist shrouded every thing, and the rain pattered on the leaves of our hut. At midnight a tree-frog ("Simook," Lepcha) broke the silence with his curious metallic clack, and others quickly joined the chorus, keeping up their strange music till morning. Like many Batrachians, this has a voice singularly unlike that of any other organised creature. The cries of beasts, birds, and insects are all explicable to our senses, and we can recognise most of them as belonging to such or such an order of animal; but the voices of many frogs are like nothing else, and allied species utter totally dissimilar noises. In some, as this, the sound is like the concussion of metals; in others, of the vibration of wires or cords; anything but the natural effects of lungs, larynx, and muscles.*

May 21.—Early this morning we proceeded upwards, our prospect more gloomy than ever. The path, which still lay up steep ridges, was very slippery, owing to the rain upon the clayey soil, and was only passable from the hold afforded by interlacing roots of trees. At 8000 feet, some enormous detached masses of micaceous gneiss rose abruptly from the ridge, they were covered with mosses and ferns, and from their summit, 7000 feet, a good view of the surrounding vegetation is obtained. The mass of the forest is formed of:—(1) Three species of oak, of which *Q. annulata?* with immense lamellated acorns, and leaves

* A very common Tasmanian species utters a sound that appears to ring in an underground vaulted chamber, beneath the feet.

sixteen inches long, is the tallest and the most abundant. —(2) Chesnut.—(3) *Laurineæ* of several species, all beautiful forest-trees, straight-boled, and umbrageous above. —(4) Magnolias.*—(5) Arborescent rhododendrons, which commence here with the *R. arboreum.* At 8000 and 9000 feet, a considerable change is found in the vegetation; the gigantic purple *Magnolia Campbellii* replacing the white; chesnut disappears, and several laurels: other kinds of maple are seen, with *Rhododendron argenteum,* and *Stauntonia,* a handsome climber, which has beautiful pendent clusters of lilac blossoms.

At 9000 feet we arrived on a long flat covered with lofty trees, chiefly purple magnolias, with a few oaks, great *Pyri* and two rhododendrons, thirty to forty feet high (*R. barbatum,* and *R. arboreum,* var. *roseum*): *Skimmia* and *Symplocos* were the common shrubs. A beautiful orchid with purple flowers (*Cœlogyne Wallichii*) grew on the trunks of all the great trees, attaining a higher elevation than most other epiphytical species, for I have seen it at 10,000 feet.

A large tick infests the small bamboo, and a more hateful insect I never encountered. The traveller cannot avoid these insects coming on his person (sometimes in great numbers) as he brushes through the forest; they get inside his dress, and insert the proboscis deeply without pain. Buried head and shoulders, and retained by a barbed lancet, the tick is only to be extracted by force, which is very painful. I

* Other trees were *Pyrus, Saurauja* (both an erect and climbing species), *Olea,* cherry, birch, alder, several maples, *Hydrangea,* one species of fig, holly, and several *Araliaceous* trees. Many species of *Magnoliaceæ* (including the genera *Magnolia, Michelia,* and *Talauma*) are found in Sikkim: *Magnolia Campbellii,* of 10,000 feet, is the most superb species known. In books on botanical geography, the magnolias are considered as most abounding in North America, east of the Rocky Mountains; but this is a great mistake, the Indian mountains and islands being the centre of this natural order.

have devised many tortures, mechanical and chemical, to induce these disgusting intruders to withdraw the proboscis, but in vain. Leeches* also swarm below 7000 feet; a small black species above 3000 feet, and a large yellow-brown solitary one below that elevation.

Our ascent to the summit was by the bed of a water-course, now a roaring torrent, from the heavy and incessant rain. A small *Anagallis* (like *tenella*), and a beautiful purple primrose, grew by its bank. The top of the mountain is another flat ridge, with depressions and broad pools. The number of additional species of plants found here was great, and all betokened a rapid approach to the alpine region of the Himalaya. In order of prevalence the trees were,—the scarlet *Rhododendron arboreum* and *barbatum*, as large bushy trees, both loaded with beautiful flowers and luxuriant foliage; *R. Falconeri*, in point of foliage the most superb of all the Himalayan species, with trunks thirty feet high, and branches bearing at their ends only leaves eighteen inches long: these are deep green above, and covered beneath with a rich brown down. Next in abundance to these were shrubs of *Skimmia Laureola*,† *Symplocos*, and Hydrangea; and there were still a few purple magnolias, very large *Pyri*, like mountain ash, and the common English

* I cannot but think that the extraordinary abundance of these *Anelides* in Sikkim may cause the death of many animals. Some marked murrains have followed very wet seasons, when the leeches appear in incredible numbers; and the disease in the cattle, described to me by the Lepchas as in the stomach, in no way differs from what leeches would produce. It is a well-known fact, that these creatures have lived for days in the fauces, nares, and stomachs of the human subject, causing dreadful sufferings, and death. I have seen the cattle feeding in places where the leeches so abounded, that fifty or sixty were frequently together on my ankles; and ponies are almost maddened by their biting the fetlocks.

† This plant has been lately introduced into English gardens, from the north-west Himalaya, and is greatly admired for its aromatic, evergreen foliage, and clusters of scarlet berries. It is a curious fact, that this plant never bears scarlet berries in Sikkim, apparently owing to the want of sun: the fruit ripens, but is of a greenish-red or purplish colour.

yew, eighteen feet in circumference, the red bark of which is used as a dye, and for staining the foreheads of Brahmins in Nepal. An erect white-flowered rose (*R. sericea*, the only species occurring in Southern Sikkim) was very abundant: its numerous inodorous flowers are pendent, apparent as a protection from the rain; and it is remarkable as being the only species having four petals instead of five.

A currant was common, always growing epiphytically on the trunks of large trees. Two or three species of Berberry, a cherry, Andromeda, *Daphne*, and maple, nearly complete, I think, the list of woody plants. Amongst the herbs were many of great interest, as a rhubarb, and *Aconitum palmatum*, which yields one of the celebrated "Bikh" poisons.* Of European genera I found *Thalictrum*, *Anemone*, *Fumaria*, violets, *Stellaria*, *Hypericum*, two geraniums, balsams, *Epilobium*, *Potentilla*, *Paris* and *Convallariæ*, one of the latter has verticillate leaves, and its root also called "bikh," is considered a very virulent poison.

Still, the absence or rarity at this elevation of several very large natural families,† which have numerous representatives at and much below the same level in the inner ranges, and on the outer of the Western Himalaya, indicate a certain peculiarity in Sikkim. On the other hand, certain tropical genera are more abundant in the temperate zone of the Sikkim mountains, and ascend much higher there than in the Western Himalaya: of this fact I have cited conspicuous examples in the palms, plantains, and tree-ferns. This

* "Bikh" is yielded by various *Aconita*. All the Sikkim kinds are called "gniong" by Lepchas and Bhoteeas, who do not distinguish them. The *A. Napellus* is abundant in the north-west Himalaya, and is perhaps as virulent a Bikh as any species.

† *Ranunculaceæ, Fumariæ, Cruciferæ, Alsineæ, Geranieæ, Leguminosæ, Potentilla, Epilobium, Crassulaceæ, Saxifrageæ, Umbelliferæ, Lonicera, Valerianeæ, Dipsaceæ,* various genera of *Compositæ, Campanulaceæ, Lobeliaceæ, Gentianeæ, Boragineæ, Srcophularineæ, Primulaceæ, Gramineæ.*

ascent and prevalence of tropical species is due to the humidity and equability of the climate in this temperate zone, and is, perhaps, the direct consequence of these conditions. An application of the same laws accounts for the extension of similar features far beyond the tropical limit in the Southern Ocean, where various natural orders, which do not cross the 30th and 40th parallels of N. latitude, are extended to the 55th of S. latitude, and found in Tasmania, New Zealand, the so-called Antarctic Islands south of that group, and at Cape Horn itself.

The rarity of Pines is perhaps the most curious feature in the botany of Tonglo, and on the outer ranges of Sikkim; for, between the level of 2,500 feet (the upper limit of *P. longifolia*) and 10,000 feet (that of the *Taxus*), there is no coniferous tree whatever in Southern Sikkim.

We encamped amongst Rhododendrons, on a spongy soil of black vegetable matter, so oozy, that it was difficult to keep the feet dry. The rain poured in torrents all the evening, and with the calm, and the wetness of the wood, prevented our enjoying a fire. Except a transient view into Nepal, a few miles west of us, nothing was to be seen, the whole mountain being wrapped in dense masses of vapour. Gusts of wind, not felt in the forest, whistled through the gnarled and naked tree-tops; and though the temperature was 50°, this wind produced cold to the feelings. Our poor Lepchas were miserably off, but always happy: under four posts and a bamboo-leaf thatch, with no covering but a single thin cotton garment, they crouched on the sodden turf, joking with the Hindoos of our party, who, though supplied with good clothing and shelter, were doleful companions.

I made a shed for my instruments under a tree; Mr. Barnes, ever active and ready, floored the tent with logs of wood, and I laid a "corduroy road" of the same to my little observatory.

During the night the rain did not abate; and the tent-roof leaked in such torrents, that we had to throw pieces of wax-cloth over our shoulders as we lay in bed. There was no improvement whatever in the weather on the following morning. Two of the Hindoos had crawled into the tent during the night, attacked with fever and ague.* The tent being too sodden to be carried, we had to remain where we were, and with abundance of novelty in the botany around, I found no difficulty in getting through the day. Observing the track of sheep, we sent two Lepchas to follow them, who returned at night from some miles west in Nepal, bringing two. The shepherds were Geroongs of Nepal, who were grazing their flocks on a grassy mountain top, from which the woods had been cleared, probably by fire. The mutton was a great boon to the Lepchas, but the Hindoos would not touch it, and several more sickening during the day, we had the tent most uncomfortably full.

During the whole of the 22nd, from 7 A.M. to 11 P.M., the thermometer never varied 6½ degrees, ranging from 47½ in the morning to 54°, its maximum, at 1 P.M., and 50¾ at night. At seven the following morning it was the same. One, sunk two feet six inches in mould and clay, stood constantly at 50¾. The dew-point was always below the temperature, at which I was surprised, for more drenching weather could not well be. The mean dew-point was 50¼, and consequent humidity, 0·973.

* It is a remarkable fact, that both the natives of the plains, under many circumstances, and the Lepchas when suffering from protracted cold and wet, take fever and ague in sharp attacks. The disease is wholly unknown amongst Europeans residing above 4000 feet, similar exposure in whom brings on rheumatism and cold.

These observations, and those of the barometer, were taken 60 feet below the summit, to which I moved the instruments on the morning of the 23rd. At a much more exposed spot the results would no doubt have been different, for a thermometer, there sunk to the same depth as that below, stood at 49¾ (or one degree colder than 60 feet lower down). My barometrical observations, taken simultaneously with those of Calcutta, give the height of Tonglo, 10,078·3 feet; Colonel Waugh's, by trigonometry, 10,079·4 feet,—a remarkable and unusual coincidence.

May 23.—We spent a few hours of alternate fog and sunshine on the top of the mountain, vainly hoping for the most modest view; our inability to obtain it was extremely disappointing, for the mountain commands a superb prospect, which I enjoyed fully in the following November, from a spot a few miles further west. The air, which was always foggy, was alternately cooled and heated, as it blew over the trees, or the open space we occupied; sometimes varying 5° and 6° in a quarter of an hour.

Having partially dried the tent in the wind, we commenced the descent, which owing to the late torrents of rain, was most fatiguing and slippery; it again commenced to drizzle at noon, nor was it till we had descended to 6000 feet that we emerged from the region of clouds. By dark we arrived at Simonbong, having descended 5000 feet, at the rate of 1000 feet an hour; and were kindly received by the Lama, who gave us his temple for the accommodation of the whole party. We were surprised at this, both because the Sikkim authorities had represented the Lamas as very averse to Europeans, and because he might well have hesitated before admitting a promiscuous horde of thirty people into a sacred building, where the little valuables on the altar, &c., were quite at our disposal. A better tribute could not well have been paid to the honesty of my Lepcha followers. Our host only begged us not to disturb his people, nor to allow the Hindoos of our party to smoke inside.

Simonbong is one of the smallest and poorest Gumpas, or temples, in Sikkim: unlike the better class, it is built of

SIMONBONG TEMPLE.

wood only. It consisted of one large room, with small sliding shutter windows, raised on a stone foundation, and roofed with shingles of wood; opposite the door a wooden altar was placed, rudely chequered with black, white, and red; to the right and left were shelves, with a few Tibetan books, wrapped in silk; a model of Symbonath temple in Nepal, a praying-cylinder,* and some implements for common purposes, bags of juniper, English wine-bottles and

* It consisted of a leathern cylinder placed upright in a frame; a projecting piece of iron strikes a little bell at each revolution, the revolution being caused by an elbowed axle and string. Within the cylinder are deposited written

glasses, with tufts of *Abies Webbiana*, rhododendron flowers, and peacock's feathers, besides various trifles, clay ornaments and offerings, and little Hindoo idols. On the altar were ranged seven little brass cups, full of water; a large conch shell, carved with the sacred lotus; a brass jug from Lhassa, of beautiful design, and a human thigh-bone, hollow, and perforated through both condyles.*

Facing the altar was a bench and a chair, and on one side a huge tambourine, with two curved iron drum-sticks. The bench was covered with bells, handsomely carved with

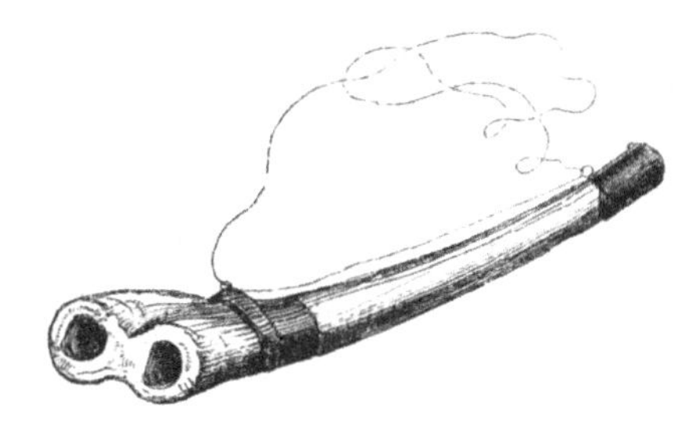

TRUMPET MADE OF A HUMAN THIGH-BONE.

idols, and censers with juniper-ashes; and on it lay the *dorge*, or double-headed thunderbolt, which the Lama holds in his hand during service. Of all these articles, the human thigh-bone is by much the most curious; it is very often that of a Lama, and is valuable in proportion to its length.† As, however, the Sikkim Lamas are burned, the relics are generally procured from Tibet, where the

prayers, and whoever pulls the string properly is considered to have repeated his prayers as often as the bell rings. Representations of these implements will be found in other parts of these volumes.

* To these are often added a double-headed rattle, or small drum, formed of two crowns of human skulls, cemented back to back; each face is then covered with parchment, and encloses some pebbles. Sometimes this instrument is provided with a handle.

† It is reported at Dorjiling, that one of the first Europeans buried at this station, being a tall man, was disinterred by the resurrectionist Bhoteeas for his *trumpet-bones.*

corpses are cut in pieces and thrown to the kites, or into the water.

Two boys usually reside in the temple, and their beds were given up to us, which being only rough planks laid on the floor, proved clean in one sense, but contrasted badly with the springy couch of bamboo the Lepcha makes, which renders carrying a mattress or aught but blankets superfluous.

May 24.—We were awakened at daylight by the discordant orisons of the Lama; these commenced by the boys beating the great tambourine, then blowing the conch-shells, and finally the trumpets and thigh-bone. Shortly the Lama entered, clad in scarlet, shorn and barefooted, wearing a small red silk mitre, a loose gown girt round the middle, and an under-garment of questionable colour, possibly once purple. He walked along, slowly muttering his prayers, to the end of the apartment, whence he took a brass bell and dorge, and, sitting down cross-legged, commenced matins, counting his beads, or ringing the bell, and uttering most dismal prayers. After various disposals of the cups, a larger bell was violently rung for some minutes, himself snapping his fingers and uttering most unearthly sounds. Finally, incense was brought, of charcoal with juniper-sprigs; it was swung about, and concluded the morning service to our great relief, for the noises were quite intolerable. Fervid as the devotions appeared, to judge by their intonation, I fear the Lama felt more curious about us than was proper under the circumstances; and when I tried to sketch him, his excitement knew no bounds; he fairly turned round on the settee, and, continuing his prayers and bell-accompaniment, appeared to be exorcising me, or some spirit within me.

After breakfast the Lama came to visit us, bringing rice,

a few vegetables, and a large bamboo-work bowl, thickly varnished with india-rubber, and waterproof, containing half-fermented millet. This mixture, called *Murwa,* is invariably offered to the traveller, either in the state of fermented grain, or more commonly in a bamboo jug, filled quite up with warm water; when the fluid, sucked through a reed, affords a refreshing drink. He gratefully accepted a few rupees and trifles which we had to spare.

Leaving Simonbong, we descended to the Little Rungeet, where the heat of the valley was very great; 80° at noon, and that of the stream 69°; the latter was an agreeable temperature for the coolies, who plunged, teeming with perspiration, into the water, catching fish with their hands. We reached Dorjiling late in the evening, again drenched with rain; our people, Hindoo and Lepcha, imprudently remaining for the night in the valley. Owing probably as much to the great exposure they had lately gone through, as to the sudden transition from a mean temperature of 50° in a bracing wind, to a hot close jungly valley at 75°, no less than seven were laid up with fever and ague.

Few excursions can afford a better idea of the general features and rich luxuriance of the Sikkim Himalaya than that to Tonglo. It is always interesting to roam with an aboriginal, and especially a mountain people, through their thinly inhabited valleys, over their grand mountains, and to dwell alone with them in their gloomy and forbidding forests, and no thinking man can do so without learning much, however slender be the means at his command for communion. A more interesting and attractive companion than the Lepcha I never lived with: cheerful, kind, and patient with a master to whom he is attached; rude but not savage, ignorant and yet intelligent; with the simple resource of a plain knife he makes his house and furnishes yours, with a speed, alacrity, and ingenuity that wile away that well-known long hour when the weary pilgrim frets for his couch. In all my dealings with these people, they proved scrupulously honest. Except for drunkenness and carelessness, I never had to complain of any of the merry troop; some of whom, bareheaded and barelegged, possessing little or nothing save a cotton garment and a long knife, followed me for many months on subsequent occasions, from the scorching plains to the everlasting snows. Ever foremost in the forest or on the bleak mountain, and ever ready to help, to carry, to encamp, collect, or cook, they cheer on the traveller by their unostentatious zeal in his service, and are spurs to his progress.

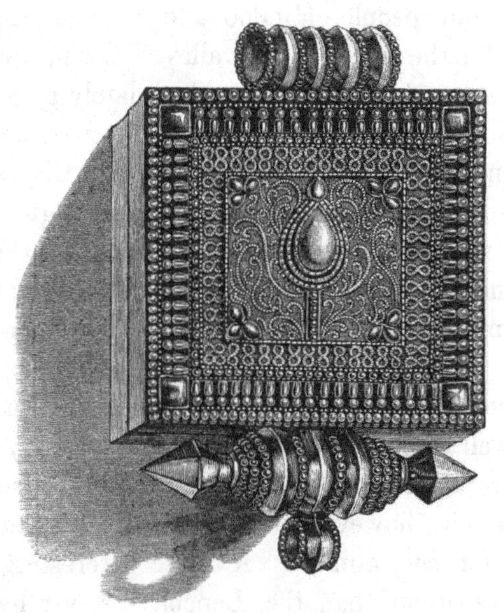

TIBETAN AMULET.

CHAPTER VIII.

Difficulty in procuring leave to enter Sikkim—Obtain permission to travel in East Nepal—Arrangements—Coolies—Stores—Servants—Personal equipment—Mode of travelling—Leave Dorjiling—Goong ridge—Behaviour of Bhotan coolies—Nepal frontier—Myong valley—Ilam—Sikkim massacre—Cultivation—Nettles—Camp at Nanki on Tonglo—Bhotan coolies run away—View of Chumulari—Nepal peaks to west—Sakkiazung—Buceros—Road to Wallanchoon—Oaks—Scarcity of water—Singular view of mountain-valleys—Encampment—My tent and its furniture—Evening occupations—Dunkotah—Crossridge of Sakkiazung—Yews—Silver-firs—View of Tambur valley—Pemmi river—Pebbly terraces—Geology—Holy springs—Enormous trees—Luculia gratissima—Khawa river, rocks of—Arrive at Tambur—Shingle and gravel terraces—Natives, indolence of—Canoe ferry—Votive offerings—Bad road—Temperature, &c.—Chingtam village, view from—Mywa river and Guola—House—Boulders—Chain-bridge—Meepo, arrival of—Fevers.

Owing to the unsatisfactory nature of our relations with the Sikkim authorities, to which I have elsewhere alluded, my endeavours to procure leave to penetrate further beyond the Dorjiling territory than Tonglo, were attended with some trouble and delay.

In the autumn of 1848, the Governor-General communicated with the Rajah, desiring him to grant me honourable and safe escort through his dominions; but this was at once met by a decided refusal, apparently admitting of no compromise. Pending further negociations, which Dr. Campbell felt sure would terminate satisfactorily, though perhaps too late for my purpose, he applied to the Nepal Rajah for permission for me to visit the Tibetan passes, west of Kinchinjunga; proposing in the meanwhile to arrange for my return through Sikkim. Through the kindness of Col. Thoresby, the Resident at that Court, and the influence of Jung Bahadoor, this request was promptly acceded to, and a guard of six Nepalese soldiers and two officers was sent to Dorjiling to conduct me to any part of the eastern districts of Nepal which I might select. I decided upon following up the Tambur, a branch of the Arun river, and exploring the two easternmost of the Nepalese passes into Tibet (Wallanchoon and Kanglachem), which would bring me as near to the central mass and loftiest part of the eastern flank of Kinchinjunga as possible.

For this expedition (which occupied three months), all the arrangements were undertaken for me by Dr. Campbell, who afforded me every facility which in his government position he could command, besides personally superintending the equipment and provisioning of my party. Taking horses or loaded animals of any kind was not expedient: the whole journey was to be performed on foot, and everything carried on men's backs. As we were to march through wholly unexplored countries, where food was only procurable at uncertain intervals, it was necessary to engage a large body of porters, some of whom should carry bags of rice for the coolies and themselves too. The difficulty of selecting these carriers, of whom thirty were required, was very great. The Lepchas, the best and most tractable, and over whom Dr. Campbell had the most direct influence, disliked employment out of Sikkim, especially in so warlike a country as Nepal: and they were besides thought unfit for the snowy regions. The Nepalese, of whom there were many residing as British subjects in Dorjiling, were mostly run-aways from their own country, and afraid of being claimed, should they return to it, by the lords of the soil. To employ Limboos, Moormis,

Hindoos, or other natives of low elevations, was out of the question; and no course appeared advisable but to engage some of the Bhotan run-aways domiciled in Dorjiling, who are accustomed to travel at all elevations, and fear nothing but a return to the country which they have abandoned as slaves, or as culprits: they are immensely powerful, and though intractable to the last degree, are generally glad to work and behave well for money. The choice, as will hereafter be seen, was unfortunate, though at the time unanimously approved.

My party mustered fifty-six persons. These consisted of myself, and one personal servant, a Portuguese half-caste, who undertook all offices, and spared me the usual train of Hindoo and Mahometan servants. My tent and equipments (for which I was greatly indebted to Mr. Hodgson), instruments, bed, box of clothes, books and papers, required a man for each. Seven more carried my papers for drying plants, and other scientific stores. The Nepalese guard had two coolies of their own. My interpreter, the coolie Sirdar (or headman), and my chief plant collector (a Lepcha), had a man each. Mr. Hodgson's bird and animal shooter, collector, and stuffer, with their ammunition and indispensables, had four more; there were besides, three Lepcha lads to climb trees and change the plant-papers, who had long been in my service in that capacity; and the party was completed by fourteen Bhotan coolies laden with food, consisting chiefly of rice with ghee, oil, capsicums, salt, and flour.

I carried myself a small barometer, a large knife and digger for plants, note-book, telescope, compass, and other instruments; whilst two or three Lepcha lads who accompanied me as satellites, carried a botanising box, thermometers, sextant and artificial horizon, measuring-tape, azimuth compass and stand, geological hammer, bottles and boxes for insects, sketch-book, &c., arranged in compartments of strong canvass bags The Nepal officer (of the rank of serjeant, I believe) always kept near me with one of his men, rendering innumerable little services. Other sepoys were distributed amongst the remainder of the party; one went ahead to prepare camping-ground, and one brought up the rear.

The course generally pursued by Himalayan travellers is to march early in the morning, and arrive at the camping-ground before or by noon, breakfasting before starting, or *en route*. I never followed this plan, because it sacrificed the mornings, which were otherwise profitably spent in collecting about camp; whereas, if I set off early, I was generally too tired with the day's march to employ in any active pursuit the rest of the daylight, which in November only lasted till 6 P.M. The men breakfasted early in the morning, I somewhat later, and all had started by 10 A.M., arriving between 4 and 6 P.M. at the next camping-ground. My tent was formed of blankets, spread over cross pieces of wood and a ridge-pole, enclosing an area of 6 to 8 feet by 4 to 6 feet. The bedstead, table, and chair were always made by my Lepchas, as described in the Tonglo excursion. The evenings I employed in writing up notes and journals, plotting maps, and ticketing the plants collected during the day's march.

I left Dorjiling at noon, on the 27th October, accompanied by Dr. Campbell, who saw me fairly off, the coolies having preceded me. Our direct route would have been over Tonglo, but the threats of the Sikkim authorities rendered it advisable to make for Nepal at once; we therefore kept west along the Goong ridge, a western prolongation of Sinchul.

On overtaking the coolies, I proceeded for six or seven miles along a zig-zag road, at about 7,500 feet elevation, through dense forests, and halted at a little hut within sight of Dorjiling. Rain and mist came on at nightfall, and though several parties of my servants arrived, none of the Bhotan coolies made their appearance, and I spent the night without food or bed, the weather being much too foggy and dark to send back to meet the missing men. They joined me late on the following day, complaining unreasonably of their loads, and without their Sirdar, who, after starting his crew, had returned to take leave of his wife and family. On the following day he appeared, and after due admonishment we started, but four miles further on were again obliged to halt for the Bhotan coolies, who were equally deaf to threats and entreaties. As they did not come up till dusk, we were obliged to encamp here, (alt. 7,400 feet) at the common source of the Balasun, which flows to the plains, and the Little Rungeet, whose course is north.

The contrast between the conduct of the Bhotan men and that of the Lepchas and Nepalese was so marked, that I seriously debated in my own mind the propriety of sending the former back to Dorjiling, but yielded to the remonstrances of their Sirdar and the Nepal guard, who represented the great difficulty we should have in replacing them, and above all, the loss of time, at this season a matter of great importance. We accordingly started again the following morning, and still keeping in a western direction, crossed the posts in the forest dividing Sikkim from Nepal, and descended into the Myong valley of the latter country, through which flows the river of that name, a tributary of the Tambur. The Myong valley is remarkably fine: it runs south-west from Tonglo, and its open character and general fertility contrast strongly with the bareness of the lower mountain spurs which flank it, and with the dense, gloomy, steep, and forest-clad gorges of Sikkim. At its lower end, about twenty miles from the frontier, is the military fort of Ilam, a celebrated stockaded post and cantonment of the Ghorkas: its position is marked by a conspicuous conical hill. The inhabitants are chiefly Brahmins, but there are also some Moormis, and a few Lepchas who escaped from Sikkim during the general massacre in 1825. Among these is a man who had formerly much influence in Sikkim; he still retains his title of Kazee,* and has had large lands assigned to him by the Nepalese Government: he sent the usual present of a kid, fowls, and eggs, and begged me to express to Dr. Campbell his desire to return to his native country, and settle at Dorjiling.

The scenery of this valley is the most beautiful I know of in the lower Himalaya, and the Cheer Pine (*P. longifolia*) is abundant, cresting the hills, which are loosely clothed with clumps of oaks and other trees, bamboos, and bracken (*Pteris*). The slopes are covered with red clay, and separate little ravines luxuriantly clothed with tropical vegetation, amongst which flow pebbly streams of transparent cool water. The villages, which are merely scattered collections of huts, are surrounded with fields of rice, buckwheat, and Indian corn, which latter the natives were now storing in little granaries, mounted on four posts, men, women, and children being all equally busy. The quantity of gigantic nettles (*Urtica heterophylla*) on the skirts of these maize fields is quite wonderful: their long white stings look most formidable, but though they sting

* This Mahometan title, by which the officers of state are known in Sikkim, is there generally pronounced Kajee.

virulently, the pain only lasts half an hour or so. These, however, with leeches, mosquitos, peepsas, and ticks, sometimes keep the traveller in a constant state of irritation.

However civilised the Hindoo may be in comparison with the Lepcha, he presents a far less attractive picture to the casual observer; he comes to your camping-ground, sits down, and stares with all his might, but offers no assistance; if he bring a present at all, he expects a return on the spot, and goes on begging till satisfied. I was amused by the cool way in which my Ghorka guard treated the village lads, when they wanted help in my service, taking them by the shoulder, pulling out their knives for them, placing them in their hands, and setting them to cut down a tree, or to chop firewood, which they seldom refused to do, when a little such douce violence was applied.

My object being to reach the Tambur, north of the great east and west mountain ridge of Sakkiazung, without crossing the innumerable feeders of the Myong and their dividing spurs, we ascended the north flank of the valley to a long spur from Tonglo, intending to follow winding ridges of that mountain to the sources of the Pemmi at the Phulloot mountains, and thence descend.

On the 3rd November I encamped on the flank of Tonglo (called Nanki in Nepal), at 9,300 feet, about 700 feet below the western summit, which is rocky, and connected by a long flat ridge with that which I had visited in the previous May. The Bhotan coolies behaved worse than ever; their conduct being in all respects typical of the turbulent, mulish race to which they belong. They had been plundering my provisions as they went along, and neither their Sirdar nor the Ghorka soldiers had the smallest authority over them. I had hired some Ghorka coolies to assist and eventually to replace them, and had made up my mind to send back the worst from the more populous banks of the Tambur, when I was relieved by their making off of their own accord. The dilemma was however awkward, as it was impossible to procure men on the top of a mountain 10,000 feet high, or to proceed towards Phulloot. No course remained but to send to Dorjiling for others, or to return to the Myong valley, and take a more circuitous route over the west end of Sakkiazung, which led through villages from which I could procure coolies day by day. I preferred the latter plan, and sent one of the soldiers to the nearest village for assistance to bring the loads down, halting a day for that purpose.

From the summit of Tonglo I enjoyed the view I had so long desired of the Snowy Himalaya, from north-east to north-west; Sikkim being on the right, Nepal on the left, and the plains of India to the southward; and I procured a set of compass bearings, of the greatest use in mapping the country. In the early morning the transparency of the atmosphere renders this view one of astonishing grandeur. Kinchinjunga bore nearly due north, a dazzling mass of snowy peaks, intersected by blue glaciers, which gleamed in the slanting rays of the rising sun, like aquamarines set in frosted silver. From this the sweep of snowed mountains to the eastward was almost continuous as far as Chola (bearing east-north-east), following a curve of 150 miles, and enclosing the whole of the northern part of Sikkim, which appeared a billowy mass of forest-clad mountains. On the north-east horizon rose the Donkia mountain (23,176 feet), and Chumulari (23,929). Though both were much more distant than the snowy ranges, being respectively eighty and ninety miles off, they raised their gigantic heads above, seeming what they really were, by far the loftiest peaks next to Kinchinjunga; and the perspective of snow is so deceptive, that though 40 to 60 miles beyond, they appeared as though almost in the same line with the ridges they overtopped. Of these mountains, Chumulari presents many attractions to the geographer, from its long disputed position, its sacred character, and the interest attached to it since Turner's mission to Tibet in 1783. It was seen and recognised by Dr. Campbell, and measured by Colonel Waugh, from Sinchul, and also from Tonglo, and was a conspicuous object in my subsequent journey to Tibet. Beyond Junnoo, one of the western peaks of Kinchinjunga, there was no continuous snowy chain; the Himalaya seemed suddenly to decline into black and rugged peaks, till in the far north-west it rose again in a white mountain mass of stupendous elevation at 80 miles distance, called, by my Nepal people, "Tsungau."* From the bearings I took of it from several positions, it is in about lat. 27° 49′ and long. 86° 24′, and is probably on the west flank of the Arun valley and river, which latter, in its course from Tibet to the plains of India, receives the waters from the west flank of Kinchinjunga, and from the east flank of the mountain in question. It is perhaps one which has been seen and measured from the Tirhoot district by some of Colonel Waugh's party, and which has been reported to be upwards of 28,000 feet in elevation; and it is the only mountain of the first class in magnitude between Gosainthan (north-east of Katmandoo) and Kinchinjunga.

* This is probably the easternmost and loftiest peak seen from Katmandoo, distant 78 miles, and estimated elevation 20,117 feet by Col. Crawford's observations. See "Hamilton's Nepal," p. 346, and plate 1.

To the west, the black ridge of Sakkiazung, bristling with pines, (*Abies Webbiana*) cut off the view of Nepal; but south-west, the Myong valley could be traced to its junction with the Tambur about thirty miles off: beyond which to the south-west and south, low hills belonging to the outer ranges of Nepal rose on the distant horizon, seventy or eighty miles off; and of these the most conspicuous were the Mahavarati which skirt the Nepal Terai. South and south-east, Sinchul and the Goong range of Sikkim intercepted the view of the plains of India, of which I had a distant peep to the south-west only.

The west top of Tonglo is very open and grassy, with occasional masses of gneiss of enormous size, but probably not in situ. The whole of this flank, and for 1000 feet down the spur to the south-west, had been cleared by fire for pasturage, and flocks of black-faced sheep were grazing. During my stay on the mountain, except in the early morning, the weather was bleak, gloomy, and very cold, with a high south-west wind. The mean temperature was 41°, extremes $\frac{53°\cdot2}{26°}$: the nights were very clear, with sharp hoar-frost; the radiating thermometer sank to 21°, the temperature at 3½ feet depth was 51°·5.

A few of the Bhotan coolies having voluntarily returned, I left Tonglo on the 5th, and descended its west flank to the Mai, a feeder of the Myong. The descent was as abrupt as that on the east face, but through less dense forest; the Sikkim side (that facing the east) being much the dampest. I encamped at dark by a small village, (Jummanoo) at 4,360 feet, having descended 5000 feet in five hours. Hence we marched eastward to the village of Sakkiazung, which we reached on the third day, crossing *en route* several spurs 4000 to 6000 feet high, from the same ridge, and as many rivers, which all fall into the Myong, and whose beds are elevated from 2,500 to 3000 feet.

Though rich and fertile, the country is scantily populated, and coolies were procured with difficulty: I therefore sent back to Dorjiling all but absolute indispensables, and on the 9th of November started up the ridge in a northerly direction, taking the road from Ilam to Wallanchoon. The ascent was gradual, through a fine forest, full of horn-bills (*Buceros*), a bird resembling the Toucan ("Dhunass" Lepcha); at 700 feet an oak (*Quercus semecarpifolia*), "Khasrou" of the Nepalese, commences, a tree which is common as far west as Kashmir, but which I never found in Sikkim, though it appears again in Bhotan.* It forms a broad-headed tree, and has a very handsome appearance; its favourite locality is on grassy open shoulders of the mountains. It was accompanied by an *Astragalus*, *Geranium*, and several other plants of the drier interior parts of Sikkim. Water is very scarce along the ridge; we walked fully eight miles without finding any, and were at length obliged to encamp at 8,350 feet by the only spring that we should be able to reach. With respect to drought, this ridge differs materially from Sikkim, where water abounds at all elevations; and the cause is obviously its position to the westward of the great ridge of Singalelah (including Tonglo) by which the S.W. currents are drained of their moisture. Here again, the east flank was much the dampest and most luxuriantly wooded.

While my men encamped on a very narrow ridge, I ascended a rocky summit, composed of great blocks of gneiss, from which I obtained a superb view to the westward. Immediately below a fearfully sudden decsent, ran

* This oak ascends in the N. W. Himalaya to the highest limit of forest (12,000 feet). No oak in Sikkim attains a greater elevation than 10,000.

the Daomy River, bounded on the opposite side by another parallel ridge of Sakkiazung, enclosing, with that on which I stood, a gulf from 6000 to 7000 feet deep, of wooded ridges, which, as it were, radiated outwards as they ascended upwards in rocky spurs to the pine-clad peaks around. To the south-west, in the extreme distance, were the boundless plains of India, upwards of 100 miles off, with the Cosi meandering through them like a silver thread.

The firmament appeared of a pale steel blue, and a broad low arch spanned the horizon, bounded by a line of little fleecy clouds (moutons); below this the sky was of a golden yellow, while in successively deeper strata, many belts or ribbons of vapour appeared to press upon the plains, the lowest of which was of a dark leaden hue, the upper more purple, and vanishing into the pale yellow above. Though well defined, there was no abrupt division between the belts, and the lowest mingled imperceptibly with the hazy horizon. Gradually the golden lines grew dim, and the blues and purples gained depth of colour; till the sun set behind the dark-blue peaked mountains in a flood of crimson and purple, sending broad beams of grey shade and purple light up to the zenith, and all around. As evening advanced, a sudden chill succeeded, and mists rapidly formed immediately below me in little isolated clouds, which coalesced and spread out like a heaving and rolling sea, leaving nothing above their surface but the ridges and spurs of the adjacent mountains. These rose like capes, promontories, and islands, of the darkest leaden hue, bristling with pines, and advancing boldly into the snowy white ocean, or starting from its bed in the strongest relief. As darkness came on, and the stars arose, a light fog gathered round me, and I quitted with reluctance one of the most impressive and magic scenes I ever beheld.

Returning to my tent, I was interested in observing how well my followers had accommodated themselves to their narrow circumstances. Their fires gleamed everywhere amongst the trees, and the people, broken up into groups of five, presented an interesting picture of native, savage, and half-civilised life. I wandered amongst them in the darkness, and watched unseen their operations; some were cooking, with their rude bronzed faces lighted up by the ruddy glow, as they peered into the pot, stirring the boiling rice with one hand, while with the other they held back their long tangled hair. Others were bringing water from the spring below, some gathering sprigs of fragrant *Artemisia* and other shrubs to form couches—some lopping branches of larger trees to screen them from nocturnal radiation; their only protection from the dew being such branches stuck in the ground, and slanting over their procumbent forms. The Bhotanese were rude and boisterous in their pursuits, constantly complaining to the Sirdars, and wrangling over their meals. The Ghorkas were sprightly, combing their raven hair, telling interminably long stories, of which money was the burthen, or singing Hindoo songs through their noses in chorus; and being neater and better dressed, and having a servant to cook their food, they seemed quite the gentlemen of the party. Still the Lepcha was the most attractive, the least restrained, and the most natural in all his actions, the simplest in his wants and appliances, with a bamboo as his water-jug, an earthen-pot as his kettle, and all manner of herbs collected during the day's march to flavour his food.

My tent was made of a blanket thrown over the limb of a tree; to this others were attached, and the whole was supported on a frame like a house. One half was occupied by my bedstead, beneath which was stowed my box of clothes, while my books and writing materials were placed under the table. The barometer hung in the most out-of-the-way corner, and my other instruments all around. A small candle was burning in a glass shade, to keep the draught and insects from the light, and I had the comfort of seeing the knife, fork, and spoon laid on a white napkin, as I entered my snug little house, and flung myself on the elastic couch to ruminate on the proceedings of the day, and speculate on those of the morrow, while waiting for my meal, which usually consisted of stewed meat and rice, with biscuits and tea. My thermometers (wet and dry bulb, and minimum) hung under a temporary canopy made of thickly plaited bamboo and leaves close to the tent, and the cooking was performed by my servant under a tree.

After dinner my occupations were to ticket and put away the plants collected during the day, write up journals, plot maps, and take observations till 10 P.M. As soon as I was in bed, one of the Nepal soldiers was accustomed to enter, spread his blanket on the ground, and sleep there as my guard. In the morning the collectors were set to change the plant-papers, while I explored the neighbourhood, and having taken observations and breakfasted, we were ready to start at 10 A.M.

Following the same ridge, after a few miles of ascent over much broken gneiss rock, the Ghorkas led me aside to the top of a knoll, 9,300 feet high, covered with stunted bushes, and commanding a splendid view to the west, of the broad, low, well cultivated valley of the Tambur, and the extensive town of Dunkotah on its banks, about twenty-five miles off; the capital of this part of Nepal, and famous for its manufactory of paper from the bark of the *Daphne*. Hence too I gained a fine view of the plains of India, including the course of the Cosi river, which, receiving the

Arun and Tambur, debouches into the Ganges opposite Colgongl (see p. 95).

A little further on we crossed the main ridge of Sakkiazung, a long flexuous chain stretching for miles to the westward from Phulloot on Singalelah, and forming the most elevated and conspicuous transverse range in this part of Nepal: its streams flow south to the Myong, and north to feeders of the Tambur. Silver firs (*Abies Webbiana*) are found on all the summits; but to my regret none occurred in our path, which led just below their limit (10,000 feet), on the southern Himalayan ranges. There were, however, a few yews, exactly like the English. The view that opened on cresting this range was again magnificent, of Kinchinjunga, the western snows of Nepal, and the valley of the Tambur winding amongst wooded and cultivated hills to a long line of black-peaked, rugged mountains, sparingly snowed, which intervene between Kinchinjunga and the great Nepal mountain before mentioned. The extremely varied colouring on the infinite number of hill-slopes that everywhere intersected the Tambur valley was very pleasing. For fully forty miles to the northward there were no lofty forest-clad mountains, nor any apparently above 4000 to 5000 feet: villages and hamlets appeared everywhere, with crops of golden mustard and purple buckwheat in full flower; yellow rice and maize, green hemp, pulse, radishes, and barley, and brown millet. Here and there deep groves of oranges, the broad-leafed banana, and sugar-cane, skirted the bottoms of the valleys, through which the streams were occasionally seen, rushing in white foam over their rocky beds. It was a goodly sight to one who had for his only standard of comparison the view from Sinchul, of the gloomy forest-clad ranges of 6000 to 10,000 feet, that intervene between that mountain and the snowy girdle of Sikkim;

though I question whether a traveller from more favoured climes would see more in this, than a thinly inhabited country, with irregular patches of poor cultivation, a vast amount of ragged forest on low hills of rather uniform height and contour, relieved by a dismal back-ground of frowning black mountains, sprinkled with snow! Kinchinjunga was again the most prominent object to the north-east, with its sister peaks of Kubra (24,005 feet), and Junnoo (25,312 feet). All these presented bare cliffs for several thousand feet below their summits, composed of white rock with a faint pink tint:—on the other hand the lofty Nepal mountain in the far west presented cliffs of black rocks. From the summit two routes to the Tambur presented themselves; one, the main road, led west and south along the ridge, and then turned north, descending to the river; the other was shorter, leading abruptly down to the Pemmi river, and thence along its banks, west to the Tambur. I chose the latter.

The descent was very abrupt on the first day, from 9,500 feet to 5000 feet, and on that following to the bed of the Pemmi, at 2000 feet; and the road was infamously bad, generally consisting of a narrow, winding, rocky path among tangled shrubs and large boulders, brambles, nettles, and thorny bushes, often in the bed of the torrent, or crossing spurs covered with forest, round whose bases it flowed. A little cultivation was occasionally met with on the narrow flat pebbly terraces which fringed the stream, usually of rice, and sometimes of the small-leaved variety of hemp (*Cannabis*), grown as a narcotic.

The rocks above 5000 feet were gneiss; below this, cliffs of very micaceous schist were met with, having a north-west strike, and being often vertical; the boulders again were always of gneiss. The streams seemed rather to occupy faults, than to have eroded courses for themselves;

their beds were invariably rocky or pebbly, and the waters white and muddy from the quantity of alumina In one little rocky dell the water gushed through a hole in a soft stratum in the gneiss; a trifling circumstance which was not lost upon the crafty Brahmins, who had cut a series of regular holes for the water, ornamented the rocks with red paint, and a row of little iron tridents of Siva, and dedicated the whole to Mahadeo.

In some spots the vegetation was exceedingly fine, and several large trees occurred: I measured a Toon (*Cedrela*) thirty feet in girth at five feet above the ground. The skirts of the forest were adorned with numerous jungle flowers, rice crops, blue *Acanthaceæ* and *Pavetta*, wild cherry-trees covered with scarlet blossoms, and trees of the purple and lilac *Bauhinia;* while *Thunbergia, Convolvulus,* and other climbers, hung in graceful festoons from the boughs, and on the dry micaceous rocks the *Luculia gratissima,* one of our common hot-house ornaments, grew in profusion, its gorgeous heads of blossoms scenting the air.

At the junction of the Pemmi and Khawa rivers, there are high rocks of mica-slate, and broad river-terraces of stratified sand and pebbles, apparently alternating with deposits of shingle. On this hot, open expanse, elevated 2250 feet, appeared many trees and plants of the Terai and plains, as pomegranate, peepul, and sal; with extensive fields of cotton, indigo, and irrigated rice.

We followed the north bank of the Khawa, which runs westerly through a gorge, between high cliffs of chlorite, containing thick beds of stratified quartz. At the angles of the river broad terraces are formed, fifteen to thirty feet above its bed, similar to those just mentioned, and planted with rows of *Acacia Serissa,* or laid out in rice fields, or sugar plantations.

I reached the east bank of the Tambur, on the 13th of November, at its junction with the Khawa, in a deep gorge. It formed a grand stream, larger than the Teesta, of a pale, sea-green, muddy colour, and flowed rapidly with a strong ripple, but no foam; it rises six feet in the rains, but ice never descends nearly so low; its breadth was sixty to eighty yards, its temperature 55° to 58°. The breadth of the foaming Khawa was twelve to fifteen yards, and its temperature 56½°. The surrounding vegetation was entirely tropical, consisting of scrubby sal trees, acacia, *Grislea*, *Emblica*, *Hibiscus*, &c.; the elevation being but 1300 feet, though the spot was twenty-five miles in a straight line from the plains. I camped at the fork of the rivers, on a fine terrace fifty feet above the water, about seventy yards long, and one hundred broad, quite flat-topped, and composed of shingle, gravel, &c., with enormous boulders of gneiss, quartz, and hornstone, much water-worn; it was girt by another broken terrace, twelve feet or so above the water, and covered with long grass and bushes.

The main road from Ilam to Wallanchoon, which I quitted on Sakkiazung, descends steeply on the opposite bank of the river, which I crossed in a canoe formed of a hollow trunk (of Toon), thirty feet long. There is considerable traffic along this road; and I was visited by numbers of natives, all Hindoos, who coolly squatted before my tent-door, and stared with their large black, vacant, lustrous eyes: they appear singularly indolent, and great beggars.

The land seems highly favoured by nature, and the population, though so scattered, is in reality considerable, the varied elevation giving a large surface; but the natives care for no more than will satisfy their immediate wants. The river swarms with fish, but they are too lazy to catch them, and they have seldom anything better to give or sell than sticks of sugar-cane, which when peeled form a refreshing morsel in these scorching marches. They have few and poor oranges, citrons, and lemons, very bad plantains, and but little else;—eggs, fowls, and milk are all scarce. Horned cattle are of course never killed by Hindoos, and it was but seldom that I could replenish my larder with a kid. Potatos are unknown, but my Sepoys often brought me large coarse radishes and legumes.

From the junction of the rivers the road led up the Tambur to Mywa Guola; about sixteen miles by the river, but fully thirty-five, as we wound, ascended, and descended, during three days' marches. We were ferried across the stream in a canoe much ruder than that of the New Zealander. I watched my party crossing by boat-loads of fifteen each; the Bhotan men hung little scraps of rags on the bushes before embarking,—the votive offerings of a Booddhist throughout central Asia;—the Lepcha, less civilised, scooped up a little water in the palm of his hand, and scattered it about, invoking the river god of his simple creed.

We always encamped upon gravelly terraces a few feet above the river, which flows in a deep gorge; its banks are very steep for 600 feet above the stream, though the mountains which flank it do not exceed 4000 to 5000 feet: this is a constant phenomenon in the Himalaya, and the roads, when low and within a few hundred feet of the river, are in consequence excessively steep and difficult; it would have been impossible to have taken ponies along that we followed, which was often not a foot broad, running along very steep cliffs, at a dizzy height above the river, and engineered with much trouble and ingenuity: often the bank was abandoned altogether, and we ascended several thousand feet to descend again. Owing to the steepness of these banks, and the reflected heat, the valley, even at this season, was excessively hot and close during the day, even when the temperature was below 70°, and tempered by a brisk breeze which rushes upwards from sunrise to sunset. The sun at this season does not, in many places, reach the bottom of these valleys until 10 A.M., and is off again by 3 P.M.; and the radiation to a clear sky is so powerful that dew frequently forms in the shade, throughout the day, and it is common at 10 A.M. to find the thermometer sink from 70° in a sheltered spot, dried by the sun, to 40° in the shade close by, where the sun has not yet penetrated. Snow never falls.

The rocks throughout this part of the river-course are mica-schists (strike north-west, dip south-west 70°, but very variable in inclination and direction); they are dry and grassy, and the vegetation wholly tropical, as is the entomology, which consists chiefly of large butterflies, *Mantis* and *Diptera*. Snowy mountains are rarely seen, and the beauty of the scenery is confined to the wooded banks of the main stream, which flows at an average inclination of fifty feet to the mile. Otters are found in the stream, and my party shot two, but could not procure them.

In one place the road ascended for 2000 feet above the river, to the village of Chingtam, situated on a lofty spur of the west bank, whence I obtained a grand view of the upper course of the river, flowing in a tremendous chasm, flanked by well-cultivated hills, and emerging fifteen miles to the northward, from black mountains of savage grandeur, whose rugged, precipitous faces were streaked with snow, and the tops of the lower ones crowned with the tabular-branched silver-fir, contrasting strongly with the tropical luxuriance around. Chingtam is an extensive village, covering an area of two miles, and surrounded with abundant culti-

J. D. H. delt. John Murray, Albemarle Street, 1854. W. L. Walton Lith.

Tambur River & Valley (East Nepal) from Chintam. (Elevn. 5,000 ft.
Looking North.

vation ; the houses, which are built in clusters, are of wood, or wattle and mud, with grass thatch. The villagers, though an indolent, staring race, are quiet and respectable; the men are handsome, the women, though less so, often good-looking. They have fine cattle, and excellent crops.

Immediately above Chingtam, the Tambur is joined by a large affluent from the west, the Mywa, which is crossed by an excellent iron bridge, formed of loops hanging from two parallel chains, along which is laid a plank of sal timber. Passing through the village, we camped on a broad terrace, from sixty to seventy feet above the junction of the rivers, whose beds are 2100 feet above the sea.

Mywa Guola (or bazaar) is a large village and mart, frequented by Nepalese and Tibetans, who bring salt, wool, gold, musk, and blankets, to exchange for rice, coral, and other commodities ; and a custom-house officer is stationed there, with a few soldiers. The houses are of wood, and well built : the public ones are large, with verandahs, and galleries of carved wood ; the workmanship is of Chinese character, and inferior to that of Katmandoo ; but in the same style, and quite unlike anything I had previously seen.

The river-terrace is in all respects similar to that at the junction of the Tambur and Khawa, but very extensive : the stones it contained were of all sizes, from a nut to huge boulders upwards of fifteen feet long, of which many strewed the surface, while others were in the bed of the river : all were of gneiss, quartz, and granite, and had doubtless been transported from great elevations, as the rocks *in situ*—both here and for several thousand feet higher up the river—were micaceous schists, dipping in various directions, and at all angles, with, however, a general strike to the north-west.

I was here overtaken by a messenger with letters from Dr. Campbell, announcing that the Sikkim Rajah had disavowed the refusal to the Governor-General's letter, and authorising me to return through any part of Sikkim I thought proper. The bearer was a Lepcha attached to the court: his dress was that of a superior person, being a scarlet jacket over a white cotton dress, the breadth of the blue stripes of which generally denotes wealth; he was accompanied by a sort of attaché, who wore a magnificent pearl and gold ear-ring, and carried his master's bow, as well as a basket on his back; while an attendant coolie bore their utensils and food. Meepo, or Teshoo (in Tibetan, Mr.), Meepo, as he was usually called, soon attached himself to me, and proved an active, useful, and intelligent companion, guide, and often collector, during many months afterwards.

The vegetation round Mywa Guola is still thoroughly tropical: the banyan is planted, and thrives tolerably, the heat being great during the day. Like the whole of the Tambur valley below 4000 feet, and especially on these flats, the climate is very malarious before and after the rains; and I was repeatedly applied to by natives suffering under attacks of fever. During the two days I halted, the mean temperature was 60° (extremes, $\frac{80°}{41°}$), that of the Tambur, 53°, and of the Mywa, 56°; each varying a few degrees (the smaller stream the most) between sunrise and 4 P.M.: the sunk thermometer was 72°.

As we should not easily be able to procure food further on, I laid in a full stock here, and distributed blankets, &c., sufficient for temporary use for all the people, dividing them into groups or messes.

CHAPTER IX.

Leave Mywa — Suspension bridge — Landslips — Vegetation — Slope of river-bed — Bees' nests — Glacial phenomena — Tibetans, clothing, ornaments, amulets, salutation, children, dogs — Last Limboo village, Taptiatok— Beautiful scenery—Tibet village of Lelyp —*Opuntia*—*Edgeworthia* —Crab-apple—Chameleon and porcupine—Praying machine—*Abies Brunoniana*— European plants—Grand scenery—Arrive at Wallanchoon—Scenery around —Trees—Tibet houses—Manis and Mendongs—Tibet household—Food— Tea-soup—Hospitality—Yaks and Zobo, uses and habits of—Bhoteeas—Yak-hair tents—Guobah of Walloong—Jhatamansi—Obstacles to proceeding— Climate and weather—Proceed—Rhododendrons, &c.—Lichens—*Poa annua* and Shepherd's purse—Tibet camp—Tuquoroma—Scenery of pass—Glaciers and snow—Summit—Plants, woolly, &c.

On the 18th November, we left Mywa Guola, and continued up the river to the village of Wallanchoon or Walloong, which was reached in six marches. The snowy peak of Junnoo (alt. 25,312 feet) forms a magnificent feature from this point, seen up the narrow gorge of the river, bearing N.N.E. about thirty miles. I crossed the Mewa, an affluent from the north, by another excellent suspension bridge. In these bridges, the principal chains are clamped to rocks on either shore, and the suspended loops occur at intervals of eight to ten feet; the single sal-plank laid on these loops swings terrifically, and the handrails not being four feet high, the sense of insecurity is very great.

The Wallanchoon road follows the west bank, but the bridge above having been carried away, we crossed by a plank, and proceeded along very steep banks of decomposed chlorite schist, much contorted, and very soapy, affording an insecure footing, especially where great landslips had occurred, which were numerous, exposing acres of a reddish and white soil of felspathic clay, sloping at an angle of 30°. Where the angle was less than 15°, rice was cultivated, and partially irrigated. The lateral streams (of a muddy opal green) had cut beds 200 feet deep in the soft earth, and were very troublesome to cross, from the crumbling cliffs on either side, and their broad swampy channels.

Five or six miles above Mywa, the valley contracts much, and the Tambur (whose bed is elevated about 3000 feet) becomes a turbulent river, shooting along its course with immense velocity, torn into foam as it lashes the spurs of rock that flank it, and the enormous boulders with which its bed is strewn.* From this elevation to 9000 feet, its sinuous track extends about thirty miles, which gives the mean fall of 200 feet to the mile, quadruple of what it is for the lower part of its course. So long as its bed is below 5000 feet, a tropical vegetation prevails in the gorge, and along the terraces, consisting of tall bamboo, *Bauhinia*, *Acacia*, *Melastoma*, &c.; but the steep mountain sides above are either bare and grassy, or cliffs with scattered shrubs and trees, and their summits are of splintered slaty gneiss, bristling with pines: those faces exposed to the south and east are invariably the driest and most grassy; while the opposite are well wooded. *Rhododendron arboreum* becomes plentiful at 5000 to 6000 feet, forming a large tree on dry clayey slopes; it is accompanied by *Indigofera*, *Andromeda*,

* In some places torrents of stone were carried down by landslips, obstructing the rivers; when in the beds of streams, they were often cemented by felspathic clay into a hard breccia of angular quartz, gneiss, and felspar nodules.

Spiræa, shrubby *Compositæ*, and very many plants absent at similar elevations on the wet outer Dorjiling ranges.

In the contracted parts of the valley, the mountains often dip to the river-bed, in precipices of gneiss, under the ledges of which wild bees build pendulous nests, looking like huge bats suspended by their wings; they are two or three feet long, and as broad at the top, whence they taper downwards: the honey is much sought for, except in spring, when it is said to be poisoned by Rhododendron flowers, just as that, eaten by the soldiers in the retreat of the Ten Thousand, was by the flowers of the *R. ponticum*.

Above these gorges are enormous accumulations of rocks, especially at the confluence of lateral valleys, where they rest upon little flats, like the river-terraces of Mywa, but wholly formed of angular shingle, flanked with beds of river-formed gravel: some of these boulders were thirty or forty yards across, and split as if they had fallen from a height; the path passing between the fragments.* At first I imagined that they had been precipitated from the mountains around; and I referred the shingle to land-shoots, which during the rains descend several thousand feet in devastating avalanches, damming up the rivers, and destroying houses, cattle, and cultivation; but though I still refer the materials of many such terraces to this cause, I consider those at the mouths of valleys to be due to ancient glacial action, especially when laden with such enormous blocks as are probably ice-transported.

A change in the population accompanies that in the natural features of the country, Tibetans replacing the

* The split fragments I was wholly unable to account for, till my attention was directed by Mr. Darwin to the observations of Charpentier and Agassiz, who refer similar ones met with in the Alps, to rocks which have fallen through crevasses in glaciers.—See "Darwin on Glaciers and Transported Boulders in North Wales." London, "Phil. Mag." xxi. p. 180.

Limboos and Khass-tribes of Nepal, who inhabit the lower region. We daily passed parties of ten or a dozen Tibetans, on their way to Mywa Guola, laden with salt; several families of these wild, black, and uncouth-looking people generally travelling together. The men are middle-sized, often tall, very square-built and muscular; they have no beard, moustache, or whiskers, the few hairs on their faces being carefully removed with tweezers. They are dressed in loose blanket robes, girt about the waist with a leather belt, in which they place their iron or brass pipes, and from which they suspend their long knives, chop-sticks, tobacco-pouch, tweezers, tinder-box, &c. The robe, boots, and cap are grey, or striped with bright colours, and they wear skull-caps, and the hair plaited into a pig-tail.

The women are dressed in long flannel petticoats and spencer, over which is thrown a sleeveless, short, striped cloak, drawn round the waist by a girdle of broad brass or silver links, to which hang their knives, scissors, needle-cases, &c., and with which they often strap their children to their backs; the hair is plaited in two tails, and the neck loaded with strings of coral and glass beads, and great lumps of amber, glass, and agate. Both sexes wear silver rings and ear-rings, set with turquoises, and square amulets upon their necks and arms, which are boxes of gold or silver, containing small idols, or the nail-parings, teeth, or other reliques of some sainted Lama, accompanied with musk, written prayers, and other charms. All are good-humoured and amiable-looking people, very square and Mongolian in countenance, with broad mouths, high cheek-bones, narrow, upturned eyes, broad, flat noses, and low foreheads. White is their natural colour, and rosy cheeks are common amongst the younger women and children, but all are begrimed with filth and smoke; added to which, they become so weather-worn from exposure to the most rigorous climate in the world, that their natural hues are rarely to be recognised. Their customary mode of saluting one another is to hold out the tongue, grin, nod, and scratch their ear; but this method entails so much ridicule in the low countries, that they do not practise it to Nepalese or strangers; most of them when meeting me, on the contrary, raised their hands to their eyes, threw themselves on the ground, and kotowed most decorously, bumping their foreheads three times on the ground; even the women did this on several occasions. On rising, they begged for a bucksheesh, which I gave in tobacco or snuff, of which they are immoderately fond. Both men and women constantly spin wool as they travel.

These motley groups of Tibetans are singularly picturesque, from the variety in their parti-coloured dresses, and their odd appearance. First comes a middle-aged

TIBET MASTIFF.

man or woman, driving a little silky black yak, grunting under his load of 260 lb. of salt, besides pots, pans, and kettles, stools, churn, and bamboo vessels, keeping up a constant rattle, and perhaps, buried amongst all, a rosy-cheeked and lipped baby, sucking a lump of cheese-curd. The main body follow in due order, and you are soon entangled amidst sheep and goats, each with its two little bags of salt: beside these, stalks the huge, grave, bull-headed mastiff, loaded like the rest, his glorious bushy tail thrown over his back in a majestic sweep, and a thick collar of scarlet wool round his neck and shoulders, setting off his long silky coat to the best advantage; he is decidedly the noblest-looking of the party, especially if a fine and pure black one, for they are often very ragged, dun-coloured, sorry beasts. He seems rather out of place, neither guarding nor keeping the party together, but he knows that neither yaks, sheep, nor goats, require his attention; all are perfectly tame, so he takes his share of work as salt-carrier by day, and watches by night as well. The children bring up the rear, laughing and chatting together; they, too, have their loads, even to the youngest that can walk alone.

The last village of the Limboos, Taptiatok, is large, and occupies a remarkable amphitheatre, apparently a lake-bed, in the course of the Tambur. After proceeding some way through a narrow gorge, along which the river foamed and roared, the sudden opening out of this broad, oval expanse, more than a mile long, was very striking: the mountains rose bare and steep, the west flank terminating in shivered masses of rock, while that on the right was more undulating, dry, and grassy: the surface was a flat gravel-bed, through which meandered the rippling stream, fringed with alder. It was a beautiful spot, the clear, cool, murmuring river, with its rapids and shallows, forcibly reminding me of trout-streams in the highlands of Scotland.

Beyond Taptiatok we again crossed the river, and

ascended over dry, grassy, or rocky spurs to Lelyp, the first Bhoteea village; it stands on a hill fully 1000 feet above the river, and commands a splendid view up the Yalloong and Kambachen valleys, which open immediately to the east, and appear as stupendous chasms in the mountains leading to the perpetual snows of Kinchin-junga. There were about fifty houses in the village, of wood and thatch, neatly fenced in with wattle, the ground between being carefully cultivated with radishes, buckwheat, wheat, and millet. I was surprised to find in one enclosure a fine healthy plant of *Opuntia*, in flower, at this latitude and elevation. A Lama, who is the head man of the place, came out to greet us, with his family and a whole troop of villagers; they were the same class of people as I have elsewhere described as Cis-nivean Tibetans, or Bhoteeas; none had ever before seen an Englishman, and I fear they formed no flattering opinion from the specimen now presented to them, as they seemed infinitely amused at my appearance, and one jolly dame clapped her hands to her sides, and laughed at my spectacles, till the hills echoed.

Elœagnus was common here, with *Edgeworthia Gardneri*,* a beautiful shrub, with globes of waxy, cowslip-coloured, deliciously scented flowers; also a wild apple, which bears a small austere fruit, like the Siberian crab. In the bed of the river rice was still cultivated by Limboos, and subtropical plants continued. I saw, too, a chameleon and a porcupine, indicating much warmth, and seeming quite foreign to the heart of these stupendous mountains. From 6000 to 7000 feet, plants of the temperate regions blend with the tropical; such as rhododendron, oak, ivy,

* A plant allied to *Daphne*, from whose bark the Nepal paper is manufactured. It was named after the eminent Indian botanist, brother of the late Miss Edgeworth.

geranium, berberry, clematis, and shrubby *Vaccinia*, which all made their appearance at Loongtoong, another Bhoteea village. Here, too, I first saw a praying machine, turned by water; it was enclosed in a little wooden house, and consisted of an upright cylinder containing a prayer, and with the words, "Om mani padmi om," (Hail to him of the Lotus and Jewel) painted on the circumference: it was placed over a stream, and made to rotate on its axis by a spindle which passed through the floor of the building into the water, and was terminated by a wheel.

Above this the road followed the west bank of the river; the latter was a furious torrent, flowing through a gorge, fringed with a sombre vegetation, damp, and dripping with moisture, and covered with long *Usnea* and pendulous mosses. The road was very rocky and difficult, sometimes leading along bluff faces of cliffs by wooden steps and single rotten planks. At 8000 feet I met with pines, whose trunks I had seen strewing the river for some miles lower down: the first that occurred was *Abies Brunoniana*, a beautiful species, which forms a stately blunt pyramid, with branches spreading like the cedar, but not so stiff, and drooping gracefully on all sides. It is unknown on the outer ranges of Sikkim, and in the interior occupies a belt about 1000 feet lower than the silver fir (*A. Webbiana*). Many sub-alpine plants occur here, as *Leycesteria*, *Thalictrum*, rose, thistles, alder, birch, ferns, berberry, holly, anemone, strawberry, raspberry, *Gnaphalium*, the alpine bamboo, and oaks. The scenery is as grand as any pictured by Salvator Rosa; a river roaring in sheets of foam, sombre woods, crags of gneiss, and tier upon tier of lofty mountains flanked and crested with groves of black firs, terminating in snow-sprinkled rocky peaks.

I now found the temperature getting rapidly cooler,

TAMBUR RIVER AT THE LOWER LIMIT OF PINES.

both that of the air, which here at 8,066 feet fell to 32° in the night, and that of the river, which was always below 40°. It was in these narrow valleys only, that I observed the return cold current rushing down the river-courses during the nights, which were usually brilliant and very cold, with copious dew: so powerful, indeed, was the radiation, that the upper blanket of my bed became coated with moisture, from the rapid abstraction of heat by the frozen tarpaulin of my tent.

The rivers here are often fringed by flats of shingle, on which grow magnificent yews and pines; some of the latter were from 120 to 150 feet high, and had been blown down, owing to their scanty hold on the soil. I measured one, *Abies Brunoniana*, twenty feet in girth. Many alpine rhododendrons occur at 9000 feet, with *Astragalus* and creeping Tamarisk. Three miles below Wallanchoon the river forks, being met by the Yangma from the north-east; they are impetuous torrents of about equal volume; the Tambur especially (here called the Walloong) is often broken into cascades, and cuts a deep gorge-like channel.

I arrived at the village of Wallanchoon on the 23rd of November. It is elevated 10,385 feet, and situated in a fine open part of the Tambur valley, differing from any part lower down in all its natural features; being broad, with a rapid but not turbulent stream, very grassy, and both the base and sides of the flanking mountains covered with luxuriant dense bushes of rhododendron, rose, berberry and juniper. Red-legged crows, hawks, wild pigeons, and finches, abounded. There was but little snow on the mountains around, which are bare and craggy above, but sloping below. Bleak and forbidding as the situation of any Himalayan village at 10,000 feet elevation must be, that of Wallanchoon is rendered the more so from the comparatively few trees; for though the silver fir and juniper are both abundant higher up the valley, they have been felled here for building materials, fuel, and export to Tibet. From the naked limbs and tall gaunt black trunks of those that remain, stringy masses of bleached lichen (*Usnea*)

WALLANCHOON VILLAGE.

many feet long, stream in the wind. Both men and women seemed fond of decorating their hair with wreaths of this lichen, which they dye yellow with leaves of *Symplocos*.

The village is very large, and occupies a flat on the east bank of the river, covered with huge boulders: the ascent to it is extremely steep, probably over an ancient moraine, though I did not recognise it as such at the time. Cresting this, the valley at once opens, and I was almost startled with the sudden change from a gloomy gorge to a broad flat and a populous village of large and good painted wooden houses, ornamented with hundreds of long poles and vertical flags, looking like the fleet of some foreign port; while a swarm of good-natured, intolerably dirty Tibetans, were kotowing to me as I advanced.

The houses crept up the base of the mountain, on the flank of which was a very large, long convent; two-storied, and painted scarlet, with a low black roof, and backed by a grove of dark junipers; while the hill-sides around were thickly studded with bushes of deep green rhododendron, scarlet berberry, and withered yellow rose. The village contained about one hundred houses, irregularly crowded together, from twenty to forty feet high, and forty to eighty feet long; each accommodating several families. All were built of upright strong pine-planks, the interstices of which were filled with yak-dung; and they sometimes rest on a low foundation wall: the door was generally at the gable end; it opened with a latch and string, and turned on a wooden pivot; the only window was a slit closed by a shutter; and the roofs were very low-pitched, covered with shingles kept down by stones. The paths were narrow and filthy; and the only public buildings besides the convents were Manis and Mendongs; of these the former are square-roofed temples, containing rows of praying-cylinders placed close together, from four to six feet high, and gaudily painted; some are turned by hand, and others by water: the latter are walls ornamented with slabs of clay and mica slate, with "Om Mani Padmi om" well carved on them in two characters, and repeated *ad infinitum*.

A Tibetan household is very slovenly; the family live higgledy-piggledy in two or more apartments, the largest of which has an open fire on the earth, or on a stone if the floor be of wood. The pots and tea-pot are earthen and copper; and these, with the bamboo churn for the brick tea, some wooden and metal spoons, bowls, and platters, comprise all the kitchen utensils.

Every one carries in the breast of his robe a little wooden cup for daily use; neatly turned from the knotted roots of maple (see p. 133). The Tibetan chiefly consumes barley, wheat, or buckwheat meal—the latter is confined to the poorer classes—with milk, butter, curd, and parched wheat; fowls, eggs, pork, and yak flesh when he can afford it, and radishes, a few potatos, legumes, and turnips in their short season. His drink is a sort of soup made from brick tea, of which a handful of leaves is churned up with salt, butter, and soda, then boiled and transferred to the tea-pot, whence it is poured scalding hot into each cup, which the good woman of the house keeps incessantly replenishing, and urging you to drain. Sometimes, but more rarely, the Tibetans make a drink by pouring boiling water over malt, as the Lepchas do over millet. A pipe of yellow mild Chinese tobacco generally follows the meal; more often, however, their tobacco is brought from the plains of India, when it is of a very inferior description. The pipe, carried in the girdle, is of brass or iron, often with an agate, amber, or bamboo mouth-piece.

Many herds of fine yaks were grazing about Wallanchoon: there were a few ponies, sheep, goats, fowls, and pigs, but very little cultivation except turnips, radishes, and potatos. The yak is a very tame, domestic animal, often handsome, and a true bison in appearance; it is invaluable to these mountaineers from its strength and hardiness, accomplishing, at a slow pace, twenty miles a day, bearing either two bags of salt or rice, or four to six planks of pine-wood slung in pairs along either flank. Their ears are generally pierced, and ornamented with a tuft of scarlet worsted; they have large and beautiful eyes, spreading horns, long silky black hair, and grand bushy tails: black is their prevailing colour, but red, dun, parti-coloured, and white are common. In winter, the flocks graze below 8000 feet, on account of the great quantity of snow above that height; in summer they find pasturage as high as 17,000 feet, consisting of grass and small tufted *Carices*, on which they browse with avidity.

The zobo, or cross between the yak and hill cow (much resembling the English cow), is but rarely seen in these mountains, though common in the North West Himalaya. The yak is used as a beast of burden; and much of the wealth of the people consists in its rich milk and curd, eaten either fresh or dried, or powdered into a kind of meal. The hair is spun into ropes, and woven into a covering for their tents, which is quite pervious to wind and rain;* from the same material are made the gauze shades for the eyes used in crossing snowy passes. The bushy tail forms the well-known "chowry" or fly-flapper of the plains of India; the bones and dung serve for fuel. The female drops one calf in April; and the young yaks are very full of gambols, tearing up and down the steep grassy and rocky slopes: their flesh is delicious, much richer and more juicy than common veal; that of the old yak is sliced and dried in the sun, forming jerked meat, which is eaten raw, the scanty proportion of fat preventing

* The latter is, however, of little consequence in the dry climate of Tibet.

its becoming very rancid, so that I found it palatable food: it is called *schat-tcheu* (dried meat). I never observed the yak to be annoyed by any insects; indeed at the elevation it inhabits, there are no large diptera, bots, or gadflies to infest it. It loves steep places, delighting to scramble among rocks, and to sun its black hide perched on the glacial boulders which strew the Wallanchoon flat, and on which these beasts always sleep. Their average value is from two to three pounds, but the price varies with the season. In autumn, when her calf is killed for food, the mother will yield no milk, unless the herdsman gives it the calf's foot to lick, or lays a stuffed skin before it, to fondle, which it does with eagerness, expressing its satisfaction by short grunts, exactly like those of a pig, a sound which replaces the low uttered by ordinary cattle. The yak, though indifferent to ice and snow and to changes of temperature, cannot endure hunger so long as the sheep, nor pick its way so well upon stony ground. Neither can it bear damp heat, for which reason it will not live in summer below 7000 feet, where liver disease carries it off after a very few years.* Lastly, the yak is ridden, especially by the fat Lamas, who find its shaggy coat warm, and its paces easy; under these circumstances it is always led. The wild yak or bison (D'hong) of central Asia, the superb progenitor of this animal, is the largest native animal of Tibet,

* Nevertheless, the yak seems to have survived the voyage to England. I find in Turner's "Tibet" (p. 189), that a bull sent by that traveller to Mr. Hastings, reached England alive, and after suffering from languor, so far recovered its health and vigour as to become the father of many calves. Turner does not state by what mother these calves were born, an important omission, as he adds that all these died but one cow, which bore a calf by an Indian bull. A painting of the yak (copied into Turner's book) by Stubbs, the animal painter, may be seen in the Museum of the Royal College of Surgeons, London. The artist is probably a little indebted to description for the appearance of its hair in a native state, for it is represented much too even in length, and reaching to too uniform a depth from the flanks.

in various parts of which country it is found ; and the Tibetans say, in reference to its size, that the liver is a load for a tame yak. The Sikkim Dewan gave Dr. Campbell and myself an animated account of the chase of this animal, which is hunted by large dogs, and shot with a blunderbuss : it is untameable and horridly fierce, falling upon you with horns and chest, and if he rasps you with his tongue, it is so rough as to scrape the flesh from the bones. The horn is used as a drinking-cup in marriage feasts, and on other grand occasions. My readers are probably familiar with Messrs. Huc and Gabet's account of a herd of these animals being frozen fast in the head-waters of the Yang-tsekiang river. There is a noble specimen in the British Museum not yet set up, and another is preparing for exhibition in the Crystal Palace at Sydenham.

The inhabitants of these frontier districts belong to two very different tribes, but all are alike called Bhoteeas (from Bhote, the proper name of Tibet), and have for many centuries been located in what is—in climate and natural features—a neutral ground between dry Tibet Proper, and the wet Himalayan gorges. They inhabit a climate too cold for either the Lepcha or Nepalese, migrating between 6000 and 15,000 feet with the seasons, always accompanied by their herds. In all respects of appearance, religion, manners, customs, and language, they are Tibetans and Lama Booddhists, but they pay tax to the Nepal and Sikkim Rajahs, to whom they render immense service by keeping up and facilitating the trade in salt, wool, musk, &c., which could hardly be conducted without their co-operation. They levy a small tax on all imports, and trade a little on their own account, but are generally poor and very indolent. In their alpine summer quarters they grow scanty crops of wheat, barley, turnips, and radishes ; and at their winter quarters, as at Loongtoong, the better classes cultivate fine crops of buck-wheat, millet, spinach, &c.; though seldom enough for their support, as in spring they are obliged to buy rice from the inhabitants of the lower regions. Equally dependent on Nepal and Tibet, they very naturally hold themselves independent of both; and I found that my roving commission from the Nepal Rajah was not respected, and the guard of Ghorkas held very cheap.

On my arrival at Wallanchoon, I was conducted to two tents, each about eight feet long, of yak's hair, striped blue and white, which had been pitched close to the village for my accommodation. Though the best that could be provided, and larger than my own, they were wretched in the extreme, being of so loose a texture that the wind blew through them: each was formed of two cloths with a long slit between them, that ran across the top, giving egress to the smoke, and ingress to the weather: they were supported on two short poles, kept to the ground by large stones, and fastened by yak's hair ropes. A fire was smoking vigorously in the centre of one, and some planks were laid at the end for my bed. A crowd of people soon came to stare and loll out their tongues at me, my party, and travelling equipage; though very civil, and only offensive in smell, they were troublesome, from their eager curiosity to see and handle everything; so that I had to place a circle of stones round the tents, whilst a soldier stood by, on the alert to keep them off. A more idle people are not to be found, except with regard to spinning, which is their constant occupation, every man and woman carrying a bundle of wool in the breast of their garments, which is spun by hand with a spindle, and wound off on two cross-pieces at its lower end. Spinning, smoking, and tea-drinking are their chief pursuits; and the women take all the active duties of the dairy and house. They live very happily together, fighting being almost unknown.

Soon after my arrival I was waited on by the Guobah (or head-man), a tall, good-looking person, dressed in a purple woollen robe, with good pearl and coral ear and finger-rings, and a broad ivory ring over the left thumb,* as a guard when using the bow; he wore a neat thick white felt cap, with the border turned up, and a silk tassel on the top; this he removed with both hands and held before him, bowing three times on entering. He was followed by a crowd, some of whom were his own people, and brought a present of a kid, fowls, rice, and eggs, and some spikenard roots (*Nardostachys Jatamansi*, a species of valerian smelling strongly of patchouli), which is a very favourite perfume. After paying some compliments, he showed me round the village. During my walk, I found that I had a good many objections to overrule before I could proceed to the Wallanchoon pass, nearly two days' journey to the northward. In the first place, the Guobah disputed the Nepal rajah's authority to pass me through his dominions; and besides the natural jealousy of these people when intruded upon, they have very good reasons for concealing the amount of revenue they raise from their position, and for keeping up the delusion that they alone can endure the excessive climate of these regions, or undergo the hardships and toil of the salt trade. My passport said nothing about the passes; my people, and especially the Ghorkas, detested the keen, cold, and cutting wind; at Mywa Guola, I had been persuaded by the Havildar to put off providing snow-boots and blankets, on the assurance that I should easily get them at Walloong, which I now

* A broad ring of this material, agate, or chalcedony, is a mark of rank here, as amongst the Man-choos, and throughout Central Asia.

found all but impossible, owing to there being no bazaar. My provisions were running short, and for the same reason I had no present hope of replenishing them. All my party had, I found, reckoned with certainty that I should have had enough of this elevation and weather by the time I reached Walloong. Some of them fell sick; the Guobah swore that the passes were full of snow, and had been impracticable since October; and the Ghorka Havildar respectfully deposed that he had no orders relative to the pass. Prompt measures were requisite, so I told all my people that I should stop the next day at Walloong, and proceed on the following on a three days' journey to the pass, with or without the Guobah's permission. To the Ghorka soldiers I said that the present they would receive, and the character they would take to their commandant, depended on their carrying out this point, which had been fully explained before starting. My servants I told that their pay and reward also depended on their implicit obedience. I took the Guobah aside and showed him troops of yaks (tethered by halters and toggles to a long rope stretched between two rocks), which had that morning arrived laden with salt from the north; I told him it was vain to try and deceive me; that my passport was ample, and that I should expect a guide, provisions, and snow-boots the next day; and that every impediment and every facility should be reported to the rajah.

During my two days' stay at Walloong, the weather was bitterly cold: as heretofore, the nights and mornings were cloudless, but by noon the whole sky became murky, the highest temperature (50°) occurring at 10 A.M. At this season the prospect from this elevation (10,385 feet), was dreary in the extreme; and the quantity of snow on the mountains, which was continually increasing, held out a dismal promise for my chance of exploring lofty uninhabited regions. All annual and deciduous vegetation had long past, and the lofty Himalayas are very poor in mosses and lichens, as compared with the European Alps, and arctic regions in general. The temperature fluctuated from 22° at sunrise, to 50° at 10 A.M.; the mean being 35°;* one night it fell to 6¼°. Throughout the day, a south wind blew strong and cold up the valley, and at sunset was replaced by a keen north blast, searching every corner, and piercing through tent and blankets. Though the sun's rays were hot for an hour or two in the morning, its genial influence was never felt in the wind. The air was never very dry, the wet-bulb thermometer standing during the day 3¾° below the dry, thus giving a mean dew-point of 30¼°. A thermometer sunk two feet stood at 44°, fully 9° above the mean temperature of the air; one exposed to the clear sky, stood, during the day, several degrees below the air in shade, and at night, from 9° to 14¾° lower. The black-bulb thermometer, in the sun, rose to 65¾° above the air, indicating upwards of 90° difference at nearly the warmest part of the day, between contiguous shaded and sunny exposures. The sky, when cloudless, was generally a cold blue or steel-grey colour, but at night the stars were large, and twinkled gloriously. The black-glass photometer indicated 10·521 inches† as the maximum intensity of sunlight; the temperature of the river close by fell to 32° during the night, and rose to 37° in the day. In my tent, the temperature

* This gives 1° Fahr. for every 309 feet of elevation, using contemporaneous observations at Calcutta, and correcting for latitude, &c.

† On three mornings the maxima occurred at between 9 and 10 A.M. They were, Nov. 24th, 10·509, Nov. 25th, 10·521. On the 25th, at Tuquoroma, I recorded 10·510. The maximum effect observed at Dorjiling (7340 feet) was 10·328, and on the plains of India 10·350. The maximum I ever recorded was in Yangma valley (15,186 feet), 10·572 at 1 P.M.

fluctuated with the state of the fire, from 26° at night to 58° when the sun beat on it; but the only choice was between cold and suffocating smoke.

After a good many conferences with the Guobah, some bullying, douce violence, persuasions, and the prescribing of pills, prayers, and charms in the shape of warm water, for the sick of the village, whereby I gained some favour, I was, on the 25th Nov., grudgingly prepared for the trip to Wallanchoon, with a guide, and some snow-boots for those of my party whom I took with me.

The path lay north-west up the valley, which became thickly wooded with silver-fir and juniper; we gradually ascended, crossing many streams from lateral gulleys, and huge masses of boulders. Evergreen rhododendrons soon replaced the firs, growing in inconceivable profusion, especially on the slopes facing the south-east, and with no other shrubs or tree-vegetation, but scattered bushes of rose, *Spiræa*, dwarf juniper, stunted birch, willow, honey-suckle, berberry, and a mountain-ash (*Pyrus*). What surprised me more than the prevalence of rhododendron bushes, was the number of species of this genus, easily recognised by the shape of their capsules, the form and woolly covering of the leaves; none were in flower, but I reaped a rich harvest of seed. At 12,000 feet the valley was wild, open, and broad, with sloping mountains clothed for 1000 feet with dark-green rhododendron bushes; the river ran rapidly, and was broken into falls here and there. Huge angular and detached masses of rock were scattered about, and to the right and left snowy peaks towered over the surrounding mountains, while amongst the latter narrow gulleys led up to blue patches of glacial ice, with trickling streams and shoots of stones. Dwarf rhododendrons with strongly-scented leaves (*R. anthopogon* and *setosum*), and abundance of a little *Andromeda*, exactly like ling, with woody stems and tufted branches, gave a heathery appearance to the hill-sides. The prevalence of lichens, common to this country and to Scotland (especially *L. geographicus*), which coloured the rocks, added an additional feature to the resemblance to Scotch Highland scenery. Along the narrow path I found the two commonest of all British weeds, a grass (*Poa annua*), and the shepherd's purse! They had evidently been imported by man and yaks, and as they do not occur in India, I could not but regard these little wanderers from the north with the deepest interest.

Such incidents as these give rise to trains of reflection in the mind of the naturalist traveller; and the farther he may be from home and friends, the more wild and desolate the country he is exploring, the greater the difficulties and dangers under which he encounters these subjects of his earliest studies in science; so much keener is the delight with which he recognises them, and the more lasting is the impression which they leave. At this moment these common weeds more vividly recal to me that wild scene than does all my journal, and remind me how I went on my way, taxing my memory for all it ever knew of the geographical distribution of the shepherd's purse, and musing on the probability of the plant having found its way thither over all Central Asia, and the ages that may have been occupied in its march.

On reaching 13,000 feet, the ground was everywhere hard and frozen, and I experienced the first symptoms of lassitude, headache, and giddiness; which however, were but slight, and only came on with severe exertion.

We encountered a group of Tibetans, encamped to leeward of an immense boulder of gneiss, against which they had raised a shelter with their salt-bags, removed from their herd of yaks, which were grazing close by. They looked miserably cold and haggard, and their little upturned eyes, much inflamed and bloodshot, testified to the hardships they had endured in their march from the salt regions: they were crouched round a small fire of juniper wood, smoking iron pipes with agate mouthpieces. A resting-house was in sight across the stream—a loose stone hut, to which we repaired. I wondered why these Tibetans had not taken possession of it, not being aware of the value they attach to a rock, on account of the great warmth which it imbibes from the sun's rays during the day, and retains at night. This invaluable property of otherwise inhospitable gneiss and granite I had afterwards many opportunities of proving; and when driven for a night's shelter to such as rude nature might afford on the bleak mountain, I have had my blankets laid beneath "the shadow of a great rock in a weary land."

The name of Dhamersala is applied, in the mountains as in the plains of India, to a house provided for the accommodation of travellers, whether it be one of the beautiful caravanserais built to gratify the piety, ostentation, or benevolence of a rajah, or such a miserable shieling of rough stone and plank as that of Tuquoroma, in which we took up our quarters, at 13,000 feet elevation. A cheerful fire soon blazed on the earthen floor, filling the room with the pungent odour of juniper, which made our eyes smart and water. The Ghorkas withdrew to one corner, and my Lepchas to a second, while one end was screened off for my couch; unluckily, the wall faced the north-east, and in that direction there was a gulley in the snowy mountains, down which the wind swept with violence, penetrating to my bed. I had calculated upon a good night's rest here, which I much needed, having been worried and unwell at Wallanchoon, owing to the Guobah's obstinacy. I had not then learnt how to treat such conduct, and just before retiring to rest had further been informed by the Havildar that the Guobah declared we should find no food on our return. To remain in these mountains without a supply was impossible, and the delay of sending to Mywa Guola would not have answered; so I long lay awake, occupied in arranging measures. The night was clear and very cold; the thermometer falling to 19° at 9 P. M., and to 12° in the night, and that by my bedside to 20°.

On the following morning (Nov. 26th) I started with a small party to visit the pass, continuing up the broad, grassy valley; much snow lay on the ground at 13,500 feet, which had fallen the previous month; and several glaciers were seen in lateral ravines at about the same elevation. After a couple of miles, we left the broad valley, which continued north-west, and struck northward up a narrow, stony, and steep gorge, crossing an immense ancient moraine at its mouth. This path, which we followed for seven or eight miles, led up to the pass, winding considerably, and keeping along the south-east exposures, which, being the most sunny, are the freest from snow. The morning was splendid, the atmosphere over the dry rocks and earth, at 14,000 feet, vibrating from the power of the sun's rays, whilst vast masses of blue glacier and fields of snow choked every gulley, and were spread over all shady places. Although, owing to the steepness and narrowness of the gorge, no view was obtained, the scenery was wild and very grand. Just below where perpetual snow descends to the path, an ugly carved head of a demon, with blood-stained cheeks and goggle-eyes, was placed in a niche of rock, and protected by a glass.

At 15,000 feet, the snow closed in on the path from all sides, whether perpetual, glacial, or only the October fall,

I could not tell; the guide declared it to be perpetual henceforward, though now deepened by the very heavy October fall; the path was cut some three feet through it. Enormous boulders of gneiss cumbered the bottom of the gorge, which gradually widened as we approached its summit; and rugged masses of black and red gneiss and mica schist pierced the snow, and stood out in dismal relief. For four miles continuously we proceeded over snow; which was much honey-combed on the surface, and treacherous from the icy streams it covered, into which we every now and then stumbled; there was scarcely a trace of vegetation, and the cold was excessive, except in the sun.

Towards the summit of the pass the snow lay very deep, and we followed the course of a small stream which cut through it, the walls of snow being breast-high on each side; the path was still frequented by yaks, of which we overtook a small party going to Tibet, laden with planks. All the party appeared alike overcome by lassitude, shortness and difficulty of breathing, a sense of weight on the stomach, giddiness and headache, with tightness across the temples.

Just below the summit was a complete bay of snow, girdled with two sharp peaks of red baked schists and gneiss, strangely contorted, and thrown up at all angles with no prevalent dip or strike, and permeated with veins of granite. The top itself, or boundary between Nepal and Tibet, is a low saddle between two rugged ridges of rock, with a cairn built on it, adorned with bits of stick and rag covered with Tibetan inscriptions. The view into Tibet was not at all distant, and was entirely of snowy mountains, piled ridge over ridge; three of these spurs must, it is said, be crossed before any descent can be made to the Chomachoo river (as the Arun is called in Tibet), on which is the frontier fort of the Tibetans, and which is reached in two or three days. There is no plain or level ground of any kind before reaching that river, of which the valley is said to be wide and flat.

Starting at 10 A.M., we did not reach the top till 3½ P.M.; we had halted nowhere, but the last few miles had been most laborious, and the three of us who gained the summit were utterly knocked up. Fortunately I carried my own barometer; it indicated 16·206 inches, giving by comparative observations with Calcutta 16,764 feet, and with Dorjiling, 16,748 feet, as the height of the pass. The thermometer stood at 18°, and the sun being now hidden behind rocks, the south-east wind was bitterly cold. Hitherto the sun had appeared as a clearly defined sparkling globe, against a dark-blue sky; but the depth of the azure blue was not so striking as I had been led to suppose, by the accounts of previous travellers, in very lofty regions. The plants gathered near the top of the pass were species of *Compositæ*, grass, and *Arenaria;* the most curious was *Saussurea gossypina*, which forms great clubs of the softest white wool, six inches to a foot high, its flowers and leaves seeming uniformly clothed with the warmest fur that nature can devise. Generally speaking, the alpine plants of the Himalaya are quite unprovided with any special protection of this kind; it is the prevalence and conspicuous nature of the exceptions that mislead, and induce the careless observer to generalise hastily from solitary instances; for the prevailing alpine genera of the Himalaya, *Arenarias*, primroses, saxifrages, fumitories, *Ranunculi*, gentians, grasses, sedges, &c., have almost uniformly naked foliage.

We descended to the foot of the pass in about two hours, darkness overtaking us by the way; the twilight, however, being prolonged by the glare of the snow. Fearing the distance to Tuquoroma might be too great to permit of our returning thither the same night, I had had a few things brought hither during the day, and finding they had arrived, we encamped under the shelter of some enormous boulders (at 13,500 feet), part of an ancient moraine, which extended some distance along the bed of the narrow valley. Except an excruciating headache, I felt no ill effects from my ascent; and after a supper of tea and biscuit, I slept soundly.

On the following morning the temperature was 28° at 6·30′ A.M., and rose to 30° when the sun appeared over the mountains at 8·15′, at which time the black bulb thermometer suddenly mounted to 112°, upwards of 80° above the temperature of the air. The sky was brilliantly clear, with a very dry, cold, north wind blowing down the snowy valley of the pass.

DEMON'S HEAD.

CHAPTER X.

Return from Wallanchoon pass—Procure a bazaar at village—Dance of Lamas—Blacking face, Tibetan custom of—Temple and convent—Leave for Kanglachem pass—Send part of party back to Dorjiling—Yangma Guola—Drunken Tibetans—Guobah of Wallanchoon—Camp at foot of Great Moraine—View from top—Geological speculations—Height of moraines—Cross dry lake-bed—Glaciers—More moraines—Terraces—Yangma temples—Jos, books and furniture—Peak of Nango—Lake—Arrive at village—Cultivation—Scenery—Potatos—State of my provisions—Pass through village—Gigantic boulders Terraces—Wild sheep—Lake-beds—Sun's power—Piles of gravel and detritus—Glaciers and moraines—Pabuk, elevation of—Moonlight scene—Return to Yangma—Temperature, &c.—Geological causes of phenomena in valley—Scenery of valley on descent.

I RETURNED to the village of Wallanchoon, after collecting all the plants I could around my camp; amongst them a common-looking dock abounded in the spots which the yaks had frequented.

The ground was covered, as with heather, with abundance of creeping dwarf juniper, *Andromeda*, and dwarf rhododendron. On arriving at the village, I refused to receive the Guobah, unless he opened a bazaar at daylight on the following morning, where my people might purchase food; and threatened to bring charges against him before his Rajah. At the same time I arranged for sending the main body of my party down the Tambur, and so back to Sikkim, whilst I should, with as few as possible, visit the Kanglachem (Tibetan) pass in the adjacent valley to the eastward, and then, crossing the Nango, Kambachen and Kanglanamo passes, reach Jongri in Sikkim, on the south flank of Kinchinjunga.

Strolling out in the afternoon I saw a dance of Lamas; they were disfigured with black paint * and covered with rags, feathers, and scarlet cloth, and they carried long poles with bells and banners attached; thus equipped, they marched through the village, every now and then halting, when they danced and gesticulated to the rude music of cymbals and horns, the bystanders applauding with shouts, crackers, and alms.

I walked up to the convents, which were long ugly buildings, several stories high, built of wood, and daubed with red and grey paint. The priests were nowhere to be found, and an old withered nun, whom I disturbed husking millet in a large wooden mortar, fled at my approach. The temple stood close by the convent, and had a broad low architrave: the walls sloped inwards, as did the lintels: the doors were black, and almost covered with a gigantic and disproportioned painting of a head, with bloody cheeks and huge teeth; it was surrounded by myriads of goggle eyes, which seemed to follow one about everywhere; and though in every respect rude, the effect was somewhat imposing. The similarly proportioned gloomy portals of Egyptian fanes naturally invite comparison; but the Tibetan temples lack the sublimity of these; and the uncomfortable creeping sensation produced by the many sleepless eyes of Boodh's numerous incarnations is very different from the awe with which we contemplate the outspread wings of the Egyptian symbol, and feel as in the presence of the God who

* I shall elsewhere have to refer to the Tibetan custom of daubing the face with black pigment to protect the skin from the excessive cold and dryness of these lofty regions; and to the ludicrous imposition that was passed on the credulity of MM. Huc and Gabet.

says," I am Osiris the Great: no man hath dared to lift my veil."

I had ascended behind the village, but returned down the "via sacra," a steep paved path flanked by mendongs or low stone dykes, into which were let rows of stone slabs, inscribed with the sacred "Om Mani Padmi om."—"Hail to him of the lotus and jewel;" an invocation of Sakkya, who is usually represented holding a lotus flower with a jewel in it.

On the following morning, a scanty supply of very dirty rice was produced, at a very high price. I had, however, so divided my party as not to require a great amount of food, intending to send most of the people back by the Tambur to Dorjiling. I kept nineteen persons in all, selecting the most willing, as it was evident the journey at this season would be one of great hardship: we took seven days' food, which was as much as they could carry. At noon, I left Wallanchoon, and mustered my party at the junction of the Tambur and Yangma, whence I dismissed the party for Dorjiling, with my collections of plants, minerals, &c., and proceeded with the chosen ones to ascend the Yangma river. The scenery was wild and very grand, our path lying through a narrow gorge, choked with pine trees, down which the river roared in a furious torrent; while the mountains on each side were crested with castellated masses of rock, and sprinkled with snow. The road was very bad, often up ladders, and along planks lashed to the faces of precipices, and overhanging the torrent, which it crossed several times by plank bridges. By dark we arrived at Yangma Guola, a collection of empty wood huts buried in the rocky forest-clad valley, and took possession of a couple. They were well built, raised on posts, with a stage and ladder at the gable end, and consisted of one good-sized apartment. Around was abundance of dock, together with three common English plants.*

The night was calm, misty, and warm ($\frac{\text{Max. } 41^\circ\ 5'}{\text{Min. } 29^\circ}$) for the elevation (9,300 feet). During the night, I was startled out of my sleep by a blaze of light, and jumping up, found myself in presence of a party of most sinister-looking, black, ragged Tibetans, armed with huge torches of pine, that filled the room with flame and pitchy smoke. I remembered their arriving just before dark, and their weapons dispelled my fears, for they came armed with bamboo jugs of Murwa beer, and were very drunk and very amiable: they grinned, nodded, kotowed, lolled out their tongues, and scratched their ears in the most seductive manner, then held out their jugs, and besought me by words and gestures to drink and be happy too. I awoke my servant (always a work of difficulty), and with some trouble ejected the visitors, happily without setting the house on fire. I heard them toppling head over heels down the stair, which I afterwards had drawn up to prevent further intrusion, and in spite of their drunken orgies, was soon lulled to sleep again by the music of the roaring river.

On the 29th November, I continued my course north up the Yangma valley, which after five miles opened considerably, the trees disappearing, and the river flowing more tranquilly, and through a broader valley, when above 11,000 feet elevation. The Guobah of Wallanchoon overtook us on the road; on his way, he said, to collect the revenues at Yangma village, but in reality to see what I was about. He owns five considerable villages, and is said to pay a tax of 6000 rupees (600*l.*) to the Rajah of Nepal: this is no doubt a great exaggeration, but the

* *Cardamine hirsuta*, *Limosella aquatica*, and *Juncus bufonius*.

revenues of such a position, near a pass frequented almost throughout the year, must be considerable. Every yak going and coming is said to pay 1*s*., and every horse 4*s*.; cattle, sheep, ponies, land, and wool are all taxed; he exports also quantities of timber to Tibet, and various articles from the plains of India. He joined my party and halted where I did, had his little Chinese rug spread, and squatted cross-legged on it, whilst his servant prepared his brick tea with salt, butter, and soda, of which he partook, snuffed, smoked, rose up, had all his traps repacked, and was off again.

We encamped at a most remarkable place: the valley was broad, with little vegetation but stunted tree-junipers: rocky snow-topped mountains rose on either side, bleak, bare, and rugged; and in front, close above my tent, was a gigantic wall of rocks, piled—as if by the Titans—completely across the valley, for about three-quarters of a mile. This striking phenomenon had excited all my curiosity on first obtaining a view of it. The path, I found, led over it, close under its west end, and wound amongst the enormous detached fragments of which it was formed, and which were often eighty feet square: all were of gneiss and schist, with abundance of granite in blocks and veins. A superb view opened from the top, revealing its nature to be a vast moraine, far below the influence of any existing glaciers, but which at some antecedent period had been thrown across by a glacier descending to 10,000 feet, from a lateral valley on the east flank. Standing on the top, and looking south, was the Yangma valley (up which I had come), gradually contracting to a defile, girdled by snow-tipped mountains, whose rocky flanks mingled with the black pine forest below. Eastward the moraine stretched south of the lateral valley, above which towered the snowy peak of Nango, tinged rosy red, and sparkling in the rays of the setting sun: blue glaciers peeped from every gulley on its side, but these were 2000 to 3000 feet above this moraine; they were small too, and their moraines were mere gravel, compared with this. Many smaller consecutive moraines, also, were evident along the bottom of that lateral valley, from this great one up to the existing glaciers. Looking up the Yangma was a flat grassy plain, hemmed in by mountains, and covered with other stupendous moraines, which rose ridge behind ridge, and cut off the view of all but the mountain tops to the north. The river meandered through the grassy plain (which appeared a mile and a half broad at the utmost, and perhaps as long), and cut through the great moraine on its eastern side, just below the junction of the stream from the glacial valley, which, at the lower part of its course, flowed over a broad steep shingle bed.

I descended to my camp, full of anxious anticipations for the morrow; while the novelty of the scene, and its striking character, the complexity of the phenomena, the lake-bed, the stupendous ice-deposited moraine, and its remoteness from any existing ice, the broad valley and open character of the country, were all marked out as so many problems suddenly conjured up for my unaided solution, and kept me awake for many hours. I had never seen a glacier or moraine on land before, but being familiar with sea ice and berg transport, from voyaging in the South Polar regions, I was strongly inclined to attribute the formation of this moraine to a period when a glacial ocean stood high on the Himalaya, made fiords of the valleys, and floated bergs laden with blocks from the lateral gulleys, which the winds and currents would deposit along certain lines. On the following morning I carried a barometer to the top of the

J.D H del[t] John Murray, Albemarle Street 1854. W L Walton, lith.

Ancient Moraine thrown across the Yangma Valley, East Nepal (Elev[n] 11,000 f[t].)

moraine, which proved to be upwards of 700 feet above the floor of the valley, and 400 above the dry lake-bed which it bounded, and to which we descended on our route up the valley. The latter was grassy and pebbly, perfectly level, and quite barren, except a very few pines at the bases of the encircling mountains, and abundance of rhododendrons, *Andromeda* and juniper on the moraines. Isolated moraines occurred along both flanks of the valley, some higher than that I have described, and a very long one was thrown nearly across from the upper end of another lateral gulley on the east side, also leading up to the glaciers of Nango. This second moraine commenced a mile and a half above the first, and abutting on the east flank of the valley, stretched nearly across, and then curving round, ran down it, parallel to and near the west flank, from which it was separated by the Yangma river: it was abruptly terminated by a conical hill of boulders, round whose base the river flowed, entering the dry lake-bed from the west, and crossing it in a south-easterly direction to the western extremity of the great moraine.

The road, on its ascent to the second moraine, passed over an immense accumulation of glacial detritus at the mouth of the second lateral valley, entirely formed of angular fragments of gneiss and granite, loosely bound together by felspathic sand. The whole was disposed in concentric ridges radiating from the mouth of the valley, and descending to the flat; these were moraines *in petto*, formed by the action of winter snow and ice upon the loose débris. A stream flowed over this débris, dividing into many branches before reaching the lake-bed, where its waters were collected, and whence it meandered southward to fall into the Yangma.

From the top of the second moraine, a very curious

ANCIENT MORAINES IN THE YANGMA VALLEY.

scene opened up the valley, of another but more stony and desolate level lake-bed, through which the Yangma (here very rapid) rushed, cutting a channel about sixty feet deep; the flanks of this second lake-bed were cut most distinctly into two principal terraces, which were again subdivided into others, so that the general appearance was that of many raised beaches, but each so broken up, that, with the exception of one on the banks of the river, none were continuous for any distance. We descended 200 feet, and crossed the valley and river obliquely in a north-west direction, to a small temple and convent which stood on a broad flat terrace under the black, precipitous, west flank: this gave me a good opportunity of examining the structure of this part of the valley, which was filled with an accumulation, probably 200 feet thick at the deepest part, of angular gravel and enormous boulders, both imbedded in the gravel, and strewed on the flat surfaces of the terraces. The latter were always broadest opposite to the lateral valleys, perfectly horizontal for the short distance that they were continuous, and very barren; there were no traces of fossils, nor could I assure myself of stratification. The accumulation was wholly glacial; and probably a lake had supervened on the melting of the great glacier and its recedence, which lake, confined by a frozen moraine, would periodically lose its waters by sudden accessions of heat melting the ice of the latter. Stratified silt, no doubt, once covered the lake bottom, and the terraces have, in succession, been denuded of it by rain and snow. These causes are now in operation amongst the stupendous glaciers of north-east Sikkim, where valleys, dammed up by moraines, exhibit lakes hemmed in between these, the base of the glacier, and the flanks of the valleys.

Yangma convents stood at the mouth of a gorge which opened upon the uppermost terrace; and the surface of the latter, here well covered with grass, was furrowed into concentric radiating ridges, which were very conspicuous from a distance. The buildings consisted of a wretched collection of stone huts, painted red, enclosed by loose stone dykes. Two shockingly dirty Lamas received me and conducted me to the temple, which had very thick walls, but was undistinguishable from the other buildings. A small door opened upon an apartment piled full of old battered gongs, drums, scraps of silk hangings, red cloth, broken praying-machines—relics much resembling those in the lumber-room of a theatre. A ladder led from this dismal hole to the upper story, which was entered by a handsomely carved and gilded door: within, all was dark, except from a little lattice-window covered with oil-paper. On one side was the library, a carved case, with a hundred gilded pigeon-holes, each holding a real or sham book, and each closed by a little square door, on which hung a bag full of amulets. In the centre of the book-case was a recess, containing a genuine Jos or Fo, graced with his Chinese attribute of very long pendulous moustaches and beard and totally wanting that air of contemplative repose which the Tibetan Lamas give to their idols. Banners were suspended around, with paintings of Lhassa, Teshoo Loombo, and various incarnations of Boodh. The books were of the usual Tibetan form, oblong squares of separate block-printed leaves of paper, made in Nepal or Bhotan from the bark of a *Daphne*, bound together by silk cords, and placed between ornamented wooden boards. On our way up the valley, we had passed some mendongs and chaits, the latter very pretty stone structures, consisting of a cube, pyramid, hemisphere, and cone placed on the top of one another, forming together the tasteful combination which appears on the cover of these volumes.

Beyond the convents the valley again contracted, and on crossing a third, but much lower, moraine, a lake opened to view, surrounded by flat terraces, and a broad gravelly shore, part of the lake being dry. To the west, the cliffs were high, black and steep: to the east a large lateral valley, filled at about 1500 feet up with blue glaciers, led (as did the other lateral valleys) to the gleaming snows of Nango; the moraine, too, here abutted on the east flank of the Yangma valley, below the mouth of the lateral one. Much snow (from the October fall) lay on the ground, and the cold was pinching in the shade; still I could not help attempting to sketch this wonderfully grand scene, especially as lakes in the Himalaya are extremely rare: the present one was about a mile long, very shallow, but broad, and as smooth as glass: it reminded me of the tarn in Glencoe. The reflected lofty peak of Nango appeared as if frozen deep down in its glassy bed, every snowy crest and ridge being rendered with perfect precision.

Nango is about 18,000 feet high; it is the next lofty mountain of the Kinchinjunga group to the west of Junnoo, and I doubt if any equally high peak occurs again for some distance further west in Nepal. Facing the Yangma valley, it presents a beautiful range of precipices of black rock, capped with a thick crust of snow: below the cliffs the snow again appears continuously and very steep, for 2000 to 3000 feet downwards, where it terminates in glaciers that descend to 14,000 feet. The steepest snow-beds appear

LOOKING ACROSS YANGMA VALLEY.

cut into vertical ridges, whence the whole snowy face is—as it were—crimped in perpendicular, closely-set, zig-zag lines, doubtless caused by the melting process, which furrows the surface of the snow into channels by which the water is carried off: the effect is very beautiful, but impossible to represent on paper, from the extreme delicacy of the shadows, and at the same time the perfect definition and precision of the outlines.

Towards the head of the lake, its bed was quite dry and gravelly, and the river formed a broad delta over it: the terraces here were perhaps 100 feet above its level, those at the lower end not nearly so much. Beyond the lake, the river became again a violent torrent, rushing in a deep chasm, till we arrived at the fork of the valley, where we once more met with numerous dry lake-beds, with terraces high up on the mountain sides.

In the afternoon we reached the village of Yangma, a miserable collection of 200 to 300 stone huts, nestling under the steep south-east flank of a lofty, flat-topped terrace, laden with gigantic glacial boulders, and projecting southward from a snowy mountain which divides the valley. We encamped on the flat under the village, amongst some stone dykes, enclosing cultivated fields. One arm of the valley runs hence N. N. E. amongst snowy mountains, and appeared quite full of moraines; the other, or continuation of the Yangma, runs W.N.W., and leads to the Kanglachem pass.

Near our camp (of which the elevation was 13,500 feet), radishes, barley, wheat, potatos, and turnips, were cultivated as summer crops, and we even saw some on the top of the terrace, 400 feet above our camp, or nearly 14,000 feet above the sea; these were grown in small fields cleared of stones, and protected by dykes.

The scenery, though dismal, (no juniper even attaining this elevation,) was full of interest and grandeur, from the number and variety of snowy peaks and glaciers all around the elevated horizon, the ancient lake-beds, now green or brown with scanty vegetation, the vast moraines, the ridges of glacial débris, the flat terraces, marking, as it were with parallel roads, the bluff sides of the mountains, the enormous boulders perched upon them, and strewed everywhere around, the little Boodhist monuments of quaint, picturesque shapes, decorated with poles and banners, the many-coloured dresses of the people, the brilliant blue of the cloudless heaven by day, the depth of its blackness by night, heightened by the light of the stars, that blaze and twinkle with a lustre unknown in less lofty regions: all these were subjects for contemplation, rendered more impressive by the stillness of the atmosphere, and the silence that reigned around. The village seemed buried in repose throughout the day: the inhabitants had already hybernated, their crops were stored, the curd made and dried, the passes closed, the soil frozen, the winter's stock of fuel housed, and the people had retired into the caverns of their half subterranean houses, to sleep, spin wool, and think of Boodh, if of anything at all, the dead, long winter through. The yaks alone can find anything to do: so long as any vegetation remains they roam and eat it, still yielding milk, which the women take morning and evening, when their shrill whistle and cries are heard for a few minutes, as they call the grunting animals. No other sounds, save the harsh roar and hollow echo of the falling rock, glacier, or snow-bed, disturbed the perfect silence of the day or night.*

I had taken three days' food to Yangma, and stayed there as long as it lasted: the rest of my provisions I had left below the first moraine, where a lateral valley leads east over the Nango pass to the Kambachen valley, which lay on the route back to Sikkim.

* Snow covers the ground at Yangma from December till April, and the falls are said to be very heavy, at times amounting to 12 feet in depth.

I was premature in complaining of my Wallanchoon tents, those provided for me at Yangma being infinitely worse, mere rags, around which I piled sods as a defence from the insidious piercing night-wind that descended from the northern glaciers in calm, but most keen, breezes. There was no food to be procured in the village, except a little watery milk, and a few small watery potatos. The latter have only very recently been introduced amongst the Tibetans, from the English garden at the Nepalese capital, I believe, and their culture has not spread in these regions further east than Kinchin-junga, but they will very soon penetrate into Tibet from Dorjiling, or eastward from Nepal. My private stock of provisions—consisting chiefly of preserved meats from my kind friend Mr. Hodgson—had fallen very low; and I here found to my dismay that of four remaining two-pound cases, provided as meat, three contained prunes, and one "*dindon aux truffes!*" Never did luxuries come more inopportunely; however the greasy French viand served for many a future meal as sauce to help me to bolt my rice, and according to the theory of chemists, to supply animal heat in these frigid regions. As for my people, they were not accustomed to much animal food; two pounds of rice, with ghee and chilis, forming their common diet under cold and fatigue. The poorer Tibetans, especially, who undergo great privation and toil, live almost wholly on barley-meal, with tea, and a very little butter and salt: this is not only the case with those amongst whom I mixed so much, but is also mentioned by MM. Huc and Gabet, as having been observed by them in other parts of Tibet.

On the 1st of December I visited the village and terrace, and proceeded to the head of the Yangma valley, in order to ascend the Kanglachem pass as far as practicable. The houses are low, built of stone, of no particular shape, and are clustered in groups against the steep face of the terrace; filthy lanes wind amongst them, so narrow, that if you are not too tall, you look into the slits of windows on either hand, by turning your head, and feel the noisome warm air in whiffs against your face. Glacial boulders lie scattered throughout the village, around and beneath the clusters of houses, from which it is sometimes difficult to distinguish the native rock. I entered one house by a narrow low door through walls four feet thick, and found myself in an apartment full of wool, juniper-wood, and dried dung for fuel: no one lived in the lower story, which was quite dark, and as I stood in it my head was in the upper, to which I ascended by a notched pole (like that in the picture of a Kamschatk house in Cook's voyage), and went into a small low room. The inmates looked half asleep, they were intolerably indolent and filthy, and were employed in spinning wool and smoking. A hole in the wall of the upper apartment led me on to the stone roof of the neighbouring house, from which I passed to the top of a glacial boulder, descending thence by rude steps to the narrow alley. Wishing to see as much as I could, I was led on a winding course through, in and out, and over the tops of the houses of the village, which alternately reminded me of a stone quarry or gravel pit, and gipsies living in old lime-kilns; and of all sorts of odd places that are turned to account as human habitations.

From the village I ascended to the top of the terrace, which is a perfectly level, sandy, triangular plain, pointing down the valley at the fork of the latter, and abutting against the flank of a steep, rocky, snow-topped mountain

DIAGRAM OF THE GLACIAL TERRACES AT THE FORK OF THE YANGMA VALLEY.

to the northward. Its length is probably half a mile from north to south, but it runs for two miles westward up the valley, gradually contracting. The surface, though level, is very uneven, being worn into hollows, and presenting ridges and hillocks of blown sand and gravel, with small black tufts of rhododendron. Enormous boulders of gneiss and granite were scattered over the surface; one of the ordinary size, which I measured, was seventy feet in girth, and fifteen feet above the ground, into which it had partly sunk. From the southern pointed end I took sketches of the opposite flanks of the valleys east and west. The river was about 400 feet below me, and flowed in a little flat lake-bed; other terraces skirted it, cut out, as it were, from the side of that I was on. On the opposite flank of the valley were several superimposed terraces, of which the highest appeared to tally with the level I occupied, and the lowest was raised very little above the river; none were continuous for any distance, but the upper one in particular, could be most conspicuously traced up and down the main valley, whilst, on looking across to the eastern valley, a much higher, but less distinctly marked one appeared on it. The road to the pass lay west-north-west up the north bank of the Yangma river, on the great terrace; for two miles it was nearly level along the gradually narrowing shelf, at times dipping into the steep gulleys formed by lateral torrents from the mountains; and as the terrace disappeared, or melted, as it were, into the rising floor of the valley, the path descended upon the lower and smaller shelf.

We came suddenly upon a flock of gigantic wild sheep, feeding on scanty tufts of dried sedge and grass; there were twenty-five of these enormous animals, of whose dimensions the term sheep gives no idea: they are very long-legged, stand as high as a calf, and have immense horns, so large that the fox is said to take up his abode in their hollows, when detached and bleaching, on the barren mountains of Tibet. Though very wild, I am sure I could easily have killed a couple had I had my gun, but I had found it necessary to reduce my party so uncompromisingly, that I could not afford a man both for my gun and instruments, and had sent the former back to Dorjiling, with Mr. Hodgson's bird-stuffers, who had broken one of theirs. Travelling without fire-arms sounds strange in India, but in these regions animal life is very rare, game is only procured with much hunting and trouble, and to come within shot of a flock of wild sheep was a contingency I never contemplated. Considering how very short we were of any food, and quite out of animal diet, I could not but bitterly regret the want of a gun, but consoled myself by reflecting that the instruments were still more urgently required to enable me to survey this extremely interesting valley. As it was, the great beasts trotted off, and turned to tantalise me by grazing within an easy stalking distance. We saw several other flocks, of thirty to forty, during the day, but never, either on this or any future occasion, within shot. The *Ovis Ammon* of Pallas stands from four to five feet high, and measures seven feet from nose to tail; it is quite a Tibetan animal, and is seldom seen below 14,000 feet, except when driven lower by snow; and I have seen it as high as 18,000 feet. The same animal, I believe, is found in Siberia, and is allied to the Big-horn of North America.

Soon after descending to the bed of the valley, which is broad and open, we came on a second dry lake-bed, a mile long, with shelving banks all round, heavily snowed on the shaded side; the river was divided into many arms, and meandered over it, and a fine glacier-bound valley opened into it from the south. There were no boulders on its surface, which was pebbly, with tufts of grass and creeping tamarisk. On the banks I observed much granite, with large mica crystals, hornstone, tourmaline, and stratified quartz, with granite veins parallel to the foliation or lamination.

A rather steep ascent of a mile, through a contracted part of the valley, led to another and smaller lake-bed, a quarter of a mile long and 100 yards broad, covered with patches of snow, and having no lateral valley opening into it: it faced the now stupendous masses of snow and ice which filled the upper part of the Yangma valley. This lake-bed (elevation,

15,186 feet) was strewed with enormous boulders; a rude stone hut stood near it, where we halted for a few minutes at 1 P.M., when the temperature was 42·2°, while the dew-point was only 20·7°.* At the same time, the black bulb thermometer, fully exposed on the snow, rose 54° above the air, and the photometer gave 10·572. Though the sun's power was so great, there was, however, no appearance of the snow melting, evaporation proceeding with too great rapidity.

Enormous piles of gravel and sand had descended upon the upper end of this lake-bed, forming shelves, terraces,

KANGLACHEM PASS.

and curving ridges, apparently consolidated by ice, and covered in many places with snow. Following the

* This indicates a very dry state of the air, the saturation-point being 0·133°; whereas, at the same hour at Calcutta it was 0·559°.

stream, we soon came to an immense moraine, which blocked up the valley, formed of angular boulders, some of which were fifty feet high. Respiration had been difficult for some time, and the guide we had taken from the village said we were some hours from the top of the pass, and could get but a little way further; we however proceeded, plunging through the snow, till on cresting the moraine a stupendous scene presented itself. A gulf of moraines, and enormous ridges of débris, lay at our feet, girdled by an amphitheatre of towering, snow-clad peaks, rising to 17,000 and 18,000 feet all around. Black scarped precipices rose on every side; deep snow-beds and blue glaciers rolled down every gulley, converging in the hollow below, and from each transporting its own materials, there ensued a complication of moraines, that presented no order to the eye. In spite of their mutual interference, however, each had raised a ridge of débris or moraine parallel to itself.

We descended with great difficulty through the soft snow that covered the moraine, to the bed of this gulf of snow and glaciers; and halted by an enormous stone, above the bed of a little lake, which was snowed all over, but surrounded by two superimposed level terraces, with sharply defined edges. The moraine formed a barrier to its now frozen waters, and it appeared to receive the drainage of many glaciers, which filtered through their gravelly ridges and moraines.

We could make no further progress; the pass lay at the distance of several hours' march, up a valley to the north, down which the glacier must have rolled that had deposited this great moraine; the pass had been closed since October, it being very lofty, and the head of this valley was far more snowy than that at Wallanchoon. We halted in the snow from 3 to 4 P. M., during which time I again took angles and observations; the height of this spot, called Pabuk, is 16,038 feet, whence the pass is probably considerably over 17,000 feet, for there was a steep ascent beyond our position. The sun sank at 3 P.M., and the thermometer immediately fell from 35° to 30¾.*

After fixing in my note and sketch books the principal features of this sublime scene, we returned down the valley: the distance to our camp being fully eight miles, night overtook us before we got half-way, but a two days' old moon guided us perfectly, a remarkable instance of the clearness of the atmosphere at these great elevations. Lassitude, giddiness, and headache came on as our exertions increased, and took away the pleasure I should otherwise have felt in contemplating by moonlight the varied phenomena, which seemed to crowd upon the restless imagination, in the different forms of mountain, glacier, moraine, lake, boulder and terrace. Happily I had noted everything on my way up, and left nothing intentionally to be done on returning. In making such excursions as this, it is above all things desirable to seize and book every object worth noticing on the way out: I always carried my note-book and pencil tied to my jacket pocket, and generally walked with them in my hand. It is impossible to begin observing too soon, or to observe too much: if the excursion is long, little is ever done on the way home; the bodily powers being mechanically exerted, the mind seeks repose, and being fevered through over-exertion, it can endure no train of thought, or be brought to bear on a subject.

During my stay at Yangma, the thermometer never rose to 50°, it fell to 14¾° at night; the ground was frozen for several inches below the surface, but at two feet depth its

* At 4 o'clock, to 29°·5, the average dew-point was 16°·3, and dryness 0·55; weight of vapour in a cubic foot, 1·33 grains.

temperature was 37½°. The black bulb thermometer rose on one occasion 84° above the surrounding air. Before leaving, I measured by angles and a base-line the elevations of the great village-terrace above the river, and that of a loftier one, on the west flank of the main valley; the former was about 400 and the latter 700 feet.

Considering this latter as the upper terrace, and concluding that it marks a water level, it is not very difficult to account for its origin. There is every reason to suppose that the flanks of the valley were once covered to the elevation of the upper terrace, with an enormous accumulation of débris; though it does not follow that the whole valley was filled by ice-action to the same depth; the effect of glaciers being to deposit moraines between themselves and the sides of the valley they fill; as also to push forward similar accumulations. Glaciers from each valley, meeting at the fork, where their depth would be 700 feet of ice, would both deposit the necessary accumulation along the flanks of the great valley, and also throw a barrier across it. The melting waters of such glaciers would accumulate in lakes, confined by the frozen earth, between the moraines and mountains. Such lakes, though on a small scale, are found at the terminations and sides of existing glaciers, and are surrounded by terraces of shingle and débris; these terraces being laid bare by the sudden drainage of the lakes during seasons of unusual warmth. To explain the phenomena of the Yangma valley, it may be necessary to demand larger lakes and deeper accumulations of débris than are now familiar to us, but the proofs of glaciers having once descended to from 8,000 to 10,000 feet in every Sikkim and east Nepal valley communicating with mountains above 16,000 feet elevation, are overwhelming, and the glaciers must, in some cases, have been fully forty miles long, and 500 feet in depth. The absence of any remains of a moraine, or of blocks of rock in the valley below the fork, is I believe, the only apparent objection to this theory; but, as I shall elsewhere have occasion to observe, the magnitude of the moraines bears no fixed proportion to that of the glacier, and at Pabuk, the steep ridges of débris, which were heaped up 200 feet high, were far more striking than the more usual form of moraine.

On my way up to Yangma I had rudely plotted the valley, and selected prominent positions for improving my plan on my return: these I now made use of, taking bearings with the azimuth compass, and angles by means of a pocket sextant. The result of my running-survey of the whole valley, from 10,000 to 16,000 feet, I have given along with a sketch-map of my routes in India, which accompanies this volume.

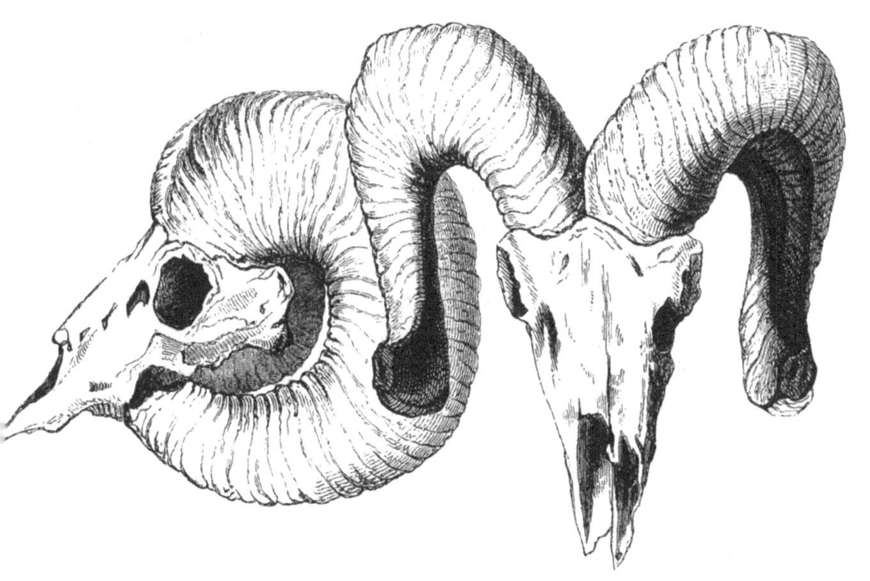

SKULLS OF OVIS AMMON.

CHAPTER XI.

Ascend to Nango mountain—Moraines—Glaciers—Vegetation—*Rhododendron Hodgsoni*—Rocks—Honey-combed surface of snow—Perpetual snow—Top of pass—View—Elevation—Geology—Distance of sound—Plants—Temperature—Scenery—Cliffs of granite and hurled boulders—Camp—Descent—Pheasants—Larch—Himalayan pines—Distribution of Deodar, note on—Tassichooding temples—Kambachen village—Cultivation—Moraines in valley, distribution of—Picturesque lake-beds, and their vegetation—Tibetan sheep and goats—*Cryptogramma crispa*—Ascent to Choonjerma pass—View of Junnoo—Rocks of its summit—Misty ocean—Nepal peaks—Top of pass—Temperature, and observations—Gorgeous sunset—Descent to Yalloong valley—Loose path—Night scenes—Musk deer.

WE passed the night a few miles below the great moraine, in a pine-wood (alt. 11,000 feet) opposite the gorge which leads to the Kambachen or Nango pass, over the south shoulder of the mountain of that name: it is situated on a ridge dividing the Yangma river from that of Kambachen, which latter falls into the Tambur opposite Lelyp.

The road crosses the Yangma (which is about fifteen feet wide), and immediately ascends steeply to the south-east, over a rocky moraine, clothed with a dense thicket of rhododendrons, mountain-ash, maples, pine, birch, juniper, &c. The ground was covered with silvery flakes of birch bark, and that of *Rhododendron Hodgsoni*, which is as delicate as tissue-paper, and of a pale flesh-colour. I had never before met with this species, and was astonished at the beauty of its foliage, which was of a beautiful bright green, with leaves sixteen inches long.

Beyond the region of trees and large shrubs the alpine rhododendrons filled the broken surface of the valley, growing with *Potentilla*, Honeysuckle, *Polygonum*, and dwarf juniper. The peak of Nango seemed to tower over the gorge, rising behind some black, splintered, rocky cliffs, sprinkled with snow; narrow defiles opened up through these cliffs to blue glaciers, and their mouths were invariably closed by beds of shingly moraines, curving outwards from either flank in concentric ridges.

Towards the base of the peak, at about 14,000 feet, the scenery is very grand; a great moraine rises suddenly to the north-west, under the principal mass of snow and ice, and barren slopes of gravel descend from it; on either side are rugged precipices; the ground is bare and stony, with patches of brown grass: and, on looking back, the valley appears very steep to the first shrubby vegetation, of dark green rhododendrons, bristling with ugly stunted pines.

We followed a valley to the south-east, so as to turn the flank of the peak; the path lying over beds of October snow at 14,000 feet, and over plashy ground, from its melting. Sometimes our way lay close to the black precipices on our right, under which the snow was deep; and we dragged ourselves along, grasping every prominence of the rock with our numbed fingers. Granite appeared in large veins in the crumpled gneiss at a great elevation, in its most beautiful and loosely-crystallised form, of pearly white prisms of felspar, glassy quartz, and milk-white flat plates of mica, with occasionally large crystals of tourmaline. Garnets were very frequent in the gneiss near the granite veins. Small rushes, grasses, and sedges formed the remaining vegetation, amongst which were the withered stalks of gentians, *Sedum*, *Arenaria*, *Silene*, and many Composite plants.

At a little below 15,000 feet, we reached enormous flat beds of snow, which were said to be perpetual, but covered deeply with the October fall. They were continuous, and like all the snow I saw at this season, the surface was honeycombed into thin plates, dipping north at a high angle; the intervening fissures were about six inches deep. A thick mist here overtook us, and this, with the great difficulty of picking our way, rendered the ascent very fatiguing. Being sanguine about obtaining a good view, I found it almost impossible to keep my temper under the aggravations of pain in the forehead, lassitude, oppression of breathing, a dense drizzling fog, a keen cold wind, a slippery footing, where I was stumbling at every few steps, and icy-cold wet feet, hands, and eyelids; the latter, odd as it sounds, I found a very disagreeable accompaniment of continued raw cold wind.

After an hour and a half's toilsome ascent, during which we made but little progress, we reached the crest, crossing a broad shelf of snow between two rocky eminences; the ridge was unsnowed a little way down the east flank; this was, in a great measure, due to the eastern exposure being the more sunny, to the prevalence of the warm and melting south-east winds that blow up the deep Kambachen valley, and to the fact that the great snow-beds on the west side are drifted accumulations.* The mist cleared

* Such enormous beds of snow in depressions, or on gentle slopes, are generally adopted as indicating the lower limit of perpetual snow. They are, however, winter accumulations, due mainly to eddies of wind, of far more snow than can be melted in the following summer, being hence perennial in the ordinary sense of the word. They pass into the state of glacier ice, and, obeying the laws that govern the motions of a viscous fluid, so admirably elucidated by Forbes ("Travels in the Alps"), they flow downwards. A careful examination of those great beds of snow in the Alps, from whose position the mean lower level of perpetual snow, in that latitude, is deduced, has convinced me that these are mainly due to accumulations of this kind, and that the true limit of perpetual snow, or that point where all that falls melts, is much higher than it is usually supposed to be.

off, and I had a partial, though limited, view. To the north the blue ice-clad peak of Nango was still 2000 feet above us, its snowy mantle falling in great sweeps and curves into glacier-bound valleys, over which the ice streamed out of sight, bounded by black aiguilles of gneiss. The Yangma valley was quite hidden, but to the eastward the view across the stupendous gorge of the Kambachen, 5000 feet below, to the waste of snow, ice, and rock, piled in confusion along the top of the range of Junnoo and Choonjerma, parallel to this but higher, was very grand indeed: this we were to cross in two days, and its appearance was such, that our guide doubted the possibility of our doing it. A third and fourth mountain mass (unseen) lay beyond this, between us and Sikkim, divided by valleys as deep as those of Yangma and Kambachen.

Having hung up my instruments, I ascended a few hundred feet to some naked rocks, to the northward; they were of much-crumpled and dislocated gneiss, thrown up at a very high angle, and striking north-west. Chlorite, schist, and quartz, in thin beds, alternated with the gneiss, and veins of granite and quartz were injected through them.

It fell calm; when the distance to which the voice was carried was very remarkable; I could distinctly hear every word spoken 300 to 400 yards off, and did not raise my voice when I asked one of the men to bring me a hammer.

The few plants about were generally small tufted *Arenarias* and woolly *Compositæ*, with a thick-rooted Umbellifer that spread its short, fleshy leaves and branches flat on the ground; the root was very aromatic, but wedged close in the rock. The temperature at 4 P.M. was 23°, and bitterly cold; the elevation, 15,770 feet; dew-point, 16°. The air was not very dry; saturation-point, 0·670°, whereas at Calcutta it was 0.498° at the same hour.

The descent was to a broad, open valley, into which the flank of Nango dipped in tremendous precipices, which reared their heads in splintered snowy peaks. At their bases were shoots of débris fully 700 feet high, sloping at a steep angle. Enormous masses of rock, detached by the action of the frost and ice from the crags, were scattered over the bottom of the valley; they had been precipitated from above, and gaining impetus in their descent, had been hurled to almost inconceivable distances from the parent cliff. All were of a very white, fine-grained crystallised granite, full of small veins of the same rock still more finely crystallised. The weathered surface of each block was black, and covered with moss and lichens; the others beautifully white, with clean, sharp-fractured edges. The material of which they were composed was so hard that I found it difficult to detach a specimen.

Darkness had already come on, and the coolies being far behind, we encamped by the light of the moon, shining through a thin fog, where we first found dwarf-juniper for fuel, at 13,500 feet. A little sleet fell during the night, which was tolerably fine, and not very cold; the minimum thermometer indicating $14\frac{1}{2}°$

Having no tent-poles, I had some difficulty in getting my blankets arranged as a shelter, which was done by making them slant from the side of a boulder, on the top of which one end was kept by heavy stones; under this roof I laid my bed, on a mass of rhododendron and juniper-twigs. The men did the same against other boulders, and lighting a huge fire opposite the mouth of my ground-nest, I sat cross-legged on the bed to eat my supper; my face scorching, and my back freezing. Rice, boiled with a few ounces of greasy *dindon aux truffes* was now my daily dinner, with chili-vinegar and tea, and I used to relish it keenly: this finished, I smoked a cigar, and wrote up my journal (in short intervals between warming myself) by the light of the fire; took observations by means of a dark-lantern; and when all this was accomplished, I went to roost.

December 5.—On looking out this morning, it was with a feeling of awe that I gazed at the stupendous ice-crowned precipices that shot up to the summit of Nango, their flanks spotted white at the places whence the gigantic masses with which I was surrounded had fallen; thence my eye wandered down their black faces to the slope of débris at the bottom, thus tracing the course which had probably been taken by that rock under whose shelter I had passed the previous night.

Meepo, the Lepcha sent by the rajah, had snared a couple of beautiful pheasants, one of which I skinned, and eat for breakfast; it is a small bird, common above 12,000 feet, but very wild; the male has two to five spurs on each of its legs, according to its age; the general colour is greenish, with a broad scarlet patch surrounding the eye; the Nepalese name is "Khalidge." The crop was distended with juniper berries, of which the flesh tasted strongly, and it was the very hardest, toughest bird I ever did eat.

We descended at first through rhododendron and juniper, then through black silver-fir (*Abies Webbiana*), and below that, near the river, we came to the Himalayan larch; a tree quite unknown, except from a notice in the journals of Mr. Griffith, who found it in Bhotan. It is a small tree, twenty to forty feet high, perfectly similar in general characters to a European larch, but with larger cones, which are erect upon the very long, pensile, whip-like branches; its leaves,—now red—were falling, and covering the rocky ground on which it grew, scattered amongst other trees. It is called "Saar" by the Lepchas and Cis-himalayan Tibetans, and "Boarga-sella" by the Nepalese, who say it is found as far west as the heads of the Cosi river: it does not inhabit Central or West Nepal, nor the North-west Himalaya. The distribution of the Himalayan pines is very remarkable. The Deodar has not been seen east of Nepal, nor the *Pinus Gerardiana*, *Cupressus torulosa*, or *Juniperus communis*. On the other hand, *Podocarpus* is confined to the east of Katmandoo. *Abies Brunoniana* does not occur west of the Gogra, nor the larch west of the Cosi, nor funereal cypress (an introduced plant, however) west of the Teesta (in Sikkim). Of the twelve * Sikkim and Bhotan *Coniferæ* (including yew, junipers, and *Podocarpus*) eight are common to the North-west Himalaya (west of Nepal), and four † are not: of the thirteen natives of the north-west provinces, again, only five ‡ are not found in Sikkim, and I have given their names below, because they show how European the absent ones are, either specifically or in affinity. I have stated that the Deodar is possibly a variety of the Cedar of Lebanon. This is now a prevalent opinion, which is strengthened by the fact that so many more Himalayan plants are now ascertained to be European than had been supposed before they were compared with European specimens; such are the yew, *Juniperus communis*, *Berberis vulgaris*, *Quercus Ballota*, *Populus alba* and *Euphratica*, &c.

* Juniper, 3; yew, *Abies Webbiana*, *Brunoniana*, and *Smithiana*: Larch, *Pinus excelsa*, and *longifolia*, and *Podocarpus neriifolia*.

† Larch, *Cupressus funebris*, *Podocarpus neriifolia*, *Abies Brunoniana*.

‡ A juniper (the European *communis*), Deodar (possibly only a variety of the Cedar of Lebanon and of Mount Atlas), *Pinus Gerardiana*, *P. excelsa*, and *Cupressus torulosa*.

The cones of the Deodar are identical with those of the Cedar of Lebanon: the Deodar has, generally longer and more pale bluish leaves and weeping branches,* but these characters seem to be unusually developed in our gardens; for several gentlemen, well acquainted with the Deodar at Simla, when asked to point it out in the Kew Gardens, have indicated the Cedar of Lebanon, and when shown the Deodar, declare that they never saw that plant in the Himalaya!

At the bottom of the valley we turned up the stream, and passing the Tassichooding convents † and temple, crossed the river—which was a furious torrent, about twelve yards wide—to the village of Kambachen, on a flat terrace a few feet above the stream. There were about a dozen houses of wood, plastered with mud and dung, scattered over a grassy plain of a few acres, fenced in, as were also a few fields, with stone dykes. The only cultivation consists of radishes, potatos, and barley: no wheat is grown, the climate being said to be too cold for it, by which is probably meant that it is foggy,—the elevation (11,380 feet)

* Since writing the above, I have seen, in the magnificent Pinetum at Dropmore, noble cedars, with the length and hue of leaf, and the pensile branches of the Deodar, and far more beautiful than that is, and as unlike the common Lebanon Cedar as possible. When it is considered from how very few wild trees (and these said to be exactly alike) the many dissimilar varieties of the *C. Libani* have been derived; the probability of this, the Cedar of Algiers, and of the Himalayas (Deodar) being all forms of one species, is greatly increased. We cannot presume to judge from the few cedars which still remain, what the habit and appearance of the tree may have been, when it covered the slopes of Libanus, and seeing how very variable *Coniferæ* are in habit, we may assume that its surviving specimens give us no information on this head. Should all three prove one, it will materially enlarge our ideas of the distribution and variation of species. The botanist will insist that the typical form of cedar is that which retains its characters best over the greatest area, namely, the Deodar; in which case the prejudice of the ignorant, and the preconceived ideas of the naturalist, must yield to the fact that the old familiar Cedar of Lebanon is an unusual variety of the Himalayan Deodar.

† These were built by the Sikkim people, when the eastern valleys of Nepal belonged to the Sikkim rajah.

being 2000 feet less than that of Yangma village, and the temperature, therefore 6° to 7° warmer; but of all the mountain gorges I have ever visited, this is by far the wildest, grandest, and most gloomy; and that man should hybernate here is indeed extraordinary, for there is no route up the valley, and all communication with Lelyp,* two marches down the river, is cut off in winter, when the houses are buried in snow, and drifts fifteen feet deep are said to be common. Standing on the little flat of Kambachen, precipices, with inaccessible patches of pine wood, appeared to the west, towering over head; while across the narrow valley wilder and less wooded crags rose in broken ridges to the glaciers of Nango. Up the valley, the view was cut off by bluff cliffs; whilst down it, the scene was most remarkable: enormous black, round backed moraines, rose, tier above tier, from a flat lake-bed, apparently hemming in the river between the lofty precipices on the east flank of the valley. These had all been deposited at the mouth of a lateral valley, opening just below the village, and descending from Junnoo, a mountain of 25,312 feet elevation, and one of the grandest of the Kinchinjunga group, whose top—though only five miles distant in a straight line—rises 13,932 feet† above the village. Few facts show more decidedly the extraordinary steepness and depth of the Kambachen valley near the village, which, though nearly 11,400 feet above the sea, lies between two mountains only eight miles apart, the one 25,312 feet high, the other (Nango), 19,000 feet.

The villagers received us very kindly, and furnished us

* Which I passed, on the Tambur, on the 21st Nov. See page 204.

† This is one of the most sudden slopes in this part of the Himalaya, the angle between the top of Junnoo and Kambachen being 2786 feet per mile, or 1 in 1·8. The slope from the top of Mont Blanc to the Chamouni valley is 2464 feet per mile, or 1 in 2·1. That from Monte Rosa top to Macugnaga greatly exceeds either.

with a guide for the Choonjerma pass, leading to the Yalloong valley, the most easterly in Nepal; but he recommended our not attempting any part of the ascent till the morrow, as it was past 1 P.M., and we should find no camping-ground for half the way up. The villagers gave us the leg of a musk deer, and some red potatos, about as big as walnuts—all they could spare from their winter-stock. With this scanty addition to our stores we started down the valley, for a few miles alternately along flat lake-beds and over moraines, till we crossed the stream from the lateral valley, and ascending a little, camped on its bank, at 11,400 feet elevation.

In the afternoon I botanized amongst the moraines, which were very numerous, and had been thrown down at right-angles to the main valley, which latter being here very narrow, and bounded by lofty precipices, must have stopped the parent glaciers, and effected the heaping of some of these moraines to at least 1000 feet above the river. The general features were modifications of those seen in the Yangma valley, but contracted into a much smaller space.

The moraines were all accumulated in a sort of delta, through which the lateral river debouched into the Kambachen, and were all deposited more or less parallel to the course of the lateral valley, but curving outwards from its mouth. The village-flat, or terrace, continued level to the first moraine, which had been thrown down on the upper or north side of the lateral valley, on whose steep flanks it abutted, and curving outwards seemed to encircle the village-flat on the south and west; where it dipped into the river. This was crossed at the height of about 100 feet, by a stony path, leading to the bed of the rapid torrent flowing through shingle and boulders, beyond which was another moraine, 250 feet high, and parallel to it a third gigantic one.

Ascending the great moraine at a place where it overhung the main river, I had a good *coup-d'œil* of the whole. The view south-east up the glacial valley—(represented in the accompanying cut)—to the snowy peaks south of Junnoo, was particularly grand, and most interesting from

ANCIENT MORAINES IN THE KAMBACHEN VALLEY.

the precision with which one great distant existing glacier was marked by two waving parallel lines of lateral moraines, which formed, as it were, a vast raised gutter, or channel, ascending from perhaps 16,000 feet elevation, till it was hidden behind a spur in the valley. With a telescope I could descry many similar smaller glaciers, with huge accumulations of shingle at their terminations; but this great one was beautifully seen by the naked eye, and formed a very curious feature in the landscape.

Between the moraines, near my tent, the soil was perfectly level, and consisted of little lake-beds strewn with gigantic boulders, and covered with hard turf of grass and sedge, and little bushes of dwarf rhododendron and prostrate juniper, as trim as if they had been clipped. Altogether these formed the most picturesque little nooks it was possible to conceive; and they exhibited the withered remains of so many kinds of primrose, gentian, anemone, potentilla, orchis, saxifrage, parnassia, campanula, and pedicularis, that in summer they must be perfect gardens of wild flowers. Around each plot of a few acres was the grand ice-transported girdle of stupendous rocks, many from 50 to 100 feet long, crested with black tabular-branched silver firs, conical deep green tree-junipers, and feathery larches; whilst amongst the blocks grew a profusion of round masses of evergreen rhododendron bushes. Beyond were stupendous frowning cliffs, beneath which the river roared like thunder; and looking up the glacial valley, the setting sun was bathing the expanse of snow in the most delicate changing tints, pink, amber, and gold.

The boulders forming the moraine were so enormous and angular, that I had great difficulty in ascending it. I saw some pheasants feeding on the black berries of the juniper, but where the large rhododendrons grew amongst the rocks I found it impossible to penetrate. The largest of the moraines is piled to upwards of 1000 feet against the south flank of the lateral valley, and stretched far up it beyond my camp, which was in a grove of silver firs. A large flock of sheep and goats, laden with salt, overtook us here on their route from Wallanchoon to Yalloong. The sheep I observed to feed on the *Rhododendron Thomsoni* and *campylocarpum*.

On the roots of one of the latter species a parasitical Broom-rape (*Orobanche*) grew abundantly; and about the moraines were more mosses, lichens, &c., than I have elsewhere seen in the loftier Himalaya, encouraged no doubt by the dampness of this grand mountain gorge, which is so hemmed in that the sun never reaches it until four or five hours after it has gilded the overhanging peaks.

December 5.—The morning was bright and clear, and we left early for the Choonjerma pass. I had hoped the route would be up the magnificent glacier-girdled valley in which we had encamped; but it lay up another, considerably south of it, and to which we crossed, ascending the rocky moraine, in the clefts of which grew abundance of a common Scotch fern, *Cryptogramma crispa!*

The clouds early commenced gathering, and it was curious to watch their rapid formation in coalescing streaks, which became first cirrhi, and then stratus, being apparently continually added to from below by the moisture-bringing southerly wind. Ascending a lofty spur, 1000 feet above the valley, against which the moraine was banked, I found it to be a distinct anticlinal axis. The pass, bearing north-west, and the valley we had descended on the previous day, rose immediately over the curved strata of quartz, topped by the glacier-crowned mountain of Nango, with four glaciers descending from its perpetual snows. The stupendous cliffs on its flanks, under which I had camped on the previous night, were very grand, but not more so than those which dipped into the chasm of the Kambachen below. Looking up the valley of the latter, was another wilderness of ice full of enormous moraines, round the bases of which the river wound.

Ascending, we reached an open grassy valley, and overtook the Tibetans who had preceded us, and who had halted here to feed their sheep. A good-looking girl of the party came to ask me for medicine for her husband's eyes, which had suffered from snow-blindness: she brought me a present of snuff, and carried a little child, stark naked, yet warm from the powerful rays of the sun, at nearly 14,000 feet elevation, in December! I prescribed for the man, and gave the mother a bright farthing to hang round the child's neck, which delighted the party. My watch was only wondered at; but a little spring measuring-tape that rolled itself up, struck them dumb, and when I threw it on the ground with the tape out, the mother shrieked and ran away, while the little savage howled after her.

Above, the path up the ascent was blocked with snow-beds, and for several miles we alternately scrambled among rocks and over slippery slopes, to the top of the first ridge, there being two to cross. The first consisted of a ridge of rocks running east and west from a superb sweep of snowy mountains to the north-west, which presented a chaotic scene of blue glacial ice and white snow, through which splintered rocks and beetling crags thrust their black heads. The view into the Kambachen gorge was magnificent, though it did not reveal the very bottom of the valley and its moraines: the black precipices of its opposite flank seemed to rise to the glaciers of Nango, fore-shortened into snow-capped precipices 5000 feet high, amongst which lay the Kambachen pass, bearing north-west by north. Lower down the valley, appeared a broad flat, called Jubla, a halting-place one stage below the village of Kambachen, on the road to Lelyp on the Tambur: it must be a remarkable geological as well as natural feature, for it appeared to jut abruptly and quite horizontally from the black cliffs of the valley.

Looking north, the conical head of Junnoo was just scattering the mists from its snowy shoulders, and standing forth to view, the most magnificent spectacle I ever beheld. It was quite close to me, bearing north-east by east, and subtending an angle of 12° 23, and is much the steepest and most conical of all the peaks of these regions. From whichever side it is viewed, it rises 9000 feet above the general mountain mass of 16,000 feet elevation, towering like a blunt cone, with a short saddle on one side, that dips in a steep cliff: it appeared as if uniformly snowed, from its rocks above 20,000 feet (like those of Kinchinjunga) being of white granite, and not contrasting with the snow. Whether the top is stratified or not, I cannot tell, but waving parallel lines are very conspicuous near it, as shown in the accompanying view.*

Looking south as evening drew on, another wonderful spectacle presented itself, similar to that which I described at Sakkiazung, but displayed here on an inconceivably grander scale, with all the effects exaggerated. I saw a sea of mist floating 3000 feet beneath me, just below the upper level of the black pines; the magnificent spurs of the snowy range which I had crossed rising out of it in rugged grandeur as promontories and peninsulas, between which the misty ocean seemed to finger up like the fiords of Norway,

* The appearance of Mont Cervin, from the Riffelberg, much reminded me of that of Junnoo, from the Choonjerma pass, the former bearing the same relation to Monte Rosa that the latter does to Kinchinjunga. Junnoo, though incomparably the more stupendous mass, not only rising 10,000 feet higher above the sea, but towering 4000 feet higher above the ridge on which it is supported, is not nearly so remarkable in outline, so sharp, or so peaked as is Mount Cervin: it is a very much grander, but far less picturesque object. The whiteness of the sides of Junnoo adds also greatly to its apparent altitude; while the strong relief in which the black cliffs of Mont Cervin protrude through its snowy mantle greatly diminish both its apparent height and distance.

J. D. H. del^t. John Murray, Albemarle Street, 1854. W. L. Walton, lith.

Junnoo 24000 f^t. from Choonjerma Pass 16,000 f^t. East Nepal.

or the salt-water lochs of the west of Scotland; whilst islets tailed off from the promontories, rising here and there out of the deceptive elements. I was so high above this mist, that it had not the billowy appearance I saw before, but was a calm unruffled ocean, boundless to the south and west, where the horizon over-arched it. A little to the north of west I discerned the most lofty group of mountains in Nepal* (mentioned at p. 185), beyond Kinchinjunga, which I believe are on the west flank of the great valley through which the Arun river enters Nepal from Tibet: they were very distant, and subtended so small an angle, that I could not measure them with the sextant and artificial horizon: their height, judging from the quantity of snow, must be prodigious.

From 4 to 5 P.M. the temperature was 24°, with a very cold wind; the elevation by the barometer was 15,260 feet, and the dew-point 10½°, giving the humidity 0·610, and the amount of vapour 1·09 grains in a cubic foot of air; the same elements at Calcutta, at the same hour, being thermometer 66½°, dew-point 60½°, humidity 0·840, and weight of vapour 5·9 grains.

I waited for an hour, examining the rocks about the pass, till the coolies should come up, but saw nothing worthy of remark, the natural history and geology being identical with those of Kambachen pass: I then bade adieu to the sublime and majestic peak of Junnoo. Thence we continued at nearly the same level for about four miles, dipping into the broad head of a snowy valley, and ascending to the second pass, which lay to the south-east.

On the left I passed a very curious isolated pillar of rock,

* Called Tsungau by the Bhoteeas. Junnoo is called Kumbo-Kurma by the Hill-men of Nepal.

amongst the wild crags to the north-east, whose bases we skirted: it resembles the Capuchin on the shoulder of Mont Blanc, as seen from the Jardin. Evening overtook us while still on the snow near the last ascent. As the sun declined, the snow at our feet reflected the most exquisitely delicate peach-bloom hue; and looking west from the top of the pass, the scenery was gorgeous beyond description, for the sun was just plunging into a sea of mist, amongst some cirrhi and stratus, all in a blaze of the ruddiest coppery hue. As it sank, the Nepal peaks to the right assumed more definite, darker, and gigantic forms, and floods of light shot across the misty ocean, bathing the landscape around me in the most wonderful and indescribable changing tints. As the luminary was vanishing, the whole horizon glowed like copper run from a smelting furnace, and when it had quite disappeared, the little inequalities of the ragged edges of the mist were lighted up and shone like a row of volcanos in the far distance. I have never before or since seen anything, which for sublimity, beauty, and marvellous effects, could compare with what I gazed on that evening from Choonjerma pass. In some of Turner's pictures I have recognized similar effects, caught and fixed by a marvellous effort of genius; such are the fleeting hues over the ice, in his "Whalers," and the ruddy fire in his "Wind, Steam, and Rain," which one almost fears to touch. Dissolving views give some idea of the magic creation and dispersion of the effects, but any combination of science and art can no more recal the scene, than it can the feelings of awe that crept over me, during the hour I spent in solitude amongst these stupendous mountains.

The moon guided us on our descent, which was to the

south, obliquely into the Yalloong valley. I was very uneasy about the coolies, who were far behind, and some of them had been frost-bitten in crossing the Kambachen pass. Still I thought the best thing was to push on, and light large fires at the first juniper we should reach. The change, on passing from off the snow to the dark earth and rock, was so bewildering, that I had great difficulty in picking my way. Suddenly we came on a flat with a small tarn, whose waters gleamed illusively in the pale moon-light: the opposite flanks of the valley were so well reflected on its gloomy surface, that we were at once brought to a stand-still on its banks: it looked like a chasm, and whether to jump across it, or go down it, or along it, was the question, so deceptive was the spectral landscape. Its true nature was, however, soon discovered, and we proceeded round it, descending. Of course there was no path, and after some perplexity amongst rocks and ravines, we reached the upper limit of wood, and halted by some bleached juniper-trees, which were soon converted into blazing fires.

I wandered away from my party to listen for the voices of the men who had lingered behind, about whom I was still more anxious, from the very great difficulty they would encounter if, as we did, they should get off the path. The moon was shining clearly in the black heavens; and its bright light, with the pale glare of the surrounding snow, obscured the milky way, and all the smaller stars; whilst the planets appeared to glow with broader orbs than elsewhere, and the great stars flashed steadily and periodically.

Deep black chasms seemed to yawn below, and cliffs rose on all sides, except down the valley, where looking across the Yalloong river, a steep range of mountains rose, seamed with torrents that were just visible like threads of silver coursing down broad landslips. It was a dead calm, and nothing broke the awful silence but the low hoarse murmur of many torrents, whose mingled voices rose and fell as if with the pulsations of the atmosphere; the undulations of which appeared thus to be marked by the ear alone. Sometimes it was the faintest possible murmur, and then it rose swelling and filling the air with sound: the effect was that of being raised from the earth's surface, and again lowered to it; or that of waters advancing and retiring. In such scenes and with such accompaniments, the mind wanders from the real to the ideal, the larger and brighter lamps of heaven lead us to imagine that we have risen from the surface of our globe and are floating through the regions of space, and that the ceaseless murmur of the waters is the Music of the Spheres.

Contemplation amid such soothing sounds and impressive scenes is very seductive, and withal very dangerous, for the temperature was at freezing-point, my feet and legs were wet through, and it was well that I was soon roused from my reveries by the monosyllabic exclamations of my coolies. They were quite knocked up, and came along grunting, and halting every minute to rest, by supporting their loads, still hanging to their backs, on their stout staves. I had still one bottle of brandy left, with which to splice the main brace. It had been repeatedly begged for in vain, and being no longer expected, was received with unfeigned joy. Fortunately with these people a little spirits goes a long way, and I kept half for future emergencies.

We camped at 13,290 feet, the air was calm and mild to the feeling, though the temperature fell to $22\frac{3}{4}°$. On the following morning we saw two musk-deer,* called "Kosturah" by the mountaineers. The musk, which hangs in a pouch near the navel of the male, is the well-known object of traffic with Bengal. This creature ranges between 8000 and 13,000 feet, on the Himalaya, often scenting the air for many hundred yards. It is a pretty grey animal, the size of a roebuck, and something resembling it, with coarse fur, short horns, and two projecting teeth from the upper jaw, said to be used in rooting up the aromatic herbs from which the Bhoteeas believe that it derives the odour of musk. This I much doubt, because the animal never frequents those very lofty regions where the herbs supposed to provide the scent are found, nor have I ever seen signs of any having been so rooted up. The *Delphinium glaciale* smells strongly and disagreeably of musk, but it is one of the most alpine plants in the world, growing at an elevation of 17,000 feet, far above the limits of the Kosturah. The female and young male are very good eating, much better than any Indian venison I ever tasted, being sweet and tender. Mr. Hodgson once kept a female alive, but it was very wild, and continued so as long as I knew it. Two of my Lepchas gave chase to these animals, and fired many arrows in vain after them: these people are fond of carrying a bow, but are very poor shots.

We descended 3000 feet to the deep valley of the Yalloong river which runs west-by-south to the Tambur, from between Junnoo and Kubra: the path was very bad, over quartz, granite, and gneiss, which cut the shoes and feet severely. The bottom of the valley, which is elevated 10,450 feet, was filled with an immense accumulation of

* There are two species of musk-deer in the Himalaya, besides the Tibetan kind, which appears identical with the Siberian animal originally described by Pallas.

angular gravel and débris of the above rocks, forming on both sides of the river a terrace 400 feet above the stream, which flowed in a furious torrent. The path led over this deposit for a good many miles, and varied exceedingly in height, in some places being evidently increased by landslips, and at others apparently by moraines.

TIBETAN CHARM-BOX.

CHAPTER XII.

Yalloong valley—Find Kanglanamo pass closed—Change route for the southward—*Picrorhiza*—View of Kubra—*Rhododendron Falconeri*—Yalloong river—Junction of gneiss and clay-slate—Cross Yalloong range—View—Descent—Yew—Vegetation—Misty weather—Tongdam village—Khabang—Tropical vegetation—Sidingbah Mountain—View of Kinchinjunga—Yangyading village—Slopes of hills, and courses of rivers—Khabili valley—Ghorkha Havildar's bad conduct—Ascend Singalelah—Plague of ticks—Short commons—Cross Islumbo pass—Boundary of Sikkim—Kulhait valley—Lingcham—Reception by Kajee—Hear of Dr. Campbell's going to meet Rajah—Views in valley—Leave for Teesta river—Tipsy Kajee—Hospitality—Murwa beer—Temples—*Acorus Calamus*—Long Mendong—Burning of dead—Superstitions—Cross Great Rungeet—Boulders, origin of—Purchase of a dog—Marshes—Lamas—Dismiss Ghorkhas—Bhoteea house—Murwa beer.

On arriving at the bottom we found a party who were travelling with sheep laden with salt; they told us that the Yalloong village, which lay up the valley on the route to the Kanglanamo pass (leading over the south shoulder of Kubra into Sikkim) was deserted, the inhabitants having retired after the October fall of snow to Yankutang, two marches down; also that the Kanglanamo pass was impracticable, being always blocked up by the October fall. I was, therefore, reluctantly obliged to abandon the plan of pursuing that route to Sikkim, and to go south, following the west flank of Singalelah to the first of the many passes over it which I might find open.

These people were very civil, and gave me a handful of the root of one of the many bitter herbs called in Bengal

"Teeta," and used as a febrifuge: the present was that of *Picrorhiza*, a plant allied to Speedwell, which grows at from 12,000 to 15,000 feet elevation, and is a powerful bitter, called "Hoonling" by the Tibetans. They had with them above 100 sheep, of a tall, long-legged, Roman-nosed breed. Each carried upwards of forty pounds of salt, done up in two leather bags, slung on either side, and secured by a band going over the chest, and another round the loins, so that they cannot slip off, when going up or down hill. These sheep are very tame, patient creatures, travelling twelve miles a day with great ease, and being indifferent to rocky or steep ground.

Looking east I had a splendid view of the broad snowy mass of Kubra, blocking up, as it were, the head of the valley with a white screen. Descending to about 10,000 feet, the *Abies Brunoniana* appeared, with fine trees of *Rhododendron Falconeri* forty feet high, and with leaves nineteen inches long! while the upper part of the valley was full of *Abies Webbiana*.

At the elevation of 9000 feet, we crossed to the east bank, and passed the junction of the gneiss and mica slate: the latter crossed the river, striking north-west, and the stream cut a dark chasm-like channel through it, foaming and dashing the spray over the splintered ridges, and the broad water-worn hog-backed masses that projected from its bed. Immense veins of granite permeated the rocks, which were crumpled in the strangest manner: isolated angular blocks of schist had been taken up by the granite in a fluid state, and remained imbedded in it.

The road made great ascents to avoid landslips, and to surmount the enormous piles of débris which encumber this valley more than any other. We encamped at 10,050 feet, on a little flat 1000 feet above the bed of the river, and on its east flank. A *Hydrangea* was the common small wood, but *Abies Webbiana* formed the forest, with great Rhododendrons. The weather was foggy, whence I judged that we were in the sea of mist I saw beneath me from the passes; the temperature, considering the elevation, was mild, 37° and 38°, which was partly due to the evolution of heat that accompanies the condensation of these vapours, the atmosphere being loaded with moisture. The thermometer fell to 28° during the night, and in the morning the ground was thickly covered with hoar-frost.

December 7.—We ascended the Yalloong ridge to a saddle 11,000 feet elevation, whence the road dips south to the gloomy gorges of the eastern feeders of the Tambur. Here we bade adieu to the grand alpine scenery, and for several days our course lay in Nepal in a southerly direction, parallel to Singalelah, and crossing every spur and river sent off by that mighty range. The latter flow towards the Tambur, and their beds for forty or fifty miles are elevated about 3000 or 4000 feet. Few of the spurs are ascended above 5000 feet, but all of them rise to 12,000 or 14,000 feet to the westward, where they join the Singalelah range.

I clambered to the top of a lofty hummock, through a dense thicket of interwoven Rhododendron bushes, the clayey soil under which was slippery from the quantity of dead leaves. I had hoped for a view of the top of Kinchinjunga, which bore north-east, but it was enveloped in clouds, as were all the snows in that direction; to the north-west, however, I obtained bearings of the principal peaks, &c., of the Yangma and Kambachen valleys. To the south and south-east, lofty, rugged and pine-clad mountains rose in confused masses, and white sheets of mist came driving up, clinging to the mountain-tops, and shrouding the landscape with extreme rapidity. The remarkable mountain of Sidingbah bore south-south-east, raising its rounded head above the clouds. I could, however, procure no other good bearing.

The descent from the Yalloong ridge to the Khabili feeders of the Tambur was very steep, and in some places almost precipitous, first through dense woods of silver fir, with *Rhod. Falconeri* and *Hodgsoni*, then through *Abies Brunoniana*, with yew (now covered with red berries) to the region of Magnolias and *Rhod. arboreum* and *barbatum*. One bush of the former was in flower, making a gorgeous show. Here also appeared the great oak with lamellated acorns, which I had not seen in the drier valleys to the westward; with many other Dorjiling trees and shrubs. A heavy mist clung to the rank luxuriant foliage, tantalizing from its obscuring all the view. Mica schist replaced the gneiss, and a thick slippery stratum of clay rendered it very difficult to keep one's footing. After so many days of bright sunshine and dry weather, I found this quiet, damp, foggy atmosphere to have a most depressing effect: there was little to interest in the meteorology, the atmospheric fluctuations being far too small; geographical discovery was at an end, and we groped our way along devious paths in wooded valleys, or ascended spurs and ridges, always clouded before noon, and clothed with heavy forest.

At 6000 feet we emerged from the mist, and found ourselves clambering down a deep gully, hemmed in by frightful rocky steeps, which exposed a fine and tolerably continuous section of schistose rocks, striking north-west, and dipping north-east, at a very high angle.

At the bottom three furious torrents met: we descended

the course of one of them, over slanting precipices, or trees lashed to the rocks, and after a most winding course our path conducted us to the village of Tarbu, high above a feeder of the Khabili river, which flows west, joining the Tambur three days' march lower down. Having no food, we had made a very long and difficult march to this place, but finding none here, proceeded on to Tonghem village on the Khabili, descending through thickets of *Rhod. arboreum* to the elevation of 5,560.

This village, or spur, called "Tonghem" by the Limboos, and "Yankutang" by the Bhoteeas, is the winter resort of the inhabitants of the upper Yalloong valley: they received us very kindly, sold us two fowls, and rice enough to last for one or two days, which was all they could spare, and gave me a good deal of information. I found that the Kanglanamo pass had been disused since the Nepal war, that it was very lofty, and always closed in October.

The night was fine, clear, and warm, but the radiation so powerful that the grass was coated with ice the following morning, though the thermometer did not fall below 33° The next day the sun rose with great power, and the vegetation reeked and steamed with the heat. Crossing the river, we first made a considerable descent, and then ascended a ridge to 5,750 feet, through a thick jungle of *Camellia*, *Eurya*, and small oak: from the top I obtained bearings of Yalloong and Choonjerma pass, and had also glimpses of the Kinchin range through a tantalizing jungle; after which a very winding and fatiguing up-and-down march southwards brought us to the village of Khabang, in the magnificent valley of the Tawa, about 800 feet above the river, and 5,500 feet above the sea.

I halted here for a day, to refresh the people, and if possible to obtain some food. I hoped, too, to find a pass into Sikkim, east over Singalelah, but was disappointed: if there had ever been one, it had been closed since the Nepal war; and there was none, for several marches further south, which would conduct us to the Iwa branch of the Khabili.

Khabang is a village of Geroongs, or shepherds, who pasture their flocks on the hills and higher valleys during summer, and bring them down to this elevation in winter: the ground was consequently infested with a tick, equal in size to that so common in the bushes, and quite as troublesome, but of a different species.

The temperature rose to 72°, and the black-bulb thermometer to 140°. Magnolias and various almost tropical trees were common, and the herbaceous vegetation was that of low elevations. Large sugar-cane (*Saccharum*), palm (*Wallichia*), and wild plantains grew near the river, and *Rhod. arboreum* was very common on dry slopes of mica-slate rocks, with the gorgeous and sweet-scented *Luculia gratissima*.

Up the valley of the Tawa the view was very grand of a magnificent rocky mountain called Sidingbah, bearing south-east by south, on a spur of the Singalelah range that runs westerly, and forms the south flank of the Tawa, and the north of the Khabili valleys. This mountain is fully 12,000 feet high, crested with rock and ragged black forest, which, on the north flank, extends to its base: to the eastward, the bare ridges of Singalelah were patched with snow, below which they too were clothed with black pines.

From the opposite side of the Tawa to Khabang (alt. 6,020 feet), I was, during our march southwards, most fortunate in obtaining a splendid view of Kinchinjunga (bearing north-east by north), with its associates, rising over the dark mass of Singalelah, its flanks showing like tier above tier of green glaciers: its distance was fully twenty-five miles, and as only about 7000 feet or 8000 feet from its summit were visible, and Kubra was foreshortened against it, its appearance was not grand; added to which, its top was round and hummocky, not broken into peaks, as when seen from the south and east. Villages and cultivation became more frequent as we proceeded southward, and our daily marches were up ridges, and down into deep valleys, with feeders from the flanks of Sidingbah to the Tambur. We passed through the village of Tchonboong, and camped at Yangyading (4,100 feet), sighted Yamroop, a large village and military post to the west of our route, crossed the Pangwa river, and reached the valley of the Khabili. During this part of the journey, I did not once see the Tambur river, though I was day after day marching only seven to ten miles distant from it, so uneven is the country. The mountains around Taptiatok, Mywa Guola, and Chingtam, were pointed out to me, but they presented no recognizable feature.

I often looked for some slope, or strike of the slopes of the spurs, in any one valley, or that should prevail through several, but could seldom trace any, except on one or two occasions, at low elevations. Looking here across the valleys, there was a tendency in the gentle slopes of the spurs to have plane faces dipping north-east, and to be bounded by a line of cliffs striking north-west, and facing the south-east. In such arrangements, the upheaved cliffs may be supposed to represent parallel lines of faults, dislocation, or rupture, but I could never trace any secondary valleys at right angles to these. There is no such uniformity of strike as to give to the rivers a zig-zag course of any regularity, or one having any apparent dependence on a prevailing arrangement of the rocks; for, though the strike of the chlorite and clay-slate at elevations below 6000 feet along its course, is certainly north-west, with a dip to north-east, the flexures of the river, as projected on the map, deviate very widely from these directions.

The valley of the Khabili is very grand, broad, open, and intersected by many streams and cultivated spurs: the road from Yamroop to Sikkim, once well frequented, runs up its north flank, and though it was long closed we determined to follow and clear it.

On the 11th of December we camped near the village of Sablakoo (4,680 feet), and procured five days' food, to last us as far as the first Sikkim village. Thence we proceeded eastward up the valley, but descending to the Iwa, an affluent of the Khabili, through a tropical vegetation of *Pinus longifolia, Phyllanthus Emblica*, dwarf date-palm, &c.

Gneiss was here the prevailing rock, uniformly dipping north-east 20°, and striking north-west. The same rock no doubt forms the mass Sidingbah, which reared its head 8000 feet above the Iwa river, by whose bed we camped at 3,780 feet. Sand-flies abounded, and were most trouble-some: troops of large monkeys were skipping about, and the whole scene was thoroughly tropical; still, the thermometer fell to 38° in the night, with heavy dew.

Though we passed numerous villages, I found unusual difficulty in getting provision, and received none of the presents so uniformly brought by the villagers to a stranger. I was not long in discovering, to my great mortification, that these were appropriated by the Ghorkha Havildar, who seemed to have profited by our many days of short allow-ance, and diverted the current of hospitality from me to himself. His coolies I saw groaning under heavy burdens, when those of my people were light; and the truth only came out when he had the impudence to attempt to impose a part of his coolies' loads on mine, to enable the former to carry more food, whilst he was pretending that he used every exertion to procure me a scanty supply of rice with my limited stock of money. I had treated this man and his soldiers with the utmost kindness, even nursing them and clothing them from my own stock of flannels, when sick and shivering amongst the snows. Though a high caste Hindoo, and one who assumed Brahmin rank, he had, I found, no objection to eat forbidden things in secret; and now that we were travelling amongst Hindoos, his caste obtained him everything, while money alone availed me. I took him roundly to task for his treachery, which caused him secretly to throw away a leg of mutton he had concealed; I also threatened to expose the humbug of his pretension to caste, but it was then too late to procure more food. Having hitherto much liked this man, and fully trusted him, I was greatly pained by his conduct.

We proceeded east for three days, up the valley, through gloomy forests of tropical trees below 5000 feet; and ascended to oaks and magnolias at 6000 feet. The path was soon obstructed, and we had to tear and cut our way, from 6000 to 10,000 feet, which took two days' very hard work. Ticks swarmed in the small bamboo jungle, and my body was covered with these loathsome insects, which got into my bed and hair, and even attached themselves to my eyelids during the night, when the constant annoyance and irritation completely banished sleep. In the daytime they penetrated my trousers, piercing to my body in many places, so that I repeatedly took off as many as twelve at one time. It is indeed marvellous how so large an insect can painlessly insert a stout barbed proboscis, which requires great force to extract it, and causes severe smarting in the operation. What the ticks feed upon in these humid forests is a perfect mystery to me, for from

6000 to 9000 feet they literally swarmed, where there was neither path nor animal life. They were, however, more tolerable than a commoner species of parasite, which I found it impossible to escape from, all classes of mountaineers being infested with it.

On the 14th, after an arduous ascent through the pathless jungle, we camped at 9,300 feet on a narrow spur, in a dense forest, amongst immense loose blocks of gneiss. The weather was foggy and rainy, and the wind cold. I ate the last supply of animal food, a miserable starved pullet, with rice and Chili vinegar; my tea, sugar, and all other superfluities having been long before exhausted.

On the following morning, we crossed the Islumbo pass over Singalelah into Sikkim, the elevation being 11,000 feet. Above our camp the trees were few and stunted, and we quickly emerged from the forest on a rocky and grassy ridge, covered with withered *Saxifrages*, *Umbelliferæ*, *Parnassia*, *Hypericum*, *&c.* There were no pines on either side of the pass; a very remarkable peculiarity of the damp mountains of Sikkim, which I have elsewhere had occasion to notice: we had left *Pinus longifolia* (a far from common tree in these valleys) at 3000 feet in the Tawa three days before, and ascended to 11,000 feet without passing a coniferous tree of any kind, except a few yews, at 9000 feet, covered with red berries.

The top of the pass was broad, grassy, and bushy with dwarf Bamboo, Rose, and Berberry, in great abundance, covered with mosses and lichens: it had been raining hard all the morning, and the vegetation was coated with ice: a dense fog obscured everything, and a violent south-east wind blew over the pass in our teeth. I collected some very curious and beautiful mosses, putting these frozen treasures into my box, in the form of exquisitely beautiful glass ornaments, or mosses frosted with silver.

A few stones marked the boundary between Nepal and Sikkim, where I halted for half an hour, and hung up my instruments: the temperature was 32°.

We descended rapidly, proceeding eastward down the broad valley of the Kulhait river, an affluent of the Great Rungeet; and as it had begun to sleet and snow hard, we continued until we reached 6,400 feet before camping.

On the following day we proceeded down the valley, and reached habitations at 4000 feet: passing many villages and much cultivation, we crossed the river, and ascended by 7 P.M., to the village of Lingcham, just below the convent of Changachelling, very tired and hungry. Bad weather had set in, and it was pitch dark and raining hard when we arrived; but the Kajee, or head man, had sent out a party with torches to conduct us, and he gave us a most hospitable reception, honoured us with a salute of musketry, and brought abundance of milk, eggs, fowls, plantains, and Murwa beer. Plenty of news was awaiting me here, and a messenger with letters was three marches further north, at Yoksun, waiting my expected return over the Kanglanamo pass. Dr. Campbell, I was told, had left Dorjiling, and was *en route* to meet the Rajah at Bhomsong on the Teesta river, where no European had ever yet been; and as the Sikkim authorities had for sixteen years steadily rejected every overture for a friendly interview, and even refused to allow the agent of the Governor-General to enter their dominions, it was evident that grave doings were pending. I knew that Dr. Campbell had long used every exertion to bring the Sikkim Rajah to a friendly conference, without having to force his way into the country for the purpose, but in vain. It will hardly be believed that though this chief's dominions were redeemed by us from the Nepalese and given back to him; though we had bound ourselves by a treaty to support him on his throne, and to defend him against the Nepalese on the west, the Bhotan people on the east, and the Tibetans on the north; and though the terms of the treaty stipulated for free intercourse, mutual protection, and friendship; the Sikkim authorities had hitherto been allowed to obstruct all intercourse, and in every way to treat the Governor-General's agent and the East India Company with contempt. An affectation of timidity, mistrust, and ignorance was assumed for the purpose of deception, and as a cloak for every insult and resistance to the terms of our treaty, and it was quoted by the Government in answer to every remonstrance on the part of their resident agent at Dorjiling.

On the following morning the Kajee waited on me with a magnificent present of a calf, a kid, fowls, eggs, rice, oranges, plantains, egg-apples, Indian corn, yams, onions, tomatos, parsley, fennel, turmeric, rancid butter, milk, and, lastly, a coolie-load of fermenting millet-seeds, wherewith to make the favourite Murwa beer. In the evening two lads arrived from Dorjiling, who had been sent a week beforehand by my kind and thoughtful friend, Mr. Hodgson, with provisions and money.

The valley of the Kulhait is one of the finest in Sikkim, and it is accordingly the site of two of the oldest and richest conventual establishments. Its length is sixteen miles, from the Islumbo pass to the Great Rungeet, for ten of which it is inhabited, the villages being invariably on long meridional spurs that project north and south from either flank; they are about 2000 feet above the river, and from 4,500 to 5000 feet above the sea. Except where these spurs project, the flanks of the valley are very steep, the mountains rising to 7000 or 8000 feet.

Looking from any spur, up or down the valley, five or six others might be seen on each side of the river, at very nearly the same average level, all presenting great uniformity of contour, namely, a gentle slope towards the centre of the valley, and then an abrupt descent to the river. They were about a quarter of a mile broad at the widest, and often narrower, and a mile or so long; some parts of their surfaces and sides were quite flat, and occasionally occupied by marshes or ponds. Cultivation is almost confined to these spurs, and is carried on both on their summits and steep flanks; between every two is a very steep gulley and water-course. The timber has long since been either wholly or partially cleared from the tops, but, to a great extent, still clothes their flanks and the intervening gorges. I have been particular in describing these spurs, because it is impossible to survey them without ascribing their comparative uniformity of level to the action of water. Similar ones are characteristic features of the valleys of Sikkim between 2000 and 8000 feet, and are rendered conspicuous by being always sites for villages and cultivation: the soil is a vegetable mould, over a deep stratum of red clay.

I am far from supposing that any geologically recent action of the sea has levelled these spurs; but as the great chain of the Himalaya has risen from the ocean, and as every part of it has been subjected to sea-action, it is quite conceivable that intervals of rest during the periods of elevation or submergence would effect their levelling. In a mountain mass so tumbled as is that of Sikkim, any level surface, or approach to it, demands study; and when, as in the Kulhait valley, we find several similar spurs with comparatively flat tops, to occupy about the same level, it is necessary to look for some levelling cause. The action of denudation is still progressing with astonishing rapidity, under an annual fall of from 100 to 150 inches of rain; but its tendency is to obliterate all such phenomena, and to give sharp, rugged outlines to these spurs, in spite of the conservative effects of vegetation.

The weather at Lingcham was gloomy, cold, and damp, with much rain and fog, and the mean temperature ($45\frac{1}{4}°$) was cold for the elevation (4,860 feet): $52\frac{1}{2}°$ was the highest temperature observed, and 39° the lowest.

A letter from Dr. Campbell reached me three days after my arrival, begging me to cross the country to the Teesta river, and meet him at Bhomsong, on its west bank, where he was awaiting my arrival. I therefore left on the 20th of December, accompanied by my friend the Kajee, who was going to pay his respects to the Rajah. He was constantly followed by a lad, carrying a bamboo of Murwa beer slung round his neck, with which he kept himself always groggy. His dress was thoroughly Lepcha, and highly picturesque, consisting of a very broad-brimmed round-crowned bamboo-platted hat, scarlet jacket, and blue-striped cloth shirt, bare feet, long knife, bow and quiver, rings and earrings, and a long pigtail. He spoke no Hindoostanee, but was very communicative through my interpreters.

Leaving the Lingcham spur, we passed steep cliffs of mica and schist, covered with brushwood and long grass, about 1000 feet above which the Changachelling convent is perched. Crossing a torrent, we came to the next village, on the spur of Kurziuk, where I was met by a deputation of women, sent by the Lamas of Changachelling, bearing enormous loads of oranges, rice, milk, butter, ghee, and the everflowing Murwa beer.

The villagers had erected a shady bower for me to rest under, of leaves and branches, and had fitted up a little bamboo stage, on which to squat cross-legged as they do, or to hang my legs from, if I preferred: after conducting me to this, the parties advanced and piled their cumbrous presents on the ground, bowed, and retired; they were succeeded by the beer-carrier, who plunged a clean drinking-tube to the bottom of the steaming bamboo jug (described at p. 175), and held it to my mouth, then placing it by my side, he bowed and withdrew. Nothing can be more fascinating than the simple manners of these kind people, who really love hospitality for its own sake, and make the stranger feel himself welcome. Just now too, the Durbar had ordered every attention to be paid me; and I hardly passed a village, however small, without receiving a present, or a cottage, where beer was not offered. This I found a most grateful beverage; and of the occasional rests under leafy screens during a hot day's march, and sips at the bamboo jug, I shall ever retain a grateful remembrance. Happily the liquor is very weak, and except by swilling, as my friend the Kajee did, it would be impossible to get fuddled by it.

At Kurziuk I was met by a most respectable Lepcha, who, as a sort of compliment, sent his son to escort us to the next village and spur of Pemiongchi, to reach which we crossed another gorge, of which the situation and features were quite similar to those of Kurziuk and Lingcham.

The Pemiongchi and Changachelling convents and temples stand a few miles apart, on the ridge forming the north flank of the Kulhait valley; and as they will be described hereafter, I now only allude to the village, which is fully 1000 feet below the convent, and large and populous.

At Pemiongchi a superior Lama met me with another

overwhelming present: he was a most jolly fat monk, shaven and girdled, and dressed in a scarlet gown: my Lepchas kotowed to him, and he blessed them by the laying on of hands.

There is a marsh on this spur, full of the common English *Acorus Calamus,* or sweet-flag, whose roots being

PEMIONGCHI GOOMPA AND CHAITS.

very aromatic, are used in griping disorders of men and cattle. Hence we descended suddenly to the Great Rungeet, which we reached at its junction with the Kulhait: the path was very steep and slippery, owing to micaceous rocks, and led along the side of an enormous Mendong,* which ran

* This remarkable structure, called the Kaysing Mendong, is 200 yards long, 10 feet high, and 6 or 8 feet broad: it is built of flat, slaty stones, and both

down the hill for several hundred yards, and had a large chait at each end, with several smaller ones at intervals. Throughout its length were innumerable inscriptions of "Om Mani Padmi om," with well carved figures of Boodh in his many incarnations, besides Lamas, &c. At the lower end was a great flat area, on which are burnt the bodies of Sikkim people of consequence: the poorer people are buried, the richer burned, and their ashes scattered or interred, but not in graves proper, of which there are none. Nor are there any signs of Lepcha interment throughout Sikkim; though chaits are erected to the memory of the departed, they have no necessary connection with the remains, and generally none at all. Corpses in Sikkim are never cut to pieces and thrown into lakes, or exposed on hills for the kites and crows to devour, as is the case in Tibet.

We passed some curious masses of crumpled chlorite slate, presenting deep canals or furrows, along which a demon once drained all the water from the Pemiongchi spur, to the great annoyance of the villagers: the Lamas, however, on choosing this as a site for their temples, easily confounded the machinations of the evil spirit, who, in the eyes of the simple Lepchas, was answerable for all the mischief.

I crossed the Great Rungeet at 1840 feet above the sea, where its bed was twenty yards in width; a rude bridge, composed of two culms of bamboo and a hand-rail, conducted me to the other side, where we camped (on the east bank) in a thick tropical jungle. In the evening

faces are covered with inscribed slates, of which there are upwards of 700, and the inscriptions, chiefly "Om Mani," &c., are in both the Uchen and Lencha Ranja characters of Tibet. A tall stone, nine feet high, covered also with inscriptions, terminates it at the lower end.

I walked down the banks of the river, which flowed in a deep gorge, cumbered with enormous boulders of granite, clay-slate, and mica-slate; the rocks *in situ* were all of the latter description, highly inclined, and much dislocated. Some of the boulders were fully ten feet in diameter, permeated and altered very much by granite veins which had evidently been injected when molten, and had taken up angular masses of the chlorite which remained, as it were, suspended in the veins.

It is not so easy to account for the present position of these blocks of granite, a rock not common at elevations below 10,000 feet. They have been transported from a considerable distance in the interior of the lofty valley to the north, and have descended not less than 8000 feet, and travelled fully fifteen miles in a straight line, or perhaps forty along the river bed. It may be supposed that moraines have transported them to 8000 feet (the lowest limit of apparent moraines), and the power of river water carried them further; if so, the rivers must have been of much greater volume formerly than they are now.

Our camp was on a gravel flat, like those of the Nepal valleys, about sixty feet above the river; its temperature was 52°, which felt cool when bathing.

From the river we proceeded west, following a steep and clayey ascent up the end of a very long spur, from the lofty mountain range called Mungbreu, dividing the Great Rungeet from the Teesta. We ascended by a narrow path, accomplishing 2,500 feet in an hour and a quarter, walking slowly but steadily, without resting; this I always found a heavy pull in a hot climate.

At about 4000 feet above the sea, the spur became more open and flat, like those of the Kulhait valley, with alternate slopes and comparative flats: from this elevation the view north, south, and west, was very fine; below us flowed the river, and a few miles up it was the conical wooded hill of Tassiding, rising abruptly from a fork of the deep river gorge, crowned with its curious temples and mendongs, and bristling with chaits: on it is the oldest monastery in Sikkim, occupying a singularly picturesque and prominent position. North of this spur, and similar to it, lay that of Raklang, with the temple and monastery of the same name, at about this elevation. In front, looking west, across the Great Rungeet, were the monasteries of Changachelling and Pemiongchi, perched aloft; and south of these were the flat-topped spurs of the Kulhait valley, with their villages, and the great mendong which I had passed on the previous day, running like a white line down the spur. To the north, beyond Tassiding, were two other monasteries, Doobdee and Sunnook, both apparently placed on the lower wooded flanks of Kinchinjunga; whilst close by was Dholing, the seventh religious establishment now in sight.

We halted at a good wooden house to refresh ourselves with Murwa beer, where I saw a woman with cancer in the face, an uncommon complaint in this country. I here bought a little black puppy, to be my future companion in Sikkim: he was of a breed between the famous Tibet mastiff and the common Sikkim hunting-dog, which is a variety of the sorry race called Pariah in the plains. Being only a few weeks old, he looked a mere bundle of black fur; and I carried him off, for he could not walk.

We camped at the village of Lingdam (alt. 5,550 feet), occupying a flat, and surrounded by extensive pools of water (for this country) containing *Acorus*, *Potamogeton*, and duckweed. Such ponds I have often met with on these terraces, and they are very remarkable, not being dammed in by any conspicuous barrier, but simply occupying depressions in the surface, from which, as I have repeatedly observed, the land dips rapidly to the valleys below.

This being the high-road from Tumloong or Sikkim Durbar (the capital, and Rajah's residence) to the numerous monasteries which I had seen, we passed many Lamas and monks on their way home from Tumloong, where they had gone to be present at the marriage of the Tupgain Lama, the eldest son of the Rajah. A dispensation having previously been procured from Lhassa, this marriage had been effected by the Lamas, in order to counteract the efforts of the Dewan, who sought to exercise an undue influence over the Rajah and his family. The Tupgain Lama having only spiritual authority, and being bound to celibacy, the temporal authority devolved on the second son, who was heir apparent of Sikkim; he, however, having died, an illegitimate son of the Rajah was favoured by the Dewan as heir apparent. The bride was brought from Tibet, and the marriage party were feasted for eighteen days at the Rajah's expense. All the Lamas whom I met were clad in red robes, with girdles, and were shaven, with bare feet and heads, or mitred; they wore rosaries of onyx, turquoise, quartz, lapis lazuli, coral, glass, amber, or wood, especially yellow berberry and sandal-wood: some had staves, and one a trident like an eel-fork, on a long staff, an emblem of the Hindoo Trinity, called Trisool Mahadeo, which represents Brahma, Siva, and Vishnu, in Hindoo; and Boodh, Dhurma, and Sunga, in Boodhist theology. All were on foot, indeed ponies are seldom used in this country; the Lamas, however, walked with becoming gravity and indifference to all around them.

The Kajee waited upon me in the evening full of importance, having just received a letter from his Rajah, which he wished to communicate to me in private; so I accompanied him to a house close by, where he was a guest, when the secret came out, that his highness was dreadfully alarmed at my coming with the two Ghorka Sepoys, whom I accordingly dismissed.

The house was of the usual Bhoteea form, of wood, well built on posts, one-storied, containing a single apartment hung round with bows, quivers, shields, baskets of rice, and cornucopias of Indian corn, the handsomest and most generous looking of all the Cerealia. The whole party were deep in a carouse on Murwa beer, and I saw the operation of making it. The millet-seed is moistened, and ferments for two days: sufficient for a day's allowance is then put into a vessel of wicker-work, lined with India-rubber to make it water-tight; and boiling water is poured on it with a ladle of gourd, from a huge iron cauldron that stands all day over the fire. The fluid, when quite fresh, tastes like negus of Cape sherry, rather sour. At this season the whole population are swilling, whether at home or travelling, and heaps of the red-brown husks are seen by the side of all the paths.

SIKKIM LAMAS WITH PRAYING CYLINDER AND DORJE; THE LATERAL FIGURES ARE MONKS OR GYLONGS.

CHAPTER XIII.

Raklang pass—Uses of nettles—Edible plants—Lepcha war—Do-mani stone—Neongong—Teesta valley—Pony, saddle, &c.—Meet Campbell—Vegetation and scenery—Presents—Visit of Dewan—Characters of Rajah and Dewan—Accounts of Tibet—Lhassa—Siling—Tricks of Dewan—Walk up Teesta—Audience of Rajah—Lamas—Kajees—Tchebu Lama, his character and position—Effects of interview—Heir-apparent—Dewan's house—Guitar—Weather—Fall of river—Tibet officers—Gigantic trees—Neongong lake—Mainom, ascent of—Vegetation—Camp on snow—Silver firs—View from top—Kinchin, &c.—Geology—Vapours—Sunset effect—Elevation—Temperature, &c.—Lamas of Neongong—Temples—Religious festival—Bamboo, flowering—Recross pass of Raklang—Numerous temples, villages, &c.—Domestic animals—Descent to Great Rungeet.

On the following morning, after receiving the usual presents from the Lamas of Dholing, and from a large posse of women belonging to the village of Barphiung, close by, we ascended the Raklang pass, which crosses the range dividing the waters of the Teesta from those of the Great Rungeet. The Kajee still kept beside me, and proved a lively companion: seeing me continually plucking and noting plants, he gave me much local information about them. He told me the uses made of the fibres of the various nettles; some being twisted for bowstrings, others as thread for sewing and weaving; while many are eaten raw and in soups, especially the numerous little succulent species. The great yellow-flowered *Begonia* was abundant, and he cut its juicy stalks to make sauce (as we do apple-sauce) for some pork which he expected to get at Bhomsong; the taste is acid and very pleasant. The large succulent fern, called *Botrychium*,* grew here plentifully; it is boiled and eaten, both here and in New Zealand. Ferns are more commonly used for food than is supposed. In Calcutta the Hindoos boil young tops of a *Polypodium* with their shrimp curries; and both in Sikkim and Nepal the watery tubers of an *Aspidium* are abundantly eaten. So also the pulp of one tree-fern affords food, but only in times of scarcity, as does that of another species in New Zealand (*Cyathea medullaris*): the pith of all is composed of a coarse sago, that is to say, of cellular tissue with starch granules.

A thick forest of Dorjiling vegetation covers the summit, which is only 6,800 feet above the sea: it is a saddle, connecting the lofty mountain of Mainom (alt. 11,000 feet) to the North, with Tendong (alt. 8,663 feet) to the south. Both these mountains are on a range which is continuous with Kinchinjunga, projecting from it down into the very heart of Sikkim. A considerable stand was made here by the Lepchas during the Nepal war in 1787; they defended the pass with their arrows for some hours, and then retired towards the Teesta, making a second stand lower down, at a place pointed out to me, where rocks on either side gave them the same advantages. The Nepalese, however, advanced to the Teesta, and then retired with little loss.

Unfortunately a thick mist and heavy rain cut off all view of the Teesta valley, and the mountains of Chola to the eastward; which I much regretted.

Descending by a very steep, slippery path, we came to a

* *Botrychium Virginicum*, Linn. This fern is eaten abundantly by the New Zealanders: its distribution is most remarkable, being found very rarely indeed in Europe, and in Norway only. It abounds in many parts of the Southern United States, the Andes of Mexico, &c., in the Himalaya mountains, Australia, and New Zealand.

fine mass of slaty gneiss, thirty feet long and thirteen feet high; not *in situ*, but lying on the mountain side: on its sloping face was carved in enormous characters, "Om Mani Padmi om;" of which letters the top-strokes afford an uncertain footing to the enthusiast who is willing to purchase a good metempsychosis by walking along the slope, with his heels or toes in their cavities. A small inscription in one corner is said to imply that this was the work of a pious monk of Raklang; and the stone is called "Do-mani," literally, "stone of prayer."

DO-MANI STONE.

The rocks and peaks of Mainom are said to overhang the descent here with grandeur; but the continued rain hid everything but a curious shivered peak, apparently of chlorite schist, which was close by, and reflected a green colour: it is of course reported to be of turquoise, and inaccessible.

Descending, the rocks became more micaceous, with broad seams of pipe-clay, originating in decomposed beds of felspathic gneiss: the natives used this to whitewash and mortar their temples.

I passed the monastery of Neongong, the monks of which were building a new temple; and came to bring me a large present. Below it is a pretty little lake, about 100 yards across, fringed with brushwood. We camped at the village of Nampok, 4,370 feet above the sea; all thoroughly sodden with rain.

During the night much snow had fallen at and above 9000 feet, but the weather cleared on the following morning, and disclosed the top of Mainom, rising close above my camp, in a series of rugged shivered peaks, crested with pines, which looked like statues of snow: to all other quarters this mountain presents a very gently sloping outline. Up the Teesta valley there was a pretty peep of snowy mountains, bearing north 35° east, of no great height.

I was met by a messenger from Dr. Campbell, who told me he was waiting breakfast; so I left my party, and, accompanied by the Kajee and Meepo, hurried down to the valley of the Rungoon (which flows east to the Teesta), through a fine forest of tropical trees; passing the villages of Broom * and Lingo, to the spur of that name; where I was met by a servant of the Sikkim Dewan's, with a pony for my use. I stared at the animal, and felt inclined to ask what he had to do here, where it was difficult enough

* On the top of the ridge above Broom, a tall stone is erected by the side of the path, covered with private marks, indicating the height of various individuals who are accustomed to measure themselves thus; there was but one mark above 5 feet 7 inches, and that was 6 inches higher. It turned out to be Campbell's, who had passed a few days before, and was thus proved to top the natives of Sikkim by a long way.

to walk up and down slippery slopes, amongst boulders of rock, heavy forest, and foaming torrents; but I was little aware of what these beasts could accomplish. The Tartar saddle was imported from Tibet, and certainly a curiosity; once—but a long time ago—it must have been very handsome; it was high-peaked, covered with shagreen and silvered ornaments, wretchedly girthed, and with great stirrups attached to short leathers. The bridle and head-gear were much too complicated for description; there were good leather, raw hide, hair-rope, and scarlet worsted all brought into use; the bit was the ordinary Asiatic one, jointed, and with two rings. I mounted on one side, and at once rolled over, saddle and all, to the other; the pony standing quite still. I preferred walking; but Dr. Campbell had begged of me to use the pony, as the Dewan had procured and sent it at great trouble: I, however, had it led till I was close to Bhomsong, when I was hoisted into the saddle and balanced on it, with my toes in the stirrups and my knees up to my breast; twice, on the steep descent to the river, my saddle and I were thrown on the pony's neck; in these awkward emergencies I was assisted by a man on each side, who supported my weight on my elbows: they seemed well accustomed to easing mounted ponies down hill without giving the rider the trouble of dismounting. Thus I entered Dr. Campbell's camp at Bhomsong, to the pride and delight of my attendants; and received a hearty welcome from my old friend, who covered me with congratulations on the successful issue of a journey which, at this season, and under such difficulties and discouragements, he had hardly thought feasible.

Dr. Campbell's tent was pitched in an orange-grove, occupying a flat on the west bank of the Teesta, close to a small enclosure of pine-apples, with a pomegranate tree in the middle. The valley is very narrow, and the vegetation wholly tropical, consisting of two species of oak, several palms, rattan-cane (screw-pine), *Pandanus*, tall grasses, and all the natives of dense hot jungles. The river is a grand feature, broad, rocky, deep, swift, and broken by enormous boulders of rock; its waters were of a pale opal green, probably from the materials of the soft micaceous rocks through which it flows.

A cane bridge crosses it,* but had been cut away (in feigned distrust of us), and the long canes were streaming from their attachments on either shore down the stream, and a triangular raft of bamboo was plying instead, drawn to and fro by means of a strong cane.

Soon after arriving I received a present from the Rajah, consisting of a brick of Tibet tea, eighty pounds of rancid yak butter, in large squares, done up in yak-hair cloth, three loads of rice, and one of Murwa for beer; rolls of bread,† fowls, eggs, dried plums, apricots, jujubes, currants, and Sultana raisins, the latter fruits purchased at Lhassa, but imported thither from western Tibet; also some trays of coarse milk-white crystallised salt, as dug in Tibet.

In the evening we were visited by the Dewan, the head and front of all our Sikkim difficulties, whose influence was paramount with the Rajah, owing to the age and infirmities of the latter, and his devotion to religion, which absorbed all his time and thoughts. The Dewan was a good-looking Tibetan, very robust, fair, muscular and well fleshed; he had a very broad Tartar face, quite free of hair; a small and beautifully formed mouth and chin, very broad cheekbones,

* Whence the name of Bhomsong Samdong, the latter word meaning bridge.

† These rolls, or rather, sticks of bread, are made in Tibet, of fine wheaten flour, and keep for a long time: they are sweet and good, but very dirtily prepared.

and a low, contracted forehead: his manners were courteous and polite, but evidently affected, in assumption of better breeding than he could in reality lay claim to. The Rajah himself was a Tibetan of just respectable extraction, a native of the Sokpo province, north of Lhassa: his Dewan was related to one of his wives, and I believe a Lhassan by birth as well as extraction, having probably also Kashmir blood in him.* Though minister, he was neither financier nor politician, but a mere plunderer of Sikkim, introducing his relations, and those whom he calls so, into the best estates in the country, and trading in great and small wares, from a Tibet pony to a tobacco pipe, wholesale and retail. Neither he nor the Rajah are considered worthy of notice by the best Tibet families or priests, or by the Chinese commissioners settled in Lhassa and Jigatzi. The latter regard Sikkim as virtually English, and are contented with knowing that its ruler has no army, and with believing that its protectors, the English, could not march an army across the Himalaya if they would.

The Dewan, trading in wares which we could supply better and cheaper, naturally regarded us with repugnance, and did everything in his power to thwart Dr. Campbell's attempts to open a friendly communication between the Sikkim and English governments. The Rajah owed everything to us, and was, I believe, really grateful; but he was a mere cipher in the hands of his minister. The priests again, while rejoicing in our proximity, were apathetic, and dreaded the more active Dewan; and the people had long given evidence of their confidence in the English. Under these circumstances it was in the hope of gaining the Rajah's own ear, and representing to him the advantages

* The Tibetans court promiscuous intercourse between their families and the Kashmir merchants who traverse their country.

of promoting an intercourse with us, and the danger of continuing to violate the terms of our treaty, that Dr. Campbell had been authorised by government to seek an interview with His Highness. At present our relations were singularly infelicitous. There was no agent on the Sikkim Rajah's part to conduct business at Dorjiling, and the Dewan insisted on sending a creature of his own, who had before been dismissed for insolence. Malefactors who escaped into Sikkim were protected, and our police interrupted in the discharge of their duties; slavery was practised; and government communications were detained for weeks and months under false pretences.

In his interviews with us the Dewan appeared to advantage: he was fond of horses and shooting, and prided himself on his hospitality. We gained much information from many conversations with him, during which politics were never touched upon. Our queries naturally referred to Tibet and its geography, especially its great feature the Yarou Tsampoo river; this he assured us was the Burrampooter of Assam, and that no one doubted it in that country. Lhassa he described as a city in the bottom of a flat-floored valley, surrounded by lofty snowy mountains: neither grapes, tea, silk, or cotton are produced near it, but in the Tartchi province of Tibet, one month's journey east of Lhassa, rice, and a coarse kind of tea are both grown. Two months' journey north-east of Lhassa is Siling, the well-known great commercial entrepôt* in west China; and there coarse silk is produced. All Tibet he described as mountainous, and an inconceivably poor country: there are no plains, save flats in the bottoms of the valleys, and the paths lead over lofty mountains. Sometimes, when the inhabitants are obliged from famine to change their habitations in winter, the old

* The entrepôt is now removed to Tang-Keou-Eul.—See Huc and Gabet.

and feeble are frozen to death, standing and resting their chins on their staves; remaining as pillars of ice, to fall only when the thaw of the ensuing spring commences.

We remained several days at Bhomsong, awaiting an interview with the Rajah, whose movements the Dewan kept shrouded in mystery. On Dr. Campbell's arrival at this river a week before, he found messengers waiting to inform him that the Rajah would meet him here; this being half way between Dorjiling and Tumloong. Thenceforward every subterfuge was resorted to by the Dewan to frustrate the meeting; and even after the arrival of the Rajah on the east bank, the Dewan communicated with Dr. Campbell by shooting across the river arrows to which were attached letters, containing every possible argument to induce him to return to Dorjiling; such as that the Rajah was sick at Tumloong, that he was gone to Tibet, that he had a religious fast and rites to perform, &c. &c.

One day we walked up the Teesta to the Rumphiup river, a torrent from Mainom mountain to the west; the path led amongst thick jungle of *Wallichia* palm, prickly rattan canes, and the *Pandanus*, or screw-pine, called "Borr," which has a straight, often forked, palm-like trunk, and an immense crown of grassy saw-edged leaves four feet long: it bears clusters of uneatable fruit as large as a man's fist, and their similarity to the pine-apple has suggested the name of "Borr" for the latter fruit also, which has for many years been cultivated in Sikkim, and yields indifferent produce. Beautiful pink balsams covered the ground, but at this season few other showy plants were in flower: the rocks were chlorite, very soft and silvery, and so curiously crumpled and contorted as to appear as though formed of scales of mica crushed together, and confusedly arranged in layers: the strike was north-west, and dip north-east from 60° to 70°.

Messengers from the Dewan overtook us at the river to announce that the Rajah was prepared and waiting to give us a reception; so we returned, and I borrowed a coat from Dr. Campbell instead of my tattered shooting-jacket; and we crossed the river on the bamboo-raft. As it is the custom on these occasions to exchange presents, I was officially supplied with some red cloth and beads: these, as well as Dr. Campbell's present, should only have been delivered during or after the audience; but our wily friend the Dewan here played us a very shabby trick; for he managed that our presents should be stealthily brought in before our appearance, thus giving to the by-standers the impression of our being tributaries to his Highness!

The audience chamber was a mere roofed shed of neat bamboo wattle, about twenty feet long: two Bhoteeas in scarlet jackets, and with bows in their hands, stood on each side of the door, and our own chairs were carried before us for our accommodation. Within was a square wicker throne, six feet high, covered with purple silk, brocaded with dragons in white and gold, and overhung by a canopy of tattered blue silk, with which material part of the walls also was covered. An oblong box (containing papers) with gilded dragons on it, was placed on the stage or throne, and behind it was perched cross-legged, an odd, black, insignificant looking old man, with twinkling upturned eyes: he was swathed in yellow silk, and wore on his head a pink silk hat with a flat broad crown, from all sides of which hung floss silk. This was the Rajah, a genuine Tibetan, about seventy years old. On some steps close by, and ranged down the apartment, were his relations, all in brocaded silk robes reaching from the throat to the ground, and girded about the waist; and wearing caps similar to that of the Rajah. Kajees, counsellors, and shaven mitred Lamas were there, to the number of twenty, all planted with their backs to the wall, mute and motionless as statues. A few spectators were huddled together at the lower end of the room, and a monk waved about an incense pot containing burning juniper and other odoriferous plants. Altogether the scene was solemn and impressive: as Campbell well expressed it, the genius of Lamaism reigned supreme.

We saluted, but received no complimentary return; our chairs were then placed, and we seated ourselves, when the Dewan came in, clad in a superb purple silk robe, worked with circular gold figures, and formally presented us. The Dewan then stood; and as the Rajah did not understand Hindoostanee, our conversation was carried on through the medium of a little bare-headed rosy-cheeked Lama, named "Tchebu," clad in a scarlet gown, who acted as interpreter. The conversation was short and constrained: Tchebu was known as a devoted servant of the Rajah and of the heir apparent; and in common with all the Lamas he hates the Dewan, and desires a friendly intercourse between Sikkim and Dorjiling. He is, further, the only servant of the Rajah capable of conversing both in Hindoo and Tibetan, and the uneasy distrustful look of the Dewan, who understands the latter language only, was very evident. He was as anxious to hurry over the interview as Dr. Campbell and Tchebu were to protract it; it was clear, therefore, that nothing satisfactory could be done under such auspices.

As a signal for departure white silk scarfs were thrown over our shoulders, according to the established custom in Tibet, Sikkim, and Bhotan; and presents were made to us of China silks, bricks of tea, woollen cloths, yaks, ponies, and salt, with worked silk purses and fans for Mrs. Campbell; after which we left. The whole scene was novel and very curious. We had had no previous idea of the extreme poverty of the Rajah, of his utter ignorance of the usages of Oriental life, and of his not having any one near to instruct him. The neglect of our salutation, and the conversion of our presents into tribute, did not arise from any ill-will: it was owing to the craft of the Dewan in taking advantage of the Rajah's ignorance of his own position and of good manners. Miserably poor, without any retinue, taking no interest in what passes in his own kingdom, subsisting on the plainest and coarsest food, passing his time in effectually abstracting his mind from the consideration of earthly things, and wrapt in contemplation, the Sikkim Rajah has arrived at great sanctity, and is all but prepared for that absorption into the essence of Boodh, which is the end and aim of all good Boodhists. The mute conduct of his Court, who looked like attendants at an inquisition, and the profound veneration expressed in every word and gesture of those who did move and speak, recalled a Pekin reception. His attendants treated him as a being of a very different nature from themselves; and well might they do so, since they believe that he will never die, but retire from the world only to re-appear under some equally sainted form.

Though productive of no immediate good, our interview had a very favourable effect on the Lamas and people, who had long wished it; and the congratulations we received thereon during the remainder of our stay in Sikkim were many and sincere. The Lamas we found universally in high spirits; they having just effected the marriage of the heir apparent, himself a Lama, said to possess much ability and prudence, and hence being very obnoxious to the

Dewan, who vehemently opposed the marriage. As, however, the minister had established his influence over the youngest, and estranged the Rajah from his eldest son, and was moreover in a fair way for ruling Sikkim himself, the Church rose in a body, procured a dispensation from Lhassa for the marriage of a priest, and thus hoped to undermine the influence of the violent and greedy stranger.

In the evening, we paid a farewell visit to the Dewan, whom we found in a bamboo wicker-work hut, neatly hung with bows, arrows, and round Lepcha shields of cane, each with a scarlet tuft of yak-hair in the middle; there were also muskets, Tibetan arms, and much horse gear; and at one end was a little altar, with cups, bells, pastiles, and images. He was robed in a fawn-coloured silk gown, lined with the softest of wool, that taken from unborn lambs: like most Tibetans, he extracts all his beard with tweezers; an operation he civilly recommended to me, accompanying the advice with the present of a neat pair of steel forceps. He aspires to be considered a man of taste, and plays the Tibetan guitar, on which he performed some airs for our amusement: the instrument is round-bodied and long-armed, with six strings placed in pairs, and probably comes from Kashmir: the Tibetan airs were simple and quite pretty, with the time well marked.

During our stay at Bhomsong, the weather was cool, considering the low elevation (1,500 feet), and very steady; the mean temperature was $52\frac{1}{4}°$, the maximum $71\frac{1}{4}°$, the minimum $42\frac{3}{4}°$. The sun set behind the lofty mountains at 3 P.M., and in the morning a thick, wet, white, dripping fog settled in the bottom of the valley, and extended to 800 or 1000 feet above the river-bed; this was probably caused by the descent of cold currents into the humid gorge: it was dissipated soon after sunrise, but formed again at sunset for a few minutes, giving place to clear starlight nights.

A thermometer sunk two feet seven inches, stood at 64°. The temperature of the water was pretty constant at 51°: from here to the plains of India the river has a nearly uniform fall of 1000 feet in sixty-nine miles, or sixteen feet to a mile: were its course straight for the same distance, the fall would be 1000 feet in forty miles, or twenty-five feet to a mile.

Dr. Campbell's object being accomplished, he was anxious to make the best use of the few days that remained before his return to Dorjiling, and we therefore arranged to ascend Mainom, and visit the principal convents in Sikkim together, after which he was to return south, whilst I should proceed north to explore the south flank of Kinchin-junga. For the first day our route was that by which I had arrived. We left on Christmas-day, accompanied by two of the Rajah's, or rather Dewan's officers, of the ranks of Dingpun and Soupun, answering to those of captain and lieutenant; the titles were, however, nominal, the Rajah having no soldiers, and these men being profoundly ignorant of the mysteries of war or drill. They were splendid specimens of Sikkim Bhoteeas (*i.e.* Tibetans, born in Sikkim, sometimes called Arrhats), tall, powerful, and well built, but insolent and bullying: the Dingpun wore the Lepcha knife, ornamented with turquoises, together with Chinese chopsticks. Near Bhomsong, Campbell pointed out a hot bath to me, which he had seen employed: it consisted of a hollowed prostrate tree trunk, the water in which was heated by throwing in hot stones with bamboo tongs. The temperature is thus raised to 114°, to which the patient submits at repeated intervals for several days, never leaving till wholly exhausted.

These baths are called "Sa-choo," literally "hot-water," in Tibetan.

We stopped to measure some splendid trees in the valley, and found the trunk of one to be forty-five feet round the buttresses, and thirty feet above them, a large size for the Himalaya: they were a species of *Terminalia* (*Pentaptera*), and called by the Lepchas "Sillok-Kun," "Kun" meaning tree.

We slept at Nampok, and the following morning commenced the ascent. On the way we passed the temple and lake of Neongong; the latter is about 400 yards round, and has no outlet. It contained two English plants, the common duckweed (*Lemna minor*), and *Potamogeton natans:* some coots were swimming in it, and having flushed a woodcock, I sent for my gun, but the Lamas implored us not to shoot, it being contrary to their creed to take life wantonly.

We left a great part of our baggage at Neongong, as we intended to return there; and took up with us bedding, food, &c., for two days. A path hence up the mountain is frequented once a year by the Lamas, who make a pilgrimage to the top for worship. The ascent was very gradual for 4000 feet. We met with snow at the level of Dorjiling (7000 feet), indicating a colder climate than at that station, where none had fallen; the vegetation was, however, similar, but not so rich, and at 8000 feet trees common also to the top of Sinchul appeared, with *R. Hodgsoni*, and the beautiful little winter-flowering primrose, *P. petiolaris*, whose stemless flowers spread like broad purple stars on the deep green foliage. Above, the path runs along the ridge of the precipices facing the south-east, and here we caught a glimpse of the great valley of the Ryott, beyond the Teesta, with Tumloong, the Rajah's residence, on its north flank, and the superb snowy peak of Chola at its head.

One of our coolies, loaded with crockery and various indispensables, had here a severe fall, and was much bruised; he however recovered himself, but not our goods.

The rocks were all of chlorite slate, which is not usual at this elevation; the strike was north-west, and dip north-east. At 9000 feet various shrubby rhododendrons prevailed, with mountain-ash, birch, and dwarf-bamboo; also *R. Falconeri*, which grew from forty to fifty feet high. The snow was deep and troublesome, so we encamped at 9,800 feet, or 800 feet below the top, in a wood of *Pyrus*, *Magnolia*, *Rhododendron*, and bamboo. As the ground was deeply covered with snow, we laid our beds on a thick layer of rhododendron twigs, bamboo, and masses of a pendent moss.

We passed a very cold night, chiefly owing to damp, the temperature falling to 24°. On the following morning we scrambled through the snow, reaching the summit after an hour's very laborious ascent, and took up our quarters in a large wooden barn-like temple (*goompa*), built on a stone platform. The summit was very broad, but the depth of the snow prevented our exploring much, and the silver firs (*Abies Webbiana*) were so tall, that no view could be obtained, except from the temple. The great peak of Kinchinjunga is in part hidden by those of Pundim and Nursing, but the panorama of snowy mountains is very grand indeed. The effect is quite deceptive; the mountains assuming the appearance of a continued chain, the distant snowy peaks being seemingly at little further distance than the nearer ones. The whole range (about twenty-two miles nearer than at Dorjiling) appeared to rise uniformly and steeply out of black pine forests, which were succeeded by the russet-brown of the rhododendron shrubs, and that again by tremendous precipices and gulleys, into which descended mighty glaciers and perpetual snows. This excessive steepness is however only apparent, being due to foreshortening.

The upper 10,000 feet of Kinchin, and the tops of Pundim, Kubra, and Junnoo, are evidently of granite, and are rounded in outline: the lower peaks again, as those of Nursing, &c., present rugged pinnacles of black and red stratified rocks, in many cases resting on white granite, to which they present a remarkable contrast. The general appearance was as if Kinchin and the whole mass of mountains clustered around it, had been up-heaved by white granite, which still forms the loftiest summits, and has raised the black stratified rocks in some places to 20,000 feet in numerous peaks and ridges. One range presented on every summit a cap of black stratified rocks of uniform inclination and dip, striking north-west, with precipitous faces to the south-west: this was clear to the naked eye, and more evident with the telescope, the range in question being only fifteen miles distant, running between Pundim and Nursing. The fact of the granite forming the greatest elevation must not be hastily attributed to that igneous rock having burst through the stratified, and been protruded beyond the latter: it is much more probable that the upheaval of the granite took place at a vast depth, and beneath an enormous pressure of stratified rocks and perhaps of the ocean; since which period the elevation of the whole mountain chain, and the denudation of the stratified rocks, has been slowly proceeding.

To what extent denudation has thus lowered the peaks we dare scarcely form a conjecture; but considering the number and variety of the beds which in some places overlie the gneiss and granite, we may reasonably conclude that many thousand feet have been removed.

It is further assumable that the stratified rocks originally took the forms of great domes, or arches. The prevailing north-west strike throughout the Himalaya vaguely indicates a general primary arrangement of the curves into waves, whose crests run north-west and south-east; an arrangement which no minor or posterior forces have wholly disturbed, though they have produced endless dislocations, and especially a want of uniformity in the amount and direction of the dip. Whether the loftiest waves were the result of one great convulsion, or of a long-continued succession of small ones, the effect would be the same, namely, that the strata over those points at which the granite penetrated the highest, would be the most dislocated, and the most exposed to wear during denudation.

We enjoyed the view of this superb scenery till noon, when the clouds which had obscured Dorjiling since morning were borne towards us by the southerly wind, rapidly closing in the landscape on all sides. At sunset they again broke, retreating from the northward, and rising from Sinchul and Dorjiling last of all, whilst a line of vapour, thrown by perspective into one narrow band, seemed to belt the Singalelah range with a white girdle, darkened to black where it crossed the snowy mountains; and it was difficult to believe that this belt did not really hang upon the ranges from twenty to thirty miles off, against which it was projected; or that its true position was comparatively close to the mountain on which we were standing, and was due to condensation around its cool, broad, flat summit.

As usual from such elevations, sunset produced many beautiful effects. The zenith was a deep blue, darkening opposite the setting sun, and paling over it into a peach colour, and that again near the horizon passing into a glowing orange-red, crossed by coppery streaks of cirrhus. Broad beams of pale light shot from the sun to the meridian, crossing the moon and the planet Venus. Far south, through gaps in the mountains, the position of the plains of India, 10,000 feet below us, was indicated by a deep leaden haze, fading upwards in gradually paler bands (of which I counted fifteen) to the clear yellow of the sunset sky. As darkness came on, the mists collected around the top of Mainom, accumulating on the windward side, and thrown off in ragged masses from the opposite.

The second night we passed here was fine, and not very cold (the mean temperature being 27°), and we kept ourselves quite warm by pine-wood fires. On the following morning the sun tinged the sky of a lurid yellow-red: to the south-west, over the plains, the belts of leaden vapour were fewer (twelve being distinguishable) and much lower than on the previous evening, appearing as if depressed on the visible horizon. Heavy masses of clouds nestled into all the valleys, and filled up the larger ones, the mountain tops rising above them like islands.

The height of our position I calculated to be 10,613 feet. Colonel Waugh had determined that of the summit by trigonometry to be 10,702 feet, which probably includes the trees which cover it, or some rocky peaks on the broad and comparatively level surface.

The mean temperature of the twenty-four hours was 32° 7 $(\frac{\text{max. } 41°\ 5}{\text{min. } 27°\ 2})$, mean dew-point 29·7, and saturation 0·82. The mercury suddenly fell below the freezing point at sunset; and from early morning the radiation was so powerful, that a thermometer exposed on snow sank to 21° 2, and stood at 25° 5, at 10 A.M. The black bulb thermometer rose to 132°, at 9 A.M. on the 27th, or 94° 2 above the temperature of the air in the shade. I did not then observe that of radiation from snow; but if, as we may assume, it was not less than on the following morning (21° 2), we shall have a difference of 148° 6 Fahr., in contiguous spots; the one exposed to the full effects of the sun, the other to that of radiation through a rarefied medium to a cloudless sky. On the 28th the black bulb thermometer, freely suspended over the snow and exposed to the sun, rose to 108°, or 78° above that of the air in the shade (32°); the radiating surface of the same snow in the shade being 21° 2, or 86° 8 colder.

Having taken a complete set of angles and panoramic sketches from the top of Mainom, with seventeen hourly observations, and collected much information from our guides, we returned on the 28th to our tents pitched by the temples at Neongong; descending 7000 feet, a very severe shake along Lepcha paths. In the evening the Lamas visited us, with presents of rice, fowls, eggs, &c., and begged subscriptions for their temple which was then building, reminding Dr. Campbell that he and the Governor-General had an ample share of their prayers, and benefited in proportion. As for me, they said, I was bound to give alms, as I surely needed praying for, seeing how I exposed myself; besides my having been the first Englishman who had visited the snows of Kinchinjunga, the holiest spot in Sikkim.

On the following morning we visited the unfinished temple. The outer walls were of slabs of stone neatly chiselled, but badly mortared with felspathic clay and pounded slate, instead of lime; the partition walls were of clay, shaped in moulds of wood; parallel planks, four feet asunder, being placed in the intended position of the walls, and left open above, the composition was placed in these boxes, a little at a time, and rammed down by the feet of many men, who walked round and round the narrow enclosure, singing, and also using rammers of heavy wood. The outer work was of good hard timber, of Magnolia ("Pendre-kun" of the Lepchas) and oak (" Sokka"). The common "Ban," or Lepcha knife, supplied the place of axe, saw, adze, and plane; and the graving work was executed with small tools, chiefly on Toon (*Cedrela*), a very soft wood (the "Simal-kun" of the Lepchas).

This being a festival day, when the natives were bringing offerings to the altar, we also visited the old temple, a small wooden building. Besides more substantial offerings, there were little cones of rice with a round wafer of butter at the top, ranged on the altar in order.* Six Lamas were at prayer, psalms, and contemplation, sitting cross-legged on two small benches that ran down the building: one was reading, with his hand and fore-finger elevated, whilst the others listened; anon they all sang hymns, repeated sacred or silly precepts to the bystanders, or joined in a chorus with boys, who struck brass cymbals, and blew straight copper trumpets six feet long, and conch-shells mounted with broad silver wings, elegantly carved with dragons. There were besides manis, or praying-cylinders, drums, gongs, books, and trumpets made of human thigh-bones, plain or mounted in silver.

* The worshippers, on entering, walk straight up to the altar, and before, or after, having deposited their gifts, they lift both hands to the forehead, fall on their knees, and touch the ground three times with both head and hands, raising the body a little between each prostration. They then advance to the head Lama, kotow similarly to him, and he blesses them, laying both hands on their heads and repeating a short formula. Sometimes the dorje is used in blessing, as the cross is in Europe, and when a mass of people request a benediction, the Lama pronounces it from the door of the temple with outstretched arms, the people all being prostrate, with their foreheads touching the ground.

Throughout Sikkim, we were roused each morning at daybreak by this wild music, the convents being so numerous that we were always within hearing of it. To me it was always deeply impressive, sounding so foreign, and awakening me so effectually to the strangeness of the wild land in which I was wandering, and of the many new and striking objects it contained. After sleep, too, during which the mind has either been at rest, or carried away to more familiar subjects, the feelings of loneliness and sometimes even of despondency, conjured up by this solemn music, were often almost oppressive.

Ascending from Neongong, we reached that pass from the Teesta to the Great Rungeet, which I had crossed on the 22nd; and this time we had a splendid view, down both the valleys, of the rivers, and the many spurs from the ridge communicating between Tendong and Mainom, with many scattered villages and patches of cultivation. Near the top I found a plant of "Praong," (a small bamboo), in full seed; this sends up many flowering branches from the root, and but few leaf-bearing ones; and after maturing its seed, and giving off suckers from the root, the parent plant dies. The fruit is a dark, long grain, like rice; it is boiled and made into cakes, or into beer, like Murwa.

Looking west from the summit, no fewer than ten monastic establishments with their temples, villages and cultivation, were at once visible, in the valley of the Great Rungeet, and in those of its tributaries; namely, Changachelling, Raklang, Dholi, Molli, Catsuperri, Dhoobdi, Sunnook, Powhungri, Pemiongchi and Tassiding, all of considerable size, and more or less remarkable in their sites, being perched on spurs or peaks at elevations varying from 3000 to 7000 feet, and commanding splendid prospects.

We encamped at Lingcham, where I had halted on

the 21st, and the weather being fine, I took bearings of all the convents and mountains around. There is much cultivation here, and many comparatively rich villages, all occupying flat-shouldered spurs from Mainom. The houses are large, and the yards are full of animals familiar to the eye but not to the ear. The cows of Sikkim, though generally resembling the English in stature, form, and colour, have humps, and grunt rather than low; and the cocks wake the morning with a prolonged howling screech, instead of the shrill crow of chanticleer.

Hence we descended north-west to the Great Rungeet, opposite Tassiding; which is one of the oldest monastic establishments in Sikkim, and one we were very anxious to visit. The descent lay through a forest of tropical trees, where small palms, vines, peppers, *Pandanus*, wild plantain, and *Pothos*, were interlaced in an impenetrable jungle, and air-plants clothed the trees.

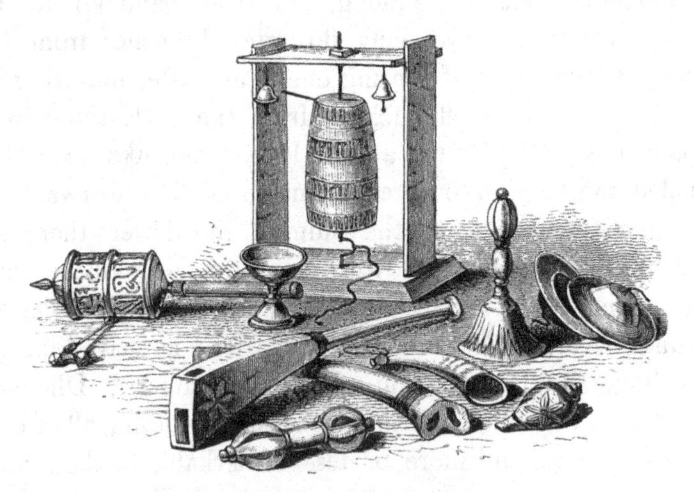

IMPLEMENTS USED IN BOODHIST TEMPLES.

Praying cylinder in stand (see p. 175); another to be carried in the hand; cymbals; bell; brass cup; three trumpets; conch; dorje.

CHAPTER XIV.

Tassiding, view of and from—Funereal cypress—Camp at Sunnook—Hot vapours —Lama's house—Temples, decorations, altars, idols, general effect—Chaits —Date of erection—Plundered by Ghorkas—Cross Ratong—Ascend to Pemiongchi—Relation of river-beds to strike of rocks—Slopes of ravines—Pemiongchi, view of—Vegetation—Elevation—Temple, decorations, &c.—Former capital of Sikkim—History of Sikkim—Nightingales—Campbell departs—Tchonpong—Edgeworthia—Cross Rungbee and Ratong—Hoar-frost on plantains—Yoksun—Walnuts—View—Funereal cypresses—Doobdi—Gigantic cypresses—Temples—Snow-fall—Sikkim, &c.—Toys.

TASSIDING hill is the steep conical termination of a long spur from a pine-clad shoulder of Kinchinjunga, called Powhungri: it divides the Great Rungeet from its main feeder, the Ratong, which rises from the south face of Kinchin. We crossed the former by a bridge formed of two bamboo stems, slung by canes from two parallel arches of stout branches lashed together.

The ascent for 2,800 feet was up a very steep, dry, zigzag path, amongst mica slate rocks (strike north-east), on which grew many tropical plants, especially the "Tukla," (*Rottlera tinctoria*), a plant which yields a brown dye. The top was a flat, curving north-west and south-east, covered with temples, chaits, and mendongs of the most picturesque forms and in elegant groups, and fringed with brushwood, wild plantains, small palms, and apple-trees. Here I saw for the first time the funereal cypress, of which some very old trees spread their weeping limbs and pensile branchlets over the buildings.* It is not wild in Sikkim, but imported there and into Bhotan from Tibet: it does not thrive well above 6000 feet elevation. It is called "Tchenden" by the Lepchas, Bhoteeas, and Tibetans, and its fragrant red wood is burnt in the temples.

GROUP OF CHAITS AT TASSIDING.

The Lamas met us on the top of the hill, bringing a noble present of fowls, vegetables and oranges, the latter most acceptable after our long and hot march. The site is admirably chosen, in the very heart of Sikkim, commanding a fine view, and having a considerable river on either side,

* I was not then aware of this tree having been introduced into England by the intrepid Mr. Fortune from China; and as I was unable to procure seeds, which are said not to ripen in Sikkim, it was a great and unexpected pleasure, on my return home, to find it alive and flourishing at Kew.

with the power of retreating behind to the convents of Sunnook and Powhungri, which are higher up on the same spur, and surrounded by forest enough to conceal an army. Considering the turbulent and warlike character of their neighbours, it is not wonderful that the monks should have chosen commanding spots, and good shelter for their indolent lives: for the same reason these monasteries secured views of one another: thus from Tassiding the great temple of Pemiongchi was seen towering 3000 feet over head, whilst to the north-west, up the course of the river, the hill-sides seemed sprinkled with monasteries.

We camped on a saddle near the village of Sunnook, at 4000 feet above the sea; and on the last day of the year we visited this most interesting monastic establishment: ascending from our camp along the ridge by a narrow path, cut here and there into steps, and passing many rocks covered with inscriptions, broken walls of mendongs, and other remains of the *via sacra* between the village and temple. At one spot we found a fissure emitting hot vapour of the temperature of 65° 5, that of the air being about 50°. It was simply a hole amongst the rocks; and near the Rungeet a similar one is said to occur, whose temperature fluctuates considerably with the season. It is very remarkable that such an isolated spring should exist on the top of a sharp ridge, 2,800 feet above the bottom of this deep valley.

The general arrangement on the summit was, first the Lamas' houses with small gardens, then three large temples raised on rudely paved platforms, and beyond these, a square walled enclosure facing the south, full of chaits and mendongs, looking like a crowded cemetery, and planted with funereal cypress (*Cupressus funebris*).

The house of the principal Lama was an oblong square,

the lower story of stone, and the upper of wood: we ascended a ladder to the upper room, which was 24 feet by 8 wattled all round, with prettily latticed windows opening upon a bamboo balcony used for drying grain, under the eaves of the broad thatched roof. The ceiling (of neat bamboo work) was hung with glorious bunches of maize, yellow, red, and brown; an altar and closed wicker cage at one end of the room held the Penates, and a few implements of worship. Chinese carpets were laid on the floor for us, and the cans of Murwa brought round.

The Lama, though one of the red sect, was dressed in a yellow flowered silk robe, but his mitre was red: he gave us much information relative to the introduction of Boodhism into Sikkim.

The three temples stand about fifty yards apart, but are not parallel to one another, although their general direction is east and west.* Each is oblong, and narrowed upwards, with the door at one end; the middle (and smallest) faces the west, the others the east: the doorways are all broad, low and deep, protected by a projecting carved portico. The walls are immensely thick, of well-masoned slaty stones; the outer surface of each slopes upwards and inwards, the inner is perpendicular. The roofs are low and thickly thatched, and project from eight to ten feet all round, to keep off the rain, being sometimes supported by long poles. There is a very low upper story, inhabited by the attendant monks and servants, accessible by a ladder at one end of the building. The main body of the temple is one large apartment, entered through a small transverse vestibule, the breadth of the temple, in which are tall cylindrical

* Timkowski, in his travels through Mongolia (i. p. 193), says, "According to the rules of Tibetan architecture, temples should face the south:" this is certainly not the rule in Sikkim, nor, so far as I could learn, in Tibet either.

praying-machines. The carving round the doors is very beautiful, and they are gaudily painted and gilded. The

DOORWAY.

northern temple is quite plain: the middle one is simply painted red, and encircled with a row of black heads, with goggle eyes and numerous teeth, on a white ground; it is said to have been originally dedicated to the evil spirits of the Lepcha creed. The southern, which contains the library, is the largest and best, and is of an irregular square shape. The inside walls and floors are plastered with clay, and painted with allegorical representations of Boodh, &c. From the vestibule the principal apartment is entered by broad folding-doors, studded with circular copper bosses, and turning on iron hinges. It is lighted by latticed windows, sometimes protected outside by a bamboo screen. Owing to the great thickness of the walls (three to four feet), a very feeble light is admitted. In the principal temple, called "Dugang," six hexagonal wooden columns, narrowed above, with peculiar broad

SOUTHERN TEMPLE.

transverse capitals, exquisitely gilded and painted, support the cross-beams of the roof, which are likewise beautifully ornamented. Sometimes a curly-maned gilt lion is placed over a column, and it is always furnished with a black bushy tail: squares, diamonds, dragons, and groups of flowers, vermilion, green, gold, azure, and white, are dispersed with great artistic taste over all the beams; the heavier masses of colour being separated by fine white lines.

The altars and idols are placed at the opposite end; and two long parallel benches, like cathedral stalls, run down the centre of the building: on these the monks sit at prayer and contemplation, the head Lama occupying a stall (often of very tasteful design) near the altar.

MIDDLE TEMPLE.

The principal Boodh, or image, is placed behind the altar under a canopy, or behind a silk screen: lesser gods, and gaily dressed and painted effigies of sainted male or female persons are ranged on either side, or placed in niches around the apartment, sometimes with separate altars before them; whilst the walls are more or less covered with paintings of monks in prayer or contemplation. The principal Boodh (Sakya Sing) sits cross-legged, with the left heel up: his left-hand always rests on his thigh, and holds the padmi or lotus and jewel, which is often a mere cup; the right-hand is either raised, with the two forefingers up, or holds the dorje, or rests on the calf of the upturned leg. Sakya has generally curled hair, Lamas have mitres, females various head-dresses; most wear immense ear-rings, and some rosaries. All are placed on rude pediments, so painted as to convey the idea of their rising out of the petals of the pink, purple, or white lotus. None are in any way disagreeable; on the

ALTAR AND IMAGES.
Central figure Akshobya, the first of the Pancha Boodha.

contrary most have a calm and pleasing expression, suggestive of contemplation.

The great or south temple contained a side altar of very elegant shape, placed before an image encircled by a glory. Flowers, juniper, peacock's feathers, pastiles, and rows of brass cups of water were the chief ornaments of the altars, besides the instruments I have elsewhere enumerated. In this temple was the library, containing several hundred books, in pigeon-holes, placed in recesses.*

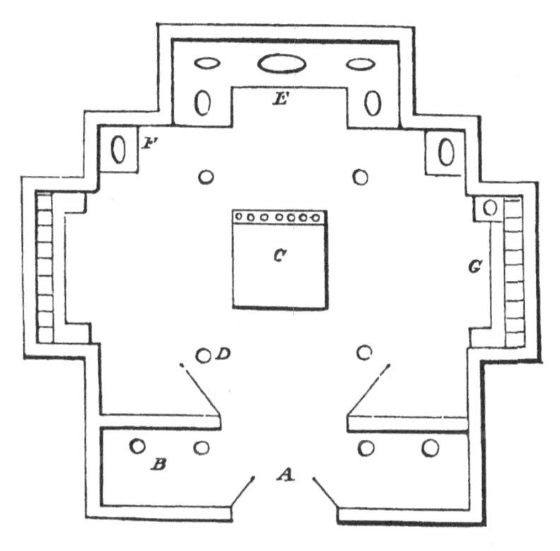

PLAN OF THE SOUTH TEMPLE.

A. entrance : B. four praying cylinders ; C. altar, with seven brass cups of water ; D. four columns ; E. and F. images ; G. library.

The effect on entering these cold and gloomy temples is very impressive ; the Dugang in particular is exquisitely ornamented and painted, and the vista from the vestibule to the principal idol, of carved and coloured pillars and beams, is very picturesque. Within, the general arrangement of the colours and gilding is felt to be harmonious and pleasing, especially from the introduction of slender white streaks between the contrasting masses of colour,

* For a particular account of the images and decorations of these temples, see Dr. Campbell's paper in "Bengal Asiatic Society's Trans.," May, 1849. The principal object of veneration amongst the Ningma or red sect of Boodhists in Sikkim and Bhotan is Goroucknath, who is always represented sitting cross-legged, holding the dorje in one hand, which is raised ; whilst the left rests in the lap and holds a cup with a jewel in it. The left arm supports a trident, whose staff pierces three sculls (a symbol of Shiva), a rosary hangs round his neck, and he wears a red mitre with a lunar crescent and sun in front.

as adopted in the Great Exhibition building of 1851. It is also well worthy of remark that the brightest colours are often used in broad masses, and when so, are always arranged chromatically, in the sequence of the rainbow's hues, and are hence never displeasing to the eye. The hues, though bright, are subdued by the imperfect light: the countenances of the images are all calm, and their expression solemn. Whichever way you turn, the eye is met by some beautiful specimen of colouring or carving, or some object of veneration. The effect is much heightened by the incense of juniper and sweet-smelling herbs which the priests burn on entering, by their grave and decorous conduct, and by the feeling of respect that is demanded by a religion which theoretically inculcates and adores virtue in the abstract, and those only amongst men who practise virtue. To the idol itself the Boodhist attaches no real importance; it is an object of reverence, not of worship, and no virtue or attribute belong to it *per se*; it is a symbol of the creed, and the adoration is paid to the holy man whom it represents.

Beyond the temples are the chaits and mendongs, scattered without much order; and I counted nearly twenty-five chaits of the same form,* between eight and thirty feet high. The largest is consecrated to the memory of the Rajah's eldest son, who, however, is not buried here. A group of these structures is, as I have often remarked, extremely picturesque, and those at Tassiding, from their

* In Sikkim the form of the cube alone is always strictly preserved; that of the pyramid and hemisphere being often much modified. The cube stands on a flight of usually three steps, and is surmounted by a low pyramid of five steps; on this is placed a swelling, urn-shaped body, which represents the hemisphere, and is surmounted by another cube. On the latter is a slender, round or angled spire (represented by a pyramid in Burma), crowned with a crescent and disc, or sun, in moon. Generally, the whole is of stone, with the exception of the spire, which is of wood, painted red.

number, variety, and size, their commanding and romantic position, and their being interspersed with weeping cypresses, are particularly so.

The Tassiding temples and convents were founded upwards of 300 years ago, by the Lamas who accompanied the first Rajah to Sikkim; and they have been continuously served by Lamas of great sanctity, many of whom have been educated at Lhassa. They were formerly very wealthy, but during the Nepal war they were plundered of all their treasures, their silver gongs and bells, their best idols, dorjes, and manis, and stripped of their ornaments; since which time Pemiongchi has been more popular. In proof of their antiquity, it was pointed out that most of the symbols and decorations were those of pure Lama Boodhism, as practised in Tibet.

Although the elevation is but 4,840 feet, the weather was cold and raw, with rain at noon, followed by thunder and lightning. These electrical disturbances are frequent about midsummer and midwinter, prevailing over many parts of India.

January 1*st*, 1849.—The morning of the new year was bright and beautiful, though much snow had fallen on the mountains; and we left Sunnook for Pemiongchi, situated on the summit of a lofty spur on the opposite side of the Ratong. We descended very steeply to the bed of the river (alt. 2,480 feet) which joins the Great Rungeet below the convents. The rocks were micaceous, dipping west and north-west 45°, and striking north and north-east, which direction prevailed for 1000 feet or so up the opposite spur. I had observed the same dip and stroke on the east flank of the Tassiding spur; but both the Ratong on its west side, and the Great Rungeet on the east, flow in channels that show no relation to either the dip or strike.

I have generally remarked in Sikkim that the channels of the rivers when cutting through or flowing at the base of bluff cliffs, are neither parallel to nor at right angles to the strike of the rocks forming the cliffs. I do not hence conclude that there is no original connection between the directions of the rivers, and the lines of fracture; but whatever may have once subsisted between the direction of the fissures and that of the strike, it is in the Sikkim Himalaya now wholly masked by shiftings, which accompanied subequent elevations and depressions.

Mr. Hopkins has mathematically demonstrated that the continued exertion of a force in raising superimposed strata would tend to produce two classes of fractures in those strata; those of the first order at right angles to the direction of the wave or ridge (or line of strike); those of the second order parallel to the strike. Supposing the force to be withdrawn after the formation of the two fractures, the result would be a ridge, or mountain chain, with diverging fissures from the summit, crossed by concentric fissures; and the courses which the rivers would take in flowing down the ridge, would successively be at right angles and parallel to the strike of the strata. Now, in the Himalaya, a prevalent strike to the north-west has been recognised in all parts of the chain, but it is everywhere interfered with by mountains presenting every other direction of strike, and by their dip never remaining constant either in amount or direction. Consequently, as might be expected, the directions of the river channels bear no apparent relation to the general strike of the rocks.

We crossed the Ratong (twenty yards broad) by a cane bridge, suspended between two rocks of green chlorite, full of veins of granite. Ascending, we passed the village of Kameti on a spur, on the face of which were strewed some enormous detached blocks of white and pink stratified quartz: the rocks *in situ* were all chlorite schist.

Looking across the valley to the flank of Mainom, the disposition of the ridges and ravines on its sides was very evident; many of the latter, throughout their westerly course, from their commencement at 10,000 feet, to their debouchure in the Great Rungeet at 2000, had a bluff, cliffy, northern flank, and a sloping southern one. The dip of the surfaces is, therefore, north-west, the exposure consequently of the villages which occupy terraces on the south flanks of the lateral valleys. The Tassiding spur presented exactly the same arrangement of its ravines, and the dip of the rocks being north-west, it follows that the planes of the sloping surfaces coincide in direction (though not in amount of inclination) with that of the dip of the subjacent strata, which is anything but a usual phenomenon in Sikkim.

The ascent to Pemiongchi continued very steep, through woods of oaks, chesnuts, and magnolias, but no tree-fern, palms, *Pothos*, or plantain, which abound at this elevation on the moister outer ranges of Sikkim. The temple (elev. 7,083 feet) is large, eighty feet long, and in excellent order, built upon the lofty terminal point of the great east and west spur that divides the Kulhait from the Ratong and Rungbee rivers; and the great Changa-chelling temple and monastery stand on another eminence of the same ridge, two miles further west.

The view of the snowy range from this temple is one of the finest in Sikkim; the eye surveying at one glance the vegetation of the Tropics and the Poles. Deep in the valleys the river-beds are but 3000 feet above the sea, and are choked with fig-trees, plantains, and palms; to these succeed laurels and magnolias, and higher up still, oaks, chesnuts, birches, &c.; there is, however, no marked line between the limits of these two last forests, which form the prevailing arboreous vegetation between 4000 and 10,000 feet, and give a lurid hue to the mountains. Pine forests succeed for 2000 feet higher, when they give place to a skirting of rhododendron and berberry. Among these appear black naked rocks, rising up in cliffs, between which are gulleys, down which the snow now (on the 1st January) descended to 12,000 feet. The mountain flanks are much more steep and rocky than those at similar heights on the outer ranges, and cataracts are very numerous, and of considerable height, though small in volume.

Pemiongchi is at the same elevation as Dorjiling, and the contrast between the shoulders of 8000 to 10,000 feet on Kinchinjunga, and those of equal height on Tendong and Tonglo, is very remarkable: looking at the latter mountains from Dorjiling, the observer sees no rock, waterfall, or pine, throughout their whole height; whereas the equally wooded flanks of these inner ranges are rocky, streaked with thread-like waterfalls, and bristling with silver firs.

This temple, the most ancient in Sikkim, is said to be 400 years old; it stands on a paved platform, and is of the same form and general character as those of Tassiding. Inside, it is most beautifully decorated, especially the beams, columns, capitals and architraves, but the designs are coarser than those of Tassiding.* The square end of every beam in the roof is ornamented either with a lotus flower or with a Tibetan character, in endless diversity

* Mr. Hodgson informed me that many of the figures and emblems in this temple are those of Tantrica Boodhism, including Shiva, Devi, and other deities usually called Brahminical; Kakotak, or the snake king, a figure terminating below in a snake, is also seen; with the tiger, elephant, and curly-maned lion.

INTERIOR OF THE TEMPLE AT PEMIONGCHI.

of colour and form, and the walls are completely covered with allegorical paintings of Lamas and saints expounding or in contemplation, with glories round their heads, mitred, and holding the dorje and jewel.

The principal image is a large and hideous figure of Sakya-thoba, in a recess under a blue silk canopy, contrasting with a calm figure of the late Rajah, wearing a cap and coronet.

Pemiongchi was once the capital of Sikkim, and called the Sikkim Durbar: the Rajah's residence was on a curious flat to the south of the temple, and a few hundred feet below it, where are the remains of (for this country) extensive walls and buildings. During the Nepal war, the Rajah was driven west across the Teesta, whilst the Ghorkas plundered Tassiding, Pemiongchi, Changachelling, and all the temples and convents to the east of that river. It was then that the famous history of Sikkim,* compiled by the Lamas of Pemiongchi, and kept at this temple, was destroyed, with the exception of a few sheets, with one of which Dr. Campbell and myself were each presented. We were told that the monks of Changachelling and those of this establishment had copied what remained, and were busy compiling from oral information, &c.: whatever value the original may have possessed, however, is irretrievably lost. A magnificent copy of the Boodhist Scriptures was destroyed at the same time; it consisted of 400 volumes, each containing several hundred sheets of Daphne paper.

The ground about the temple was snowed; and we descended a few hundred feet, to encamp in a most picturesque grove, among chaits and inscribed stones, with

* This remarkable and beautiful manuscript was written on thick oblong sheets of Tibet paper, painted black to resist decay, and the letters were yellow and gold. The Nepalese soldiers wantonly employed the sheets to roof the sheds they erected, as a protection from the weather.

a peep of the temples above. Nightingales warbled deliciously night and morning, which rather surprised us, as the minimum thermometer fell to 27·8°, and the ground next day was covered with hoar-frost; the elevation being 6,580 feet. These birds migrate hither in October and November, lingering in the Himalayan valleys till the cold of early spring drives them further south, to the plains of India, whence they return north in March and April.

On the 2nd of January I parted from my friend, who was obliged to hurry to the great annual fair at Titalya. I regretted much being unable to accompany Dr. Campbell to this scene of his disinterested labours, especially as the Nawab of Moorshedabad was to be present, one of the few wealthy native princes of Bengal who still keep a court worth seeing; but I was more anxious to continue my explorations northward till the latest moment: I however accompanied him for a short distance on his way towards Dorjiling. We passed the old Durbar, called Phieungoong ("Bamboo-hill," so named from the abundance of a small bamboo, "Phieung.") The buildings, now in ruins, occupy a little marshy flat, hemmed in by slate rocks, and covered with brambles and *Andromeda* bushes. A wall, a bastion, and an arched gateway, are the only traces of fortifications; they are clothed with mosses, lichens, and ferns.

A steep zigzag path, descending amongst long grass and scarlet rhododendrons, leads to the Kaysing Mendong.* Here I bade adieu to Dr. Campbell, and toiled up the hill, feeling very lonely. The zest with which he had entered into all my pursuits, and the aid he had afforded me, together with the charm that always attends companionship with one who enjoys every incident of travel, had so attracted me to him that I found it difficult to recover my spirits.

* Described at page 287.

It is quite impossible for any one who cannot from experience realise the solitary wandering life I had been leading for months, to appreciate the desolate feeling that follows the parting from one who has heightened every enjoyment, and taken far more than his share of every annoyance and discomfort: the few days we had spent together appeared then, and still, as months.

On my return to Pemiongchi I spent the remainder of the day sketching in the great temple, gossiping with the Lamas, and drinking salted and buttered tea-soup, which I had begun to like, when the butter was not rancid.

My route hence was to be along the south flank of Kinchinjunga, north to Jongri, which lay about four or five marches off, on the road to the long deserted pass of Kanglanamo, by which I had intended entering Sikkim from Nepal, when I found the route up the Yalloong valley impracticable. The village and ruined convents of Yoksun lay near the route, and the temples of Doobdi, Catsuperri and Molli, on the Ratong river.

I descended to the village of Tchonpong (alt. 4,980 feet), where I was detained a day to obtain rice, of which I required ten days' supply for twenty-five people. On the way I passed groves of the paper-yielding *Edgeworthia Gardneri:* it bears round heads of fragrant, beautiful, yellow flowers, and would be a valuable acquisition to an English conservatory.

From Tchonpong we descended to the bed of the Rungbee (alt. 3,160 feet), an affluent of the Ratong, flowing in a deep gulley with precipitous sides of mica schist full of garnets, dipping west and north-west 45°: it was spanned by a bridge of two loose bamboo culms, about fifteen yards long, laid across without handrails; after wet sand had been thrown on it the bare-footed coolies crossed easily enough, but I, having shoes on, required a hand to steady me. From this point we crossed a lofty spur to the Ratong (alt. 3000 feet), where we encamped, the coolies being unable to proceed further on such very bad roads. This river descends from the snows of Kinchin, and consequently retains the low temperature 42°, being fully 7° colder than the Rungbee, which at an elevation of but 3000 feet appears very remarkable: it must however be observed that scarcely anywhere does the sun penetrate to the bottom of its valley.

We encamped on a gravelly flat, fifty feet above the river, strewn with water-worn boulders, and so densely covered with tall *Artemisiæ*, gigantic grasses, bamboo, plantain, fern, and acacia, that we had to clear a space in the jungle, which exhaled a rank heavy smell.

Hoar-frost formed copiously in the night, and though above the sun's rays were very powerful, they did not reach this spot till 7·30 A.M., the frost remaining in the shade till nearly 9 A.M.; and this on plantains, and other inhabitants of hot-houses in England.

Hence I ascended to Yoksun, one of the most curious and picturesque spots in Sikkim, and the last inhabited place towards Kinchinjunga. The path was excessively steep and rocky for the first mile or two, and then alternately steep and flat. Mixed with many tropical trees, were walnuts of the common English variety; a tree, which, though planted here, is wild near Dorjiling, where it bears a full-sized fruit, as hard as a hickory-nut: those I gathered in this place were similar, whereas in Bhotan the cultivated nut is larger, thin-shelled, and the kernel is easily removed. We ascended one slope, of an angle of 36° 30′, which was covered with light black mould, and had been recently cleared by fire: we found millet now cultivated on it. From the top the view of the Ratong valley was very fine: to the north lay Yoksun, appearing from this height to occupy a flat, two miles long and one broad, girdled by steep mountains to the north and east, dipping very suddenly 2,200 feet to the Ratong on the west. To the right was a lofty hill, crowned with the large temple and convents of Doobdi, shadowed by beautiful weeping cypresses, and backed by lofty pine-clad mountains. Northward, the gorge of the Ratong opened as a gloomy defile, above which rose partially snowed mountains, which shut out Kinchinjunga. To the west, massive pine-clad mountains rose steeply; while the little hamlet of Lathiang occupied a remarkable shelf overhanging the river, appearing inaccessible except by ropes from above. South-west, the long spurs of Molli and Catsuperri, each crowned with convents or temples, descended from Singalelah; and parallel to them on the south, but much longer and more lofty, was the great mountain range north of the Kulhait, with the temples and convents of Pemiongchi, and Changachelling, towering in the air. The latter range dips suddenly to the Great Rungeet, where Tassiding, with its chaits and cypresses, closed the view. The day was half cloud, half sunshine; and the various effects of light and shade, now bringing out one or other of the villages and temples, now casting the deep valleys into darker gloom, was wonderfully fine.

Yoksun was the earliest civilised corner of Sikkim, and derived its name (which signifies in Lepcha "three chiefs") from having been the residence of three Lamas of great influence, who were the means of introducing the first Tibetan sovereign into the country. At present it boasts of but little cultivation, and a scattered population, inhabiting a few hamlets, 5,500 feet above the sea: beautiful lanes and paths wind everywhere over the gentle slopes, and through the copsewood that has replaced the timber-trees of a former period. Mendongs and chaits are very numerous, some of great size; and there are also the ruins of two very large temples, near which are some magnificent weeping cypresses, eighty feet high. These fine trees are landmarks from all parts of the flat; they form irregular cones of pale bright green, with naked gnarled tops, the branches weep gracefully, but not like the picture in Macartney's Embassy to China, whence originated the famous willow-pattern of our crockery. The ultimate branchlets are very slender and pendulous; my Lepcha boys used to make elegant chaplets of them, binding the withes with scarlet worsted. The trunk is quite erect, smooth, cylindrical, and pine-like; it harbours no moss, but air-plants, Orchids, and ferns, nestle on the limbs, and pendulous lichens, like our beard-moss, wave from the branches.

In the evening I ascended to Doobdi. The path was broad, and skilfully conducted up a very steep slope covered with forest: the top, which is 6,470 feet above the sea, and nearly 1000 above Yoksun, is a broad partially paved platform, on which stand two temples, surrounded by beautiful cypresses: one of these trees (perhaps the oldest in Sikkim) measured sixteen and a half feet in girth, at five feet from the ground, and was apparently ninety feet high: it was not pyramidal, the top branches being dead and broken, and the lower limbs spreading; they were loaded with masses of white-flowered Cœlogynes, and Vacciniums. The younger trees were pyramidal.

I was received by a monk of low degree, who made many apologies for the absence of his superior, who had been ordered an eight years' penance and seclusion from the world, of which only three had passed. On inquiry,

TEMPLE AND WEEPING CYPRESS.

I learnt the reason for this; the holy father having found himself surrounded by a family, to which there would have been no objection, had he previously obtained a dispensation. As, however, he had omitted this preliminary, and was able to atone by prayer and payment, he had been condemned to do penance; probably at his own suggestion, as the seclusion will give him sanctity, and eventually lead to his promotion, when his error shall have been forgotten.

Both temples are remarkable for their heavily ornamented, two-storied porticos, which occupy nearly the whole of one end. The interior decorations are in a ruinous condition, and evidently very old; they have no Hindoo emblems.

The head Lama sent me a present of dried peaches, with a bag of walnuts, called "Koal-kun" by the Lepchas, and "Taga-sching" by the Bhoteeas; the two terminations alike signifying "tree."

The view of Yoksun from this height was very singular: it had the appearance of an enormous deposit banked up against a spur to the south, and mountains to the east, and apparently levelled by the action of water: this deposit seemed as though, having once completely filled the valley of the Ratong, that river had cut a gorge 2000 feet deep between it and the opposite mountain.

Although the elevation is so low, snow falls abundantly at Doobdi in winter; I was assured that it has been known of the depth of five feet, a statement I consider doubtful; the quantity is, however, certainly greater than at equal heights about Dorjiling, no doubt owing to its proximity to Kinchinjunga.

I was amused here by watching a child playing with a popgun, made of bamboo, similar to that of quill, with which most English children are familiar, which propels pellets by means of a spring-trigger made of the upper part of the quill. It is easy to conclude such resemblances between the familiar toys of different countries to be accidental, but I question their being really so. On the plains of India, men may often be seen for hours together, flying what with us are children's kites; and I procured a jews'-harp from Tibet. These are not the toys of savages, but the amusements of people more than half-civilised, and with whom we have had indirect communication from the earliest ages. The Lepchas play at quoits, using slate for the purpose, and at the Highland games of "putting the stone" and "drawing the stone." Chess, dice, draughts, Punch, hockey, and battledore and shuttlecock, are all Indo-Chinese or Tartarian; and no one familiar with the wonderful instances of similarity between the monasteries, ritual, ceremonies, attributes, vestments, and other paraphernalia of the eastern and western churches, can fail to acknowledge the importance of recording even the most trifling analogies or similarities between the manners and customs of the young as well as of the old.

CHAPTER XV.

Leave Yoksun for Kinchinjunga—Ascend Ratong valley—Salt-smuggling over Ratong—Landslips—Plants—Buckeem—Blocks of gneiss—Mon Lepcha—View—Weather—View from Gubroo—Kinchinjunga, tops of—Pundim cliff—Nursing—Vegetation of Himalaya—Coup d'œil of Jongri—Route to Yalloong—Arduous route of salt-traders from Tibet—Kinchin, ascent of—Lichens—Surfaces sculptured by snow and ice—Weather at Jongri—Snow—Shades for eyes.

I LEFT Yoksun on an expedition to Kinchinjunga on the 7th of January. It was evident that at this season I could not attain any height; but I was most anxious to reach the lower limit of that mass of perpetual snow which descends in one continuous sweep from 28,000 to 15,000 feet, and radiates from the summit of Kinchin, along every spur and shoulder for ten to fifteen miles, towards each point of the compass.

The route lay for the first mile over the Yoksun flat, and then wound along the almost precipitous east flank of the Ratong, 1000 feet above its bed, leading through thick forest. It was often difficult, crossing torrents by culms of bamboo, and leading up precipices by notched poles and roots of trees. I wondered what could have induced the frequenting of such a route to Nepal, when there were so many better ones over Singalelah, till I found from my guide that he had habitually smuggled salt over this pass to avoid the oppressive duty levelled by the Dewan on all imports from Tibet by the eastern passes: he further told me that it took five days to reach Yalloong in Nepal from Yoksun, on the third of which the Kanglanamo pass is crossed, which is open from April to November, but is always heavily snowed. Owing to this duty, and the remoteness of the eastern passes, the people on the west side of the Great Rungeet were compelled to pay an enormous sum for salt; and the Lamas of Changachelling and Pemiongchi petitioned Dr. Campbell to use his influence with the Nepal Court to have the Kanglanamo pass re-opened, and the power of trading with the Tibetans of Wallanchoon, Yangma, and Kambachen, restored to them: the pass having been closed since the Nepalese war, to prevent the Sikkim people from kidnapping children and slaves, as was alleged to be their custom.*

We passed some immense landslips, which had swept the forest into the torrent, and exposed white banks of angular detritus of gneiss and granite: we crossed one 200 yards long, by a narrow treacherous path, on a slope of 35°: the subjacent gneiss was nearly vertical, striking north-east. We camped at 6,670 feet, amongst a vegetation I little expected to find so close to the snows of Kinchin; it consisted of oak, maple, birch, laurel, rhododendron, white *Daphne*, jessamine, *Arum*, *Begonia*, *Cyrtandraceæ*, pepper, fig, *Menispermum*, wild cinnamon, *Scitamineæ*, several epiphytic orchids, vines, and ferns in great abundance.

On the following day, I proceeded north-west up the Ratong river, here a furious torrent; which we crossed,

* An accusation in which there was probably some truth; for the Sikkim Dingpun, who guided Dr. Campbell and myself to Mainom, Tassiding, &c., since kidnapped, or caused to be abducted, a girl of Brahmin parents, from the Mai valley of Nepal, a transaction which cost him some 300 rupees. The Nepal Durbar was naturally furious, the more so as the Dingpun had no caste, and was therefore abhorred by all Brahmins. Restitution was demanded through Dr. Campbell, who caused the incensed Dingpun to give up his paramour and her jewels. He vowed vengeance against Dr. Campbell, and found means to gratify it, as I shall hereafter show.

and then ascended a very steep mountain called "Mon Lepcha." Immense detached masses of gneiss, full of coarse garnets, lay on the slope, some of which were curiously marked with a series of deep holes, large enough to put one's fist in, and said to be the footprints of the sacred cow. They appeared to me to have been caused by the roots of trees, which spread over the rocks in these humid regions, and wear channels in the hardest material, especially when they follow the direction of its lamination or stratification.

I encamped at a place called Buckeem (alt. 8,650 ft.), in a forest of *Abies Brunoniana* and *Webbiana*, yew, oak, various rhododendrons, and small bamboo. Snow lay in patches at 8000 feet, and the night was cold and clear. On the following morning I continued the ascent, alternately up steeps and along perfectly level shelves, on which were occasionally frozen pools, surrounded with dwarf juniper and rhododendrons. Across one I observed the track of a yak in the snow; it presented two ridges, probably from the long hair of this animal, which trails on the ground, sweeping the snow from the centre of its path. At 11,000 feet the snow lay deep and soft in the woods of silver fir, and the coolies waded through it with difficulty.

Enormous fractured boulders of gneiss were frequent over the whole of Mon Lepcha, from 7000 to 11,000 feet: they were of the same material as the rock *in situ*, and as unaccountable in their origin as the loose blocks on Dorjiling and Sinchul spurs at similar elevations, often cresting narrow ridges. I measured one angular detached block, forty feet high, resting on a steep narrow shoulder of the spur, in a position to which it was impossible it could have rolled; and it is equally difficult to suppose that glacial ice deposited it 4000 feet above the bottom of the gorge, except we conclude the valley to have been filled with ice to that depth. A glance at the map will show that Mon Lepcha is remarkably situated, opposite the face of Kinchin-junga, and at the great bend of the Ratong. Had that valley ever been filled with water during a glacial period, Mon Lepcha would have formed a promontory, and many floating bergs from Kinchin would have been stranded on its flank: but I nowhere observed these rocks to be of so fine a granite as I believe the upper rocks of Kinchin to be, and I consequently cannot advance even that far-fetched solution with much plausibility.

As I ascended, the rocks became more granitic, with large crystals of mica. The summit was another broad bare flat, elevated 13,080 feet, and fringed by a copse of rose, berberry, and very alpine rhododendrons: the Himalayan heather (*Andromeda fastigiata*) grew abundantly here, affording us good fuel.

The toilsome ascent through the soft snow and brushwood delayed the coolies, who scarcely accomplished five miles in the day. Some of them having come up by dark, I prepared to camp on the mountain-top, strewing thick masses of *Andromeda* and moss (which latter hung in great tufts from the bushes) on the snow; my blankets had not arrived, but there was no prospect of a snow-storm.

The sun was powerful when I reached the summit, and I was so warm that I walked about barefoot on the frozen snow without inconvenience, preferring it to continuing in wet stockings: the temperature at the time was 29½°, with a brisk south-east moist wind, and the dew point 22° 8.

The night was magnificent, brilliant starlight, with a pale mist over the mountains: the thermometer fell to 15½° at 7½ P.M., and one laid upon wood with its bulb freely exposed, sank to 7½°: the snow sparkled with broad flakes of hoar-frost in the full moon, which was so bright, that I recorded my observations by its light. Owing to the extreme cold of radiation, I passed a very uncomfortable night. The minimum thermometer fell to 1° in shade.* The sky was clear; and every rock, leaf, twig, blade of grass, and the snow itself, were covered with broad rhomboidal plates of hoar-frost, nearly one-third of an inch across: while the metal scale of the thermometer instantaneously blistered my tongue. As the sun rose, the light reflected from these myriads of facets had a splendid effect.

Before sunrise the atmosphere was still, and all but cloudless. To the south-east were visible the plains of India, at least 140 miles distant; where, as usual, horizontal layers of leaden purple vapour obscured the horizon: behind these the sun rose majestically, instantly dispersing them, while a thin haze spread over all the intervening mountains, from its slanting beams reaching me through otherwise imperceptible vapours: these, as the sun mounted higher, again became invisible, though still giving that transparency to the atmosphere and brilliant definition of the distances, so characteristic of a damp, yet clear day.

Mon Lepcha commands a most extensive view of Sikkim, southward to Dorjiling. At my feet lay the great and profound valley of the Ratong, a dark gulf of vegetation. Looking northward, the eye followed that river to the summit of Kinchinjunga (distant eighteen miles), which fronts the beholder as Mont Blanc does when seen from the mountains on the opposite side of the valley of Chamouni. To the east are the immense precipices and

* At sunrise the temperature was 11½°; that of grass, cleared on the previous day from snow, and exposed to the sky, 6½°; that on wool, 2° 2; and that on the surface of the snow, 0° 7.

glaciers of Pundim, and on the west those of Kubra, forming great supporters to the stupendous mountain between them. Mon Lepcha itself is a spur running south-east from the Kubra shoulder: it is very open, and covered with rounded hills for several miles further north, terminating in a conspicuous conical black hummock * called Gubroo, of 15,000 feet elevation, which presents a black cliff to the south.

Kinchinjunga rises in three heads, of nearly equal height,† which form a line running north-west. It exposes many white or grey rocks, bare of snow, and disposed in strata ‡ sloping to the west; the colour of all which above 20,000 feet, and the rounded knobbed form of the summit, suggest a granitic formation. Lofty snowed ridges project from Kubra into the Ratong valley, presenting black precipices of stratified rocks to the southward. Pundim has a very grand appearance; being eight miles distant, and nearly 9000 feet above Mon Lepcha, it subtends an angle of 12°; while Kinchin top, though 15,000 feet higher than Mon Lepcha, being eighteen miles distant, rises only 9° 30′ above the true horizon: these angular heights are too small to give much grandeur and

* This I have been told is the true Kubra; and the great snowy mountain behind it, which I here, in conformity with the Dorjiling nomenclature, call Kubra, has no name, being considered a part of Kinchin.

† The eastern and western tops are respectively 27,826 and 28,177 feet above the level of the sea.

‡ I am aware that the word strata is inappropriate here; the appearance of stratification or bedding, if it indicate any structure of the rock, being, I cannot doubt, due to that action which gives parallel cleavage planes to granite in many parts of the world, and to which the so-called lamination or foliation of slate and gneiss is supposed by many geologists to be due. It is not usual to find this structure so uniformly and conspicuously developed through large masses of granite, as it appeared to me to be on the sides of Kinchinjunga and on the top of Junnoo, as seen from the Choonjerma pass (p. 264, plate); but it is sometimes very conspicuous, and nowhere more than in the descent of the Grimsel towards Meyringen, where the granite on the east flank of that magnificent gorge seems cleft into parallel nearly vertical strata.

apparent elevation to mountains, however lofty; nor would they do so in this case, were it not that the Ratong valley which intervenes, is seen to be several thousand feet lower, and many degrees below the real horizon.

KINCHINJUNGA AND PUNDIM FROM MON LEPCHA.

Pundim has a tremendous precipice to the south, which, to judge from its bareness of snow, must be nearly perpendicular; and it presented a superb geological section. The height of this precipice I found by angles with a pocket sextant to be upwards of 3,400 feet, and that of its top to be 21,300 above the sea, and consequently only 715 feet less than that of the summit of Pundim itself (which is 22,015 feet). This cliff is of black stratified rocks, sloping to the west, and probably striking north-west; permeated from top to bottom by veins of white granite, disposed in zigzag lines, which produce a contortion of the gneiss, and give it a marbled appearance. The same structure may be seen in miniature on the transported blocks which abound in the Sikkim rivers; where veins of finely grained granite are forced in all directions through the gneiss, and form parallel seams or beds between the laminæ of that rock, united by transverse seams, and crumpling up the gneiss itself, like the crushed leaves of a book. The summit of Pundim itself is all of white rock, rounded in shape, and forming a cap to the gneiss, which weathers into precipices.

A succession of ridges, 14,000 to 18,000 feet high, presented a line of precipices running south from Pundim for several miles: immense granite veins are exposed on their surfaces, and they are capped by stratified rocks, sloping to the east, and apparently striking to the north-west, which, being black, contrast strongly with the white granite beneath them: these ridges, instead of being round-topped, are broken into splintered crags, behind which rises the beautiful conical peak of Nursing, 19,139 feet above the sea, eight miles distant, and subtending an angle of 8° 30′.

At the foot of these precipices was a very conspicuous series of lofty moraines, round whose bases the Ratong wound; these appeared of much the same height, rising several hundred feet above the valley: they were comparatively level-topped, and had steep shelving rounded sides.

I have been thus particular in describing the upper Ratong valley, because it drains the south face of the loftiest mountain on the globe; and I have introduced angular heights, and been precise in my details, because the vagueness with which all terms are usually applied to the apparent altitude and steepness of mountains and precipices, is apt to give false impressions. It is essential to attend to such points where scenery of real interest and importance is to be described. It is customary to speak of peaks as towering in the air, which yet subtend an angle of very few degrees; of almost precipitous ascents, which, when measured, are found to be slopes of 18° or 20°; and of cliffs as steep and stupendous, which are inclined at a very moderate angle.

The effect of perspective is as often to deceive in details as to give truth to general impressions; and those accessories are sometimes wanting in nature, which, when supplied by art, give truth to the landscape. Thus, a streak of clouds adds height to a peak which should appear lofty, but which scarcely rises above the true horizon; and a belt of mist will sunder two snowy mountains which, though at very different distances, for want of a play of light and shade on their dazzling surfaces, and from the extreme transparency of the air in lofty regions, appear to be at the same distance from the observer.

The view to the southward from Mon Lepcha, including the country between the sea-like plains of India and the loftiest mountain on the globe, is very grand, and neither wanting in variety nor in beauty. From the deep valleys choked with tropical luxuriance to the scanty yak pasturage on the heights above, seems but a step at the first *coup-d'œil*, but resolves itself on a closer inspection into five belts: 1, palm and plantain; 2, oak and laurel; 3, pine; 4, rhododendron and grass; and 5, rock and snow. From the bed of the Ratong, in which grow palms with screw-pine and plantain, it is only seven miles in a direct line to the perpetual ice. From the plains of India, or outer Himalaya, one may behold snowy peaks rise in the distance behind a foreground of tropical forest; here, on the contrary, all the intermediate phases of vegetation are seen at a glance. Except in the Himalaya this is no common phenomenon, and is owing to the very remarkable depth of the river-beds. That part of the valley of the Ratong where tropical vegetation ceases, is but 4000 feet above the sea, and though fully fifty miles as the crow flies (and perhaps 200 by the windings of the river) from the plains of India, is only eight in a straight line (and forty by the windings) from the snows which feed that river. In other words, the descent is so rapid, that in eight miles the Ratong waters every variety of vegetation, from the lichen of the poles to the palm of the tropics; whilst throughout the remainder of its mountain course, it falls from 4000 to 300 feet, flowing amongst tropical scenery, through a valley whose flanks rise from 5000 to 12,000 feet above its bed.

From Mon Lepcha we proceeded north-west towards Jongri, along a very open rounded bare mountain, covered with enormous boulders of gneiss, of which the subjacent rock is also composed. The soil is a thick clay full of angular stones, everywhere scooped out into little depressions which are the dry beds of pools, and are often strewed with a thin layer of pebbles. Black tufts of alpine aromatic rhododendrons of two kinds (*R. anthopogon* and *setosum*), with dwarf juniper, comprised all the conspicuous vegetation at this season.

After a two hours' walk, keeping at 13,000 feet elevation, we sighted Jongri.* There were two stone huts on the bleak face of the spur, scarcely distinguishable at the distance of half a mile from the great blocks around

* I am assured by Capt. Sherwill, who, in 1852, proceeded along and surveyed the Nepal frontier beyond this point to Gubroo, that this is not Jongri, but Yangpoong. The difficulty of getting precise information, especially as to the names of seldom-visited spots, is very great. I was often deceived myself, undesignedly, I am sure, on the part of my informants; but in this case I have Dr. Campbell's assurance, who has kindly investigated the subject, that there is no mistake on my part. Captain Sherwill has also kindly communicated to me a map of the head waters of the Rungbee, Yungya, and Yalloong rivers, of which, being more correct than my own, I have gladly availed myself for my map. Gubroo, he informs me, is 15,000 feet in altitude, and dips in a precipice 1000 feet high, facing Kubra, which prevented his exploring further north.

them. To the north Gubroo rose in dismal grandeur, backed by the dazzling snows of Kubra, which now seemed quite near, its lofty top (alt. 24,005 feet) being only eight miles distant. Much snow lay on the ground in patches, and there were few remains of herbaceous vegetation; those I recognised were chiefly of poppy, *Potentilla*, gentian, geranium, fritillary, *Umbelliferæ*, grass, and sedges.

On our arrival at the huts the weather was still fine, with a strong north-west wind, which meeting the warm moist current from the Ratong valley, caused much precipitation of vapour. As I hoped to be able to visit the surrounding glaciers from this spot, I made arrangements for a stay of some days: giving up the only habitable hut to my people, I spread my blankets in a slope from its roof to the ground, building a little stone dyke round the skirts of my dwelling, and a fire-place in front.

Hence to Yalloong in Nepal, by the Kanglanamo pass, is two days' march: the route crosses the Singalelah range at an elevation of about 15,000 feet, south of Kubra, and north of a mountain that forms a conspicuous feature south-west from Jongri, as a crest of black fingered peaks, tipped with snow.

It is difficult to conceive the amount of labour expended upon every pound of salt imported into this part of Sikkim from Tibet, and as an enumeration of the chief features of the routes it must follow, will give some idea of what the circuit of the loftiest mountain in the globe involves, I shall briefly allude to them; premising that the circuit of Mont Blanc may be easily accomplished in four days. The shortest route to Yoksun (the first village south of Kinchin) from the nearest Tibetan village north of that mountain, involves a detour of one-third of the circumference of Kinchin. It is evident that the most direct way must be that nearest the mountain-top, and therefore that which reaches the highest accessible elevation on its shoulders, and which, at the same time, dips into the shallowest valleys between those shoulders. The actual distance in a straight line is about fifty miles, from Yoksun to the mart at or near Tashirukpa.

The marches between them are as follows:—

1. To Yalloong two days; crossing Kanglanamo pass, 15,000 feet high.

3. To foot of Choonjerma pass, descending to 10,000 feet.

4. Cross Choonjerma pass, 15,260 feet, and proceed to Kambachen, 11,400 feet.

5. Cross Nango pass, 15,770, and camp on Yangma river, 11,000 feet.

6. Ascend to foot of Kanglachem pass, and camp at 15,000 feet.

7. Cross Kanglachem pass, probably 16,500 feet; and

8—10. It is said to be three marches hence to the Tibetan custom-house, and that two more snowy passes are crossed.

This allows no day of rest, and gives only five miles—as the crow flies—to be accomplished each day, but I assume fully fourteen of road distance; the labour spent in which would accomplish fully thirty over good roads. Four snowed passes at least are crossed, all above 15,000 feet, and after the first day the path does not descend below 10,000 feet. By this route about one-third of the circuit of Kinchinjunga is accomplished. Supposing the circuit were to be completed by the shortest practicable route, that is, keeping as near the summit as possible, the average time required for a man with his load would be upwards of a month.

To reach Tashirukpa by the eastern route from Yoksun, being a journey of about twenty-five days, requires a long detour to the southward and eastward, and afterwards the ascent of the Teesta valley, to Kongra Lama, and so north to the Tibetan Arun.

My first operation after encamping and arranging my instruments, was to sink the ground thermometer; but the earth being frozen for sixteen inches, it took four men several hours' work with hammer and chisel, to penetrate so deep. There was much vegetable matter for the first eight or ten inches, and below that a fine red clay. I spent the afternoon, which was fine, in botanising. When the sun shone, the smell of the two rhododendrons was oppressive, especially as a little exertion at this elevation brings on headache. There were few mosses; but crustaceous lichens were numerous, and nearly all of them of Scotch, Alpine, European, and Arctic kinds. The names of these, given by the classical Linnæus and Wàhlenberg, tell in some cases of their birth-places, in others of their hardihood, their lurid colours and weather-beaten aspects; such as *tristis*, *gelida*, *glacialis*, *arctica*, *alpina*, *saxatilis*, *polaris*, *frigida*, and numerous others equally familiar to the Scotch botanist. I recognised many as natives of the wild mountains of Cape Horn, and the rocks of the stormy Antarctic ocean; since visiting which regions I had not gathered them. The lichen called *geographicus* was most abundant, and is found to indicate a certain degree of cold in every latitude; descending to the level of the sea in latitude 52° north, and 50° south, but in lower latitudes only to be seen on mountains. It flourishes at 10,000 feet on the Himalaya, ascending thence to 18,000 feet. Its name, however, was not intended to indicate its wide range, but the curious maplike patterns which its yellow crust forms on the rocks.

Of the blocks of gneiss scattered over the Jongri spur, many are twenty feet in diameter. The ridge slopes gently south-west to the Choroong river, and more steeply north-east to the Ratong, facing Kinchin: it rises so very gradually to a peaked mountain between Jongri and Kubra, that it is not possible to account for the transport and deposit of these boulders by glaciers of the ordinary form, viz., by a stream of ice following the course of a valley; and we are forced to speculate upon the possibility of ice having capped the whole spur, and moved downwards, transporting blocks from the prominences on various parts of the spur.

The cutting up of the whole surface of this rounded mountain into little pools, now dry, of all sizes, from ten to about one hundred yards in circumference, is a very striking phenomenon. The streams flow in shallow transverse valleys, each passing through a succession of such pools, accompanying a step-like character of the general surface. The beds are stony, becoming more so where they enter the pools, upon several of the larger of which I observed curving ridges of large stones, radiating outwards on to their beds from either margin of the entering stream: more generally large stones were deposited opposite every embouchure.

This superficial sculpturing must have been a very recent operation; and the transport of the heavy stones opposite the entrance of the streams has been effected by ice, and perhaps by snow; just as the arctic ice strews the shores of the Polar ocean with rocks.

The weather had been threatening all day, northern and westerly currents contending aloft with the south-east trade-wind of Sikkim, and meeting in strife over the great upper valley of the Ratong. Stately masses of white cumuli wheeled round that gulf of glaciers, partially dissipating in an occasional snow-storm, but on the whole gradually accumulating.

On my arrival the thermometer was 32°, with a powerful sun shining, and it fell to 28° at 4 P.M., when the north wind set in. At sunset the moon rose through angry masses of woolly cirrus; its broad full orb threw a flood of yellow light over the serried tops south of Pundim; thence advancing obliquely towards Nursing, "it stood tip-toe" for a few minutes on that beautiful pyramid of snow, whence it seemed to take flight and mount majestically into mid-air, illuminating Kinchin, Pundim, and Kubra.

I sat at the entrance of my gipsy-like hut, anxiously watching the weather, and absorbed in admiration of the moonrise, from which my thoughts were soon diverted by its fading light as it entered a dense mass of mare's-tail cirrus. It was very cold, and the stillness was oppressive. I had been urged not to attempt such an ascent in January, my provisions were scanty, firewood only to be obtained from some distance, the open undulating surface of Jongri was particularly exposed to heavy snow-drifts, and the path was, at the best, a scarcely perceptible track. I followed every change of the wind, every fluctuation of the barometer and thermometer, each accession of humidity, and the courses of the clouds aloft. At 7 P.M., the wind suddenly shifted to the west, and the thermometer instantly rose from 20° to 30°. After 8 P.M., the temperature fell again, and the wind drew round from west by south to north-east, when the fog cleared off. The barometer rose no more than it usually does towards 10 P.M., and though it clouded again, with the temperature at 17°, the wind seemed steady, and I went to bed with a relieved mind.

Jan. 10.—During the night the temperature fell to 11° 2, and at 6 A.M. was 19° 8, falling again to 17° soon after. Though clouds were rapidly coming up from the west and south-west, the wind remained northerly till 8 A.M., when it shifted to south-west, and the temperature rose to 25°. As it continued fine, with the barometer high, I ventured on a walk towards Gubroo, carefully taking bearings of my position. I found a good many plants in a rocky valley close to that mountain, which I in vain attempted to ascend. The air was 30°, with a strong and damp south-west wind, and the cold was so piercing, that two lads who were with me, although walking fast, became benumbed, and could not return without assistance. At 11 A.M., a thick fog obliged us to retrace our steps: it was followed by snow in soft round pellets like sago, that swept across the hard ground. During the afternoon it snowed unceasingly, the wind repeatedly veering round the compass, always from west to east by south, and so by north to west again. The flakes were large, soft, and moist with the south wind, and small, hard, and dry with the north. Glimpses of blue sky were constantly seen to the south, under the gloomy canopy above, but they augured no change. As darkness came on, the temperature fell to 15°, and it snowed very hard; at 6 P.M., it was 11°, but rose afterwards to 18°

The night was very cold and wintry: I sat for some hours behind a blanket screen (which had to be shifted every few minutes) at my tent-door, keeping up a sulky fire, and peering through the snow for signs of improvement, but in vain. The clouds were not dense, for the moon's light was distinct, shining on the glittering snow-flakes that fell relentlessly: my anxiety was great, and I could not help censuring myself severely for exposing a party to so great danger at such a season. I found comfort in the belief that no idle curiosity had prompted me, and that with a good motive and a strong prestige of success, one can surmount a host of difficulties. Still the snow fell; and my heart sank, as my fire declined, and the flakes sputtered on the blackening embers; my little puppy, who had gambolled all day amongst the drifting white pellets, now whined, and crouched under my thick woollen cloak; the inconstant searching wind drifted the snow into the tent, whose roof so bagged in with the accumulation that I had to support it with sticks, and dreaded being smothered, if the weight should cause it to sink upon my bed during my sleep. The increasing cold drove me, however, to my blankets, and taking the precaution of stretching a tripod stand over my head, so as to leave a breathing hole, by supporting the roof if it fell in, I slept soundly, with my dog at my feet.

At sunrise the following morning the sky was clear, with a light north wind; about two feet of snow had fallen, the drifts were deep, and all trace of the path obliterated. The minimum thermometer had fallen to 3° 7, the temperature rose to 27° at 9 A.M., after which the wind fell, and with it the thermometer to 18°. Soon, however, southerly breezes set in, bringing up heavy masses of clouds.

My light-hearted companions cheerfully prepared to leave the ground; they took their appointed loads without a murmur, and sought protection for their eyes from the glare of the newly fallen snow, some with as much of my crape veil as I could spare, others with shades of brown paper, or of hair from the yaks' tails, whilst a few had spectacle-shades of woven hair; and the Lepchas loosened their pigtails, and combed their long hair over their eyes and faces. It is from fresh-fallen snow alone that much inconvenience is felt; owing, I suppose, to the light reflected from the myriads of facets which the crystals of snow present. I have never suffered inconvenience in crossing beds of old snow, or glaciers with weathered surfaces, which absorb a great deal of light, and reflect comparatively little, and that little coloured green or blue.

The descent was very laborious, especially through the several miles of bush and rock which lie below the summit: so that, although we started at 10 A. M., it was dark by the time we reached Buckeem, where we found two lame coolies, whom we had left on our way up, and who were keeping up a glorious fire for our reception.

MAITRYA, THE SIXTH OR COMING BOODH.

CHAPTER XVI.

Ratong river below Mon Lepcha—Ferns—Vegetation of Yoksun, tropical—*Araliaceæ*, fodder for cattle—Rice-paper plant—Geology of Yoksun—Lake—Old temples—Funereal cypresses—Gigantic chait—Altars—Songboom—Weather—Catsuperri—Velocity of Ratong—Worship at Catsuperri lake—Scenery—Willow—Lamas and ecclesiastical establishments of Sikkim—Tengling—Changachelling temples and monks—Portrait of myself on walls—Block of mica-schist—Lingcham Kajee asks for spectacles—Hee-hill—Arrive at Little Rungeet—At Dorjiling—Its deserted and wintry appearance.

On the following day we marched to Yoksun: the weather was fair, though it was evidently snowing on the mountains above. I halted at the Ratong river, at the foot of Mon Lepcha, where I found its elevation to be 7,150 feet; its edges were frozen, and the temperature of the water 36°; it is here a furious torrent flowing between gneiss rocks which dip south-south-east, and is flanked by flat-topped beds of boulders, gravel and sand, twelve to fourteen feet thick. Its vegetation resembles that of Dorjiling, but is more alpine, owing no doubt to the proximity of Kinchinjunga. The magnificent *Rhododendron argenteum* was growing on its banks. On the other hand, I was surprised to see a beautiful fern (a *Trichomanes*, very like the Irish one) which is not found at Dorjiling. The same day, at about the same elevation, I gathered sixty species of fern, many of very tropical forms.* No doubt the range of such genera is extended in proportion to the extreme damp and equable

* They consisted of the above-mentioned *Trichomanes*, three *Hymenophyllæ*, *Vittaria*, *Pleopeltis*, and *Marattia*, together with several *Selaginellas*.

climate, here, as about Dorjiling. Tree-ferns are however absent, and neither plantains, epiphytical *Orchideæ*, nor palms, are so abundant, or ascend so high as on the outer ranges. About Yoksun itself, which occupies a very warm sheltered flat, many tropical genera occur, such as tall bamboos of two kinds, grasses allied to the sugar-cane, scarlet *Erythrina*, and various *Araliaceæ*, amongst which was one species whose pith was of so curious a structure,that I had no hesitation in considering the then unknown Chinese substance called rice-paper to belong to a closely allied plant.*

The natives collect the leaves of many Aralias as fodder for cattle, for which purpose they are of the greatest service in a country where grass for pasture is so scarce; this is the more remarkable, since they belong to the natural family of ivy, which is usually poisonous; the use of this food, however, gives a peculiar taste to the butter. In other parts of Sikkim, fig-leaves are used for the same purpose, and branches of a bird-cherry (*Prunus*), a plant also of a very poisonous family, abounding in prussic acid.

We were received with great kindness by the villagers of Yoksun, who had awaited our return with some anxiety, and on hearing of our approach had collected large supplies of food; amongst other things were tares (called by the Lepchas "Kullai"), yams ("Book"), and a bread made by bruising together damp maize and rice into tough thin cakes ("Katch-ung tapha"). The Lamas of Doobdi were especially civil, having a favour to ask, which was that I would intercede with Dr. Campbell to procure the permission of the Nepalese

* The Chinese rice-paper has long been known to be cut from cylinders of pith which has always a central hollow chamber, divided into compartments by septa or excessively thin plates. It is only within the last few months that my supposition has been confirmed, by my father's receiving from China, after many years of correspondence, specimens of the rice-paper plant itself, which very closely resemble, in botanical characters, as well as in outward appearance of size and habit, the Sikkim plant.

to reopen the Kanglanamo pass, and thus give some occupation to their herds of yaks, which were now wandering idly about.

I botanized for two days on the Yoksun flat, searching for evidence of lacustrine strata or moraines, being more than ever convinced by the views I had obtained of this place from Mon Lepcha, that its uniformity of surface was due to water action. It is certainly the most level area of its size that I know of in Sikkim, though situated in one of the deepest valleys, and surrounded on almost all sides by very steep mountains; and it is far above the flat gravel terraces of the present river-beds. I searched the surface of the flat for gravel beds in vain, for though it abounds in depressions that must have formerly been lake-beds, and are now marshes in the rainy season, these were all floored with clay. Along the western edge, where the descent is very steep for 1800 feet to the Ratong, I found no traces of stratified deposits, though the spurs which projected from it were often flattened at top. The only existing lake has sloping clay banks, covered with spongy vegetable mould; it has no permanent affluent or outlet, its present drainage being subterranean, or more probably by evaporation; but there is an old water channel several feet above its level. It is eighty to a hundred yards across, and nearly circular; its depth three or four feet, increased to fifteen or sixteen in the rains; like all similar pools in Sikkim, it contains little or no animal life at this season, and I searched in vain for shells, insects, or frogs. All around were great blocks of gneiss, some fully twelve feet square.

The situation of this lake is very romantic, buried in a tall forest of oaks and laurels, and fringed by wild camellia shrubs; the latter are not the leafy, deep green, large-blossomed plants of our greenhouses, but twiggy bushes with small scattered leaves, and little yellowish flowers like those of the tea-plant. The massive walls of a ruined temple rise close to the water, which looks like the still moat of a castle: beside it are some grand old funereal cypresses, with ragged scattered branches below, where they struggle for light in the dense forest, but raising their heads aloft as bright green pyramids.

After some difficulty I found the remains of a broad path that divided into two; one of them led to a second ruined temple, fully a mile off, and the other I followed to a grove, in which was a gigantic chait; it was a beautiful lane throughout, bordered with bamboo, brambles, gay-flowered

ALTAR AND SONG-BOOM AT YOKSUN.

Melastomaceæ like hedge-roses, and scarlet *Erythrina*: there were many old mendongs and chaits on the way, which I was always careful to leave on the right hand in passing, such being the rule among Boodhists, the same which ordains that the praying-cylinder or "Mani" be made to revolve in a direction against the sun's motion.

This great chait is the largest in Sikkim; it is called "Nirbogong," and appears to be fully forty feet high; facing it is a stone altar about fifteen feet long and four broad, and behind this again is a very curious erection called "Song-boom," used for burning juniper as incense; it resembles a small smelting furnace, and consists of an elongated conical stone building eight feet high, raised on a single block; it is hollow, and divided into three stories or chambers; in the lower of which is a door, by which fuel is placed inside, and the smoke ascending through holes in the upper slabs, escapes by lateral openings from the top compartment. These structures are said to be common in Tibet, but I saw no other in Sikkim.

During my stay at Yoksun, the weather was very cold, especially at night, considering the elevation (5,600 feet): the mean temperature was 39°, the extremes being 19° 2 and 60°; and even at 8 A.M. the thermometer, laid on the frosty grass, stood at 20°; temperatures which are rare at Dorjiling, 1500 feet higher. I could not but regard with surprise such half tropical genera as perennial-leaved vines, *Saccharum*, *Erythrina*, large bamboos, *Osbeckia* and cultivated millet, resisting such low temperatures.*

On the 14th January I left Yoksun for the lake and temples of Catsuperri, the former of which is by much the largest in Sikkim. After a steep descent of 1800 feet, we reached the Ratong, where its bed is only 3,790 feet above

* This is no doubt due to the temperature of the soil being always high: I did not sink a thermometer at Yoksun, but from observations taken at similar elevations, the temperature of the earth, at three feet depth, may be assumed to be 55°.

the sea; it is here a turbulent stream, twelve yards across, with the usual features of gravel terraces, huge boulders of gneiss and some of the same rock *in situ*, striking north-east. Some idea of its velocity may be formed from the descent it makes from the foot of Mon Lepcha, where the elevation of its bed was 7,150 feet, giving a fall of 3,350 feet in only ten miles.

Hence I ascended a very steep spur, through tropical vegetation, now become so familiar to me that I used to count the number of species belonging to the different large natural orders, as I went along. I gathered only thirty-five ferns at these low elevations, in the same space as produces from fifty to sixty in the more equable and humid regions of 6000 feet; grasses on the other hand were much more numerous. The view of the flat of Yoksun from Lung-schung village, opposite to it, and on about the same level, is curious; as is that of the hamlet of Lathiang on the same side, which I have before noticed as being placed on a very singular flat shelf above the Ratong, and is overhung by rocks.

Ascending very steeply for several thousand feet, we reached a hollow on the Catsuperri spur, beyond which the lake lies buried in a deep forest. A Lama from the adjacent temple accompanied us, and I found my people affecting great solemnity as they approached its sacred bounds; they incessantly muttered "Om mani," &c., kotowed to trees and stones, and hung bits of rag on the bushes. A pretence of opposing our progress was made by the priest, who of course wanted money; this I did not appear to notice, and after a steep descent, we were soon on the shores of what is, for Sikkim, a grand sheet of water, (6,040 feet above the sea), without any apparent outlet: it may be from three to five hundred yards across in the rains, but was much less now, and was bordered by a broad marsh of bog moss (*Sphagnum*), in which were abundance of *Azolla*, colouring the waters red, and sedges. Along the banks were bushes of *Rhododendron barbatum* and *Berberis insignis*,* but the mass of the vegetation was similar to that of Dorjiling.

We crossed the marsh to the edge of the lake by a rude paved way of decaying logs, through which we often plunged up to our knees. The Lama had come provided with a piece of bark, shaped like a boat, some juniper incense and a match-box, with which he made a fire, and put it in the boat, which he then launched on the lake as a votive offering to the presiding deity. It was a dead calm, but the impetus he gave to the bark shot it far across the lake, whose surface was soon covered with a thick cloud of white smoke. Taking a rupee from me the priest then waved his arm aloft, and pretended to throw the money into the water, singing snatches of prayers in Tibetan, and at times shrieking at the top of his voice to the Dryad who claims these woods and waters as his own. There was neither bird, beast, nor insect to be seen, and the scenery was as impressive to me, as the effect of the simple service was upon my people, who prayed with redoubled fervour, and hung more rags on the bushes.

I need hardly say that this invocation of the gods of the woods and waters forms no part of Lama worship; but the Lepchas are but half Boodhists; in their hearts they dread the demons of the grove, the lake, the snowy mountain and the torrent, and the crafty Lama takes advantage of this, modifies his practices to suit their requirements, and is content with the formal recognition of the spiritual supremacy of the church. This is most remarkably shown in their acknowledgment of the day on which offerings had been made from time immemorial by the pagan Lepchas to

* This magnificent new species has not been introduced into England; it forms a large bush, with deep-green leaves seven inches long, and bunches of yellow flowers.

the genius of Kinchinjunga, by holding it as a festival of the church throughout Sikkim.*

The two Catsuperri temples occupy a spur 445 feet above the lake, and 6,485 feet above the sea; they are poor, and only remarkable for a miserable weeping-willow tree planted near them, said to have been brought from Lhassa. The monks were very civil to me, and offered amongst other things a present of excellent honey. One was an intelligent man, and gave me much information: he told me that there were upwards of twenty religious establishments in Sikkim, containing more than 1000 priests. These have various claims upon the devout: thus, Tassiding, Doobdi, Changachelling, and Pemiongchi, are celebrated for their antiquity, and the latter also for being the residence of the head Lama; Catsuperri for its lake; Raklang for its size, &c. All are under one spiritual head, who is the Tupgain Lama, or eldest son of the Rajah; and who resides at the Phadong convent, near Tumloong: the Lama of Pemiongchi is, however, the most highly respected, on account of his age, position, and sanctity. Advancement in the hierarchy is dependent chiefly on interest, but indirectly on works also; pilgrimages to Lhassa and Teshoo Loombo are the highest of

* On that occasion an invocation to the mountain is chanted by priests and people in chorus. Like the Lama's address to the genius of Catsuperri lake, its meaning, if it ever had any, is not now apparent. It runs thus:—

"Kanchin-jinga, Pemi Kadup
Guetche Tangla, Dursha tember
Zu jinga Pemsum Serkiem
Dischze Kubra Kanchin tong."

This was written for me by Dr. Campbell, who, like myself, has vainly sought its solution; it is probably a mixture of Tibetan and Lepcha, both as much corrupted as the celebrated "Om mani padmi hoom," which is universally pronounced by Lepchas "Menny pemmy hoom." This reminds me that I never got a solution of this sentence from a Lama, of whatever rank or learning; and it was only after incessant inquiry, during a residence of many years in Nepal, that Mr. Hodgson at last procured the interpretation, or rather paraphrase: "Hail to him (Sakya) of the lotus and the jewel," which is very much the same as M. Klaproth and other authorities have given.

these, and it is clearly the interest of the supreme pontiffs of those ecclesiastical capitals to encourage such, and to intimate to the Sikkim authorities, the claims those who perform them have for preferment. Dispensations for petty offences are granted to Lamas of low degree and monks, by those of higher station, but crimes against the church are invariably referred to Tibet, and decided there.

The election to the Sikkim Lamaseries is generally conducted on the principle of self-government, but Pemiongchi and some others are often served by Lamas appointed from Tibet, or ordained there, at some of the great convents. I never heard of an instance of any Sikkim Lama arriving at such sanctity as to be considered immortal, and to reappear after death in another individual, nor is there any election of infants. All are of the Ningma, Dookpa, or Shammar sect, and are distinguished by their red mitres; they were once dominant throughout Tibet, but after many wars * with the yellow-caps, they were driven from that country, and took refuge principally in the Himalaya. The Bhotan or Dhurma †

* The following account of the early war between the red and the yellow-mitred Lamas was given me by Tchebu Lama:—For twenty-five generations the red-caps (Dookpa or Ningma) prevailed in Tibet, when they split into two sects, who contended for supreme power; the Lama of Phado, who headed the dissenters, and adopted a yellow mitre, being favoured by the Emperor of China, to whom reference was made. A persecution of the red Lamas followed, who were caught by the yellow-caps, and their mitres plunged into dyeing vats kept always ready at the Lamaseries. The Dookpa, however, still held Teshoo Loombo, and applied to the Sokpo (North Tibet) Lamas for aid, who bringing horses and camels, easily prevailed over the Gelookpa or yellow sect, but afterwards treacherously went over to them, and joined them in an attack on Teshoo Loombo, which was plundered and occupied by the Gelookpas. The Dookpa thereafter took refuge in Sikkim and Bhotan, whence the Bhotan Rajah became their spiritual chief under the name of Dhurma Rajah, and is now the representative of that creed. Gooroucknath is still the Dookpa's favourite spiritual deity of the older creed, which is, however, no longer in the ascendant. The Dalai Lama of Teshoo Loombo is a Gelookpa, as is the Rimbochay Lama, and the Potala Lama of Lhassa, according to Tchebu Lama, but Turner ("Travels in Tibet," p. 315) says the contrary; the Gelookpa consider Sakya Thoba (or Tsongkaba) alias Mahamouni, as their great avatar.

† Bhotan is generally known as the Dhurma country. See page 136, in note.

Rajah became the spiritual head of this sect, and, as is well known, disputes the temporal government also of his country with the Deva Rajah, who is the hereditary temporal monarch, and never claims spiritual jurisdiction. I am indebted to Dr. Campbell for a copy and translation of the Dhurma Rajah's great seal, containing the attributes of his spirituality, a copy of which I have appended to the end of this chapter.

The internal organisation of the different monastic establishments is very simple. The head or Teshoo Lama * rules supreme; then come the monks and various orders of priests, and then those who are candidates for orders, and dependents, both lay-brothers and slaves: there are a few nunneries in Sikkim, and the nuns are all relatives or connections of the Rajah, his sister is amongst them. During the greater part of the year, all lead a more or less idle life; the dependents being the most occupied in carrying wood and water, cultivating the land, &c.

The lay-brothers are often skilful workmen, and are sometimes lent or hired out as labourers, especially as housebuilders and decorators. No tax of any kind is levied on the church, which is frequently very rich in land, flocks, and herds, and in contributions from the people: land is sometimes granted by the Rajah, but is oftener purchased by the priests, or willed, or given by the proprietor. The services, to which I have already alluded, are very irregularly performed; in most temples only on festival days, which correspond to the Tibetan ones so admirably described in MM. Huc and Gabet's narrative; in a few, however, service is performed daily, especially in such as stand near frequented roads, and hence reap the richest harvest.

* I have been informed by letters from Dr. Campbell that the Pemiongchi Lama is about to remove the religious capital of Sikkim to Dorjiling, and build there a grand temple and monastery; this will be attractive to visitors, and afford the means of extending our knowledge of East Tibet.

Like all the natives of Tibet and Sikkim, the priests are intolerably filthy; in some cases so far carrying out their doctrines as not even to kill the vermin with which they swarm. All are nominally bound to chastity, but exemptions in favour of Lamas of wealth, rank, or power, are granted by the supreme pontiffs, both in Tibet and Sikkim. I constantly found swarms of children about the Lamaseries, who were invariably called nephews and nieces.

Descending from the Catsuperri temples, I encamped at the village of Tengling (elevation 5,257 feet), where I was waited upon by a bevy of forty women, Lepchas and Sikkim Bhoteeas, accompanied by their children, and bringing presents of fowls, rice and vegetables, and apologising for the absence of their male relatives, who were gone to carry tribute to the Rajah. Thence I marched to Changachelling, first descending to the Tengling river, which divides the Catsuperri from the Molli ridge, and which I crossed.

Tree-ferns here advance further north than in any other part of Sikkim. I did not visit the Molli temples, but crossed the spur of that name, to the Rungbee river, whose bed is 3,300 feet above the sea; thence I ascended upwards of 3,500 feet to the Changachelling temples, passing Tchongpong village. The ridge on which both Pemiongchi and Changachelling are built, is excessively narrow at top; it is traversed by a "via sacra," connecting these two establishments; this is a pretty wooded walk, passing mendongs and chaits hoary with lichens and mosses; to the north the snows of Kinchinjunga are seen glimmering between the trunks of oaks, laurels, and rhododendrons, while to the south the Sinchul and Dorjiling spurs shut out the view of the plains of India.

Changachelling temples and chaits crown a beautiful

rocky eminence on the ridge, their roofs, cones and spires peeping through groves of bamboo, rhododendrons, and arbutus; the ascent is by broad flights of steps cut in the mica-slate rocks, up which shaven and girdled monks, with rosaries and long red gowns, were dragging loads of bamboo stems, that produced a curious rattling noise. At the summit there is a fine temple, with the ruins of several others, and of many houses: the greater part of the principal temple, which is two-storied and divided into several compartments, is occupied by families. The monks were busy repairing the part devoted to worship, which consists of a large chamber and vestibule of the usual form: the outside walls are daubed red, with a pigment of burnt felspathic clay, which is dug hard by Some were painting the vestibule with colours brought from Lhassa, where they had been trained to the art. Amongst other figures was one playing on a guitar, a very common symbol in the vestibules of Sikkim temples: I also saw an angel playing on the flute, and a snake-king offering fruit to a figure in the water, who was grasping a serpent. Amongst the figures I was struck by that of an Englishman, whom, to my amusement, and the limner's great delight, I recognised as myself. I was depicted in a flowered silk coat instead of a tartan shooting jacket, my shoes were turned up at the toes, and I had on spectacles and a tartar cap, and was writing notes in a book. On one side a snake-king was politely handing me fruit, and on the other a horrible demon was writhing.

A crowd had collected to see whether I should recognise myself, and when I did so, the merriment was extreme. They begged me to send them a supply of vermilion, gold-leaf, and brushes; our so called camel's-hair pencils being much superior to theirs, which are made of marmot's hair.

I was then conducted to a house, where I found salted and buttered tea and Murwa beer smoking in hospitable preparation. As usual, the house was of wood, and the inhabited apartments above the low basement story were approached by an outside ladder, like a Swiss cottage: within were two rooms floored with earth; the inner was small, and opened on a verandah that faced Kinchinjunga, whence the keen wind whistled through the apartment.

The head Lama, my jolly fat friend of the 20th of December, came to breakfast with me, followed by several children, nephews and nieces he said; but they were uncommonly like him for such a distant relationship, and he seemed extremely fond of them, and much pleased when I stuffed them with sugar.

Changachelling hill is remarkable for having on its summit an immense tabular mass of chlorite slate, resting apparently horizontally on variously inclined rocks of the same: it is quite flat-topped, ten to twelve yards each way, and the sides are squared by art; the country people attribute its presence here to a miracle.

The view of the Kinchin range from this spot being one of the finest in Sikkim, and the place itself being visible from Dorjiling, I took a very careful series of bearings, which, with those obtained at Pemiongchi, were of the utmost use in improving my map, which was gradually progressing. To my disappointment I found that neither priest nor people knew the name of a single snowy mountain. I also asked in vain for some interpretation of the lines I have quoted at p. 365; they said they were Lepcha worship, and that they only used them for the gratification of the people, on the day of the great festival of Kinchinjunga.

Hence I descended to the Kulhait river, on my route back to Dorjiling, visiting my very hospitable tippling friend, the Kajee of Lingcham, on the way down: he humbly begged me to get him a pair of spectacles, for no other object than to look wise, as he had the eyes of a hawk; he told me that mine drew down universal respect in Sikkim, and that I had been drawn with them on, in the temple at Changachelling; and that a pair would not only wonderfully become him, but afford him the most pleasing recollections of myself. Happily I had the means of gratifying him, and have since been told that he wears them on state occasions.

I encamped by the river, 3,160 feet above the sea, amongst figs and plantains, on a broad terrace of pebbles, boulders and sand, ten feet above the stream; the rocks in the latter were covered with a red conferva. The sand on the banks was disposed in layers, alternately white and red, the white being quartz, and the red pulverised garnets. The arranging of these sand-bands by the water must be due to the different specific gravities of the garnet and quartz; the former being lighter, is lifted by the current on to the surface of the quartz, and left there when the waters retire.

On the next day I ascended Hee hill, crossed it at an elevation of 7,290 feet, and camped on the opposite side at 6,680 feet, in a dense forest. The next march was still southward to the little Rungeet guard-house, below Dorjiling spur, which I reached after a fatiguing walk amidst torrents of rain. The banks of the little Rungeet river, which is only 1,670 feet above the sea, are very flat and low, with broad terraces of pebbles and shingle, upon which are huge gneiss boulders, fully 200 feet above the stream.

On the 19th of January, I ascended the Tukvor spur to Dorjiling, and received a most hospitable welcome from my friend Mr. Muller, now almost the only European inhabitant of the place; Mr. Hodgson having gone down on a shooting excursion in the Terai, and Dr. Campbell being on duty on the Bhotan frontier. The place looked what it really was —wholly deserted. The rain I had experienced in the valley, had here been snow, and the appearance of the broad snowed patches clear of trees, and of the many houses without smoke or inhabitant, and the tall scattered trees with black bark and all but naked branches, was dismal in the extreme. The effect was heightened by an occasional Hindoo, who flitted here and there along the road, crouching and shivering, with white cotton garments and bare legs.

The delight of my Lepcha attendants at finding themselves safely at home again, knew no bounds; and their parents waited on me with presents, and other tokens of their goodwill and gratitude. I had no lack of volunteers for a similar excursion in the following season, though with their usual fickleness, more than half failed me, long before the time arrived for putting their zeal to the proof.

I am indebted to Dr. Campbell for the accompanying impression and description of the seal of the Dhurma Rajah, or sovereign pontiff of Bhotan, and spiritual head of the whole sect of the Dookpa, or red-mitred Lama Boodhists. The translations were made by Aden Tchebu Lama, who accompanied us into Sikkim in 1849, and I believe they are quite correct. The Tibetan characters run from left to right.

The seal of the Dhurma Rajah is divided into a centre portion and sixteen rays. In the centre is the word Dookyin, which means "The Dookpa Creed;" around the "Dookyin" are sixteen similar letters, meaning "I," or "I am." The sixteen radial compartments contain his titles and attributes, thus, commencing from the centre erect one, and passing round from left to right:—

1. I am the Spiritual and Temporal Chief of the Realm.
2. The Defender of the Faith.
3. Equal to Saruswati in learning.
4. Chief of all the Boodhs.
5. Head expounder of the Shasters.
6. Caster out of devils.
7. The most learned in the Holy Laws.
8. An Avatar of God (or, by God's will).
9. Absolver of sins.
10. I am above all the Lamas of the Dookpa Creed.
11. I am of the best of all Religions—the Dookpa.
12. The punisher of unbelievers.
13. Unequalled in expounding the Shasters.
14. Unequalled in holiness and wisdom.
15. The head (or fountain) of all Religious Knowledge.
16. The Enemy of all false Avatars.

CHAPTER XVII.

EXCURSION TO TERAI.

Dispatch collections—Acorns — Heat — Punkabaree—Bees—Vegetation—Haze—Titalya—Earthquake—Proceed to Nepal frontier—Terai, geology of—Physical features of Himalayan valleys—Elephants, purchase of, &c.—River-beds—Mechi river—Return to Titalya—Leave for Teesta—Climate of plains—Jeelpigoree—Cooches—Alteration in the appearance of country by fires, &c.—Grasses—Bamboos—Cottages—Rajah of Cooch Behar—Condition of people—Hooli festival—Ascend Teesta—Canoes — Cranes — Forest — Baikant-pore — Rummai—Religion—Plants at foot of mountains—Exit of Teesta—Canoe voyage down to Rangamally—English genera of plants—Birds—Beautiful scenery — Botanizing on elephants—Willow—Siligoree—Cross Terai—Geology—Iron—Loharghur—Coal and sandstone beds—Mechi fisherman—Hailstorm—Ascent to Khersiong—To Dorjiling—Vegetation—Geology—Folded quartz-beds—Spheres of feldspar—Lime deposits.

HAVING arranged the collections (amounting to eighty loads) made during 1848, they were conveyed by coolies to the foot of the hills, where carts were provided to carry them five days' journey to the Mahanuddy river, which flows into the Ganges, whence they were transported by water to Calcutta.

On the 27th of February, I left Dorjiling to join Mr. Hodgson, at Titalya on the plains. The weather was raw, cold, and threatening: snow lay here and there at 7000 feet, and all vegetation was very backward, and wore a wintry garb. The laurels, maples, and deciduous-leaved oaks, hydrangea and cherry, were leafless, but the abundance of chesnuts and evergreen oaks, rhododendrons, *Aucuba*, *Limonia*, and other shrubs, kept the forest well clothed. The oaks had borne a very unusual number of acorns during the last season, which were now falling, and strewing the road in some places so abundantly, that it was hardly safe to ride down hill.

The plains of Bengal were all but obscured by a dense haze, partly owing to a peculiar state of the atmosphere that prevails in the dry months, and partly to the fires raging in the Terai forest, from which white wreaths of smoke ascended, stretching obliquely for miles to the eastward, and filling the air with black particles of grass-stems, carried 4000 feet aloft by the heated ascending currents that impinge against the flanks of the mountains.

In the tropical region the air was scented with the white blossoms of the *Vitex Agnus-castus*, which grew in profusion by the road-side; but the forest, which had looked so gigantic on my arrival at the mountains the previous year, appeared small after the far more lofty and bulky oaks and pines of the upper regions of the Himalaya.

The evening was sultry and close, the heated surface of the earth seemed to load the surrounding atmosphere with warm vapours, and the sensation, as compared with the cool pure air of Dorjiling, was that of entering a confined tropical harbour after a long sea-voyage.

I slept in the little bungalow of Punkabaree, and was wakened next morning by sounds to which I had long been a stranger, the voices of innumerable birds, and the humming of great bees that bore large holes for their dwellings in the beams and rafters of houses: never before had I been so forcibly struck with the absence of animal life in the regions of the upper Himalaya.

Breakfasting early, I pursued my way in the so-called cool of the morning, but this was neither bright nor fresh; the night having been hazy, there had been no terrestrial radiation, and the earth was dusty and parched; while the sun rose through a murky yellowish atmosphere with ill-defined orb. Thick clouds of smoke pressed upon the plains, and the faint easterly wind wafted large flakes of grass charcoal sluggishly through the air.

Vegetation was in great beauty, though past its winter prime. The tropical forest of India has two flowering seasons; one in summer, of the majority of plants; and the other in winter, of *Acanthaceæ*, *Bauhinia*, *Dillenia*, *Bombax*, &c. Of these the former are abundant, and render the jungle gay with large and delicate white, red, and purple blossoms. Coarse, ill-favoured vultures wheeled through the air, languid Bengalees had replaced the active mountaineers, jackal-like curs of low degree teemed at every village, and ran howling away from the onslaught of my mountain dog; and the tropics, with all their beauty of flower and genial warmth, looked as forbidding and unwholesome as they felt oppressive to a frame that had so long breathed the fresh mountain air.

Mounted on a stout pony, I enjoyed my scamper of sixteen miles over the wooded plains and undulating gravelly slopes of the Terai, intervening between the foot of the mountains and Siligoree bungalow, where I rested for an hour. In the afternoon I rode on leisurely to Titalya, sixteen miles further, along the banks of the Mahanuddy, the atmosphere being so densely hazy, that objects a few miles off were invisible, and the sun quite concealed, though its light was so powerful that no part of the sky could be steadily gazed upon. This state of the air is very curious, and has met with various attempts at explanation,*

* Dr. M'Lelland ("Calcutta Journal of Natural History," vol. i., p. 52), attributes the haze of the atmosphere during the north-west winds of this season, wholly to suspended earthy particles. But the haze is present even in the calmest weather, and extreme dryness is in all parts of the world usually accompanied by an obscure horizon. Captain Campbell (" Calcutta Journal of Natural History," vol. ii., p. 44)

all unsatisfactory to me: it accompanies great heat, dryness, and elasticity of the suspended vapours, and is not affected by wind. During the afternoon the latter blew with violence, but being hot and dry, brought no relief to my still unacclimated frame. My pony alone enjoyed the freedom of the boundless plains, and the gallop or trot being fatiguing in the heat, I tried in vain to keep him at a walk; his spirits did not last long, however, for he flagged after a few days' tropical heat. My little dog had run thirty miles the day before, exclusive of all the detours he had made for his own enjoyment, and he flagged so much after twenty more this day, that I had to take him on my saddle-bow, where, after licking his hot swollen feet, he fell fast asleep, in spite of the motion.

After leaving the wooded Terai at Siligoree, trees became scarce, and clumps of bamboos were the prevalent features; these, with an occasional banyan, peepul, or betel-nut palm near the villages, were the only breaks on the distant horizon. A powerfully scented *Clerodendron*, and an *Osbeckia* gay with blossoms like dog-roses, were abundant; the former especially under trees, where the seeds are dropped by birds.

At Titalya bungalow, I received a hearty welcome from Mr. Hodgson, and congratulations on the success of my Nepal journey, which afforded a theme for many conversations.

In the evening we had three sharp jerking shocks of an earthquake in quick succession, at 9·8 P.M., appearing to come up from the southward: they were accompanied by a hollow rumbling sound like that of a waggon passing over a wooden bridge. The shock was felt strongly at Dorjiling, and registered by Mr. Muller at 9·10 P.M.: we had

also objects to Dr. M'Clelland's theory, citing those parts of Southern India which are least likely to be visited by dust-storms, as possessing an equally hazy atmosphere; and further denies its being influenced by the hygrometric state of the atmosphere.

accurately adjusted our watches (chronometers) the previous morning, and the motion may therefore fairly be assumed to have been transmitted northwards through the intervening distance of forty miles, in two minutes. Both Mr. Muller and Mr. Hodgson had noted a much more severe shock at 6·10 P.M. the previous evening, which I, who was walking down the mountain, did not experience; this caused a good deal of damage at Dorjiling, in cracking well-built walls. Earthquakes are frequent all along the Himalaya, and are felt far in Tibet; they are, however, most common towards the eastern and western extremities of India; owing in the former case to the proximity of the volcanic forces in the bay of Bengal. Cutch and Scinde, as is well known, have suffered severely on many occasions, and in several of them the motion has been propagated through Affghanistan and Little Tibet, to the heart of Central Asia.*

On the morning of the 1st of March, Dr. Campbell arrived at the bungalow, from his tour of inspection along the frontier of Bhotan and the Rungpore district; and we accompanied him hence along the British and Sikkim frontier, as far west as the Mechi river, which bounds Nepal on the east.

Terai is a name loosely applied to a tract of country at the very foot of the Himalaya: it is Persian, and signifies damp. Politically, the Terai generally belongs to the hill-states beyond it; geographically, it should appertain to the plains of India; and geologically, it is a sort of neutral country, being composed neither of the alluvium of the plains, nor of the rocks of the hills, but for the most part of alternating beds of sand, gravel, and boulders brought from the mountains. Botanically it is readily defined as the region of forest-trees; amongst which the Sal, the most valuable

* See "Wood's Travels to the Oxus."

of Indian timber, is conspicuous in most parts, though not now in Sikkim, where it has been destroyed. The Terai soil is generally light, dry, and gravelly (such as the Sal always prefers), and varies in breadth, from ten miles along the Sikkim frontier, to thirty and more on the Nepalese. In the latter country it is called the Morung, and supplies Sal and Sissoo timber for the Calcutta market, the logs being floated down the Konki and Cosi rivers to the Ganges. The gravel-beds extend uninterruptedly upon the plains for fully twenty miles south of the Sikkim mountains, the gravel becoming smaller as the distance increases, and large blocks of stone not being found beyond a few miles from the rocks of the Himalaya itself, even in the beds of rivers, however large and rapid. Throughout its breadth this formation is conspicuously cut into flat-topped terraces, flanking the spurs of the mountains, at elevations varying from 250 to nearly 1000 feet above the sea. These terraces are of various breadth and length, the smallest lying uppermost, and the broadest flanking the rivers below. The isolated hills beyond are also flat-topped and terraced. This deposit contains no fossils; and its general appearance and mineral constituents are the only evidence of its origin, which is no doubt due to a retiring ocean that washed the base of the Sikkim Himalaya, received the contents of its rivers, and, wearing away its bluff spurs, spread a talus upwards of 1000 feet thick along its shores. It is not at first sight evident whether the terracing is due to periodic retirements of the ocean, or to the levelling effects of rivers that have cut channels through the deposit. In many places, especially along the banks of the great streams, the gravel is smaller, obscurely interstratified with sand, and the flattened pebbles over-lap rudely, in a manner characteristic of the effects of running water; but such is not the case with the main body of the deposit, which is unstratified, and much coarser.

The alluvium of the Gangetic valley is both interstratified with the gravel, and passes into it, and was no doubt deposited in deep water, whilst the coarser matter * was accumulating at the foot of the mountains.

This view is self-evident, and has occurred, I believe, to almost every observer, at whatever part of the base of the Himalaya he may have studied this deposit. Its position, above the sandstones of the Sewalik range in the north-west Himalaya, and those of Sikkim, which appear to be modern fossiliferous rocks, indicates its being geologically of recent formation; but it still remains a subject of the utmost importance to discover the extent and nature of the ocean to whose agency it is referred. I have elsewhere remarked that the alluvium of the Gangetic valley may to a great degree be the measure of the denudation which the Himalaya has suffered along its Indian watershed. It was, no doubt, during the gradual rise of that chain from the ocean, that the gravel and alluvium were deposited; and in the terraces and alternation of these, there is evidence that there have been many subsidences and elevations of the coast-line, during which the gravel has suffered greatly from denudation.

I have never looked at the Sikkim Himalaya from the plains without comparing its bold spurs enclosing sinuous river gorges, to the weather-beaten front of a mountainous coast; and in following any of its great rivers, the scenery

* This, too, is non-fossiliferous, and is of unknown depth, except at Calcutta, where the sand and clay beds have been bored through, to the depth of 120 feet, below which the first pebbles were met with. Whence these pebbles were derived is a curious problem. The great Himalayan rivers convey pebbles but a very few miles from the mountains on to the plains of India; and there is no rock *in situ* above the surface, within many miles of Calcutta, in any direction.

of its deep valleys no less strikingly resembles that of such narrow arms of the sea (or fiords) as characterize every mountainous coast, of whatever geological formation: such as the west coast of Scotland and Norway, of South Chili and Fuegia, of New Zealand and Tasmania. There are too in these Himalayan valleys, at all elevations below 6000 feet, terraced pebble-beds, rising in some cases eighty feet above the rivers, which I believe could only have been deposited by them when they debouched into deep water; and both these, and the beds of the rivers, are strewed, down to 1000 feet, with masses of rock. Such accumulations and transported blocks are seen on the raised beaches of our narrow Scottish salt water lochs, exposed by the rising of the land, and they are yet forming of immense thickness on many coasts by the joint action of tides and streams.

I have described meeting with ancient moraines in every Himalayan valley I ascended, at or about 7000 or 8000 feet elevation, proving, that at one period, the glaciers descended fully so much below the position they now occupy: this can only be explained by a change of climate,* or by a depression of the mountain mass equal to 8000 feet, since the formation of these moraines.

The country about Titalya looks desert, from that want of trees and cultivation, so characteristic of the upper level throughout this part of the plains, which is covered with

* Such a change of temperature, without any depression or elevation of the mountains, has been thought by Capt. R. Strachey ("Journal of Geological Society"), an able Himalayan observer, to be the necessary consequence of an ocean at the foot of these mountains; for the amount of perpetual snow, and consequent descent of the glaciers, increasing indirectly in proportion to the humidity of the climate, and the snow-fall, he conjectured that the proximity of the ocean would prodigiously increase such a deposition of snow.—To me, this argument appears inconclusive; for the first effect of such a vast body of water would be to raise the temperature of winter; and as it is the rain, rather than the sun of summer, which removes the Sikkim snow, so would an increase of this rain elevate, rather than depress, the level of perpetual snow.

short, poor pasture-grass. The bungalow stands close to the Mahanuddy, on a low hill, cut into an escarpment twenty feet high, which exposes a section of river-laid sand and gravel, alternating with thick beds of rounded pebbles.

Shortly after Dr. Campbell's arrival, the meadows about the bungalow presented a singular appearance, being dotted over with elephants, brought for purchase by Government. It was curious to watch the arrival of these enormous animals, which were visible nearly two miles across the flat plains; nor less interesting was it to observe the wonderful docility of these giants of the animal kingdom, often only guided by naked boys, perched on their necks, scolding, swearing, and enforcing their orders with the iron goad. There appeared as many tricks in elephant-dealers as in horse-jockeys, and of many animals brought, but few were purchased. Government limits the price to about 75*l*., and the height to the shoulder must not be under seven feet, which, incredible as it appears, may be estimated within a fraction as being three times the circumference of the forefoot. The pedigree is closely inquired into, the hoofs are examined for cracks, the teeth for age, and many other points attended to.

The Sikkim frontier, from the Mahanuddy westward to the Mechi, is marked out by a row of tall posts. The country is undulating; and though fully 400 miles from the ocean, and not sixty from the top of the loftiest mountain on the globe, its average level is not 300 feet above that of the sea. The upper levels are gravelly, and loosely covered with scattered thorny jujube bushes, occasionally tenanted by the *Florican*, which scours these downs like a bustard. Sometimes a solitary fig, or a thorny acacia, breaks the horizon, and there are a few gnarled trees of the scarlet *Butea frondosa*.

On our route I had a good opportunity of examining the line of junction between the alluvial plains that stretch south to the Ganges, and the gravel deposit flanking the hills. The rivers always cut broad channels with scarped terraced sides, and their low banks are very fertile, from the mud annually spread by the ever-shifting streams that meander within their limits; there are, however, few shrubs and no trees. The houses, which are very few and scattered, are built on the gravelly soil above, the lower level being very malarious.

Thirty miles south of the mountains, numerous isolated flat-topped hills, formed of stratified gravel and sand with large water-worn pebbles, rise from 80 to 200 feet above the mean level, which is about 250 feet above the sea; these, too, have always scarped sides, and the channels of small streams completely encircle them.

At this season few insects but grasshoppers are to be seen, even mosquitos being rare. Birds, however, abound, and we noticed the common sparrow, hoopoe, water-wagtail, skylark, osprey, and several egrets.

We arrived on the third day at the Mechi river, to the west of which the Nepal Terai (or Morung) begins, whose belt of Sal forest loomed on the horizon, so raised by refraction as to be visible as a dark line, from the distance of many miles. It is, however, very poor, all the large trees having been removed. We rode for several miles into it, and found the soil dry and hard, but supporting a prodigious undergrowth of gigantic harsh grasses that reached to our heads, though we were mounted on elephants. Besides Sal there was abundance of *Butea*, *Diospyros*, *Terminalia*, and *Symplocos*, with the dwarf *Phœnix* palm, and occasionally *Cycas*. Tigers, wild elephants, and the rhinoceros, are said to be found here; but we saw none.

The old and new Mechi rivers are several miles apart, but flow in the same depression, a low swamp many miles broad, which is grazed at this season, and cultivated during the rains. The grass is very rich, partly owing to the moisture of the climate, and partly to the retiring waters of the rivers; both circumstances being the effects of proximity to the Himalaya. Hence cattle (buffalos and the common humped cow of India) are driven from the banks of the Ganges 300 miles to these feeding grounds, for the use of which a trifling tax is levied on each animal. The cattle are very carelessly herded, and many are carried off by tigers.

Having returned to Titalya, Mr. Hodgson and I set off in an eastern direction for the Teesta river, whose embouchure from the mountains to the plains I was anxious to visit. Though the weather is hot, and oppressively so in the middle of the day, there are few climates more delicious than that of these grassy savannahs from December to March. We always started soon after daybreak on ponies, and enjoyed a twelve to sixteen miles' gallop in the cool of the morning before breakfast, which we found prepared on our arrival at a tent sent on ahead the night before. The road led across an open country, or followed paths through interminable rice-fields, now dry and dusty. On poor soil a white-flowered *Leucas* monopolized the space, like our charlock and poppy: it was apparently a pest to the agriculturist, covering the surface in some places like a sprinkling of snow. Sometimes the river-beds exposed fourteen feet of pure stratified sand, with only an inch of vegetable soil above.

At this season the mornings are very hazy, with the thermometer at sunrise 60°; one laid on grass during the night falling 7° below that temperature: dew forms, but never copiously: by 10 A.M. the temperature has risen to 75°, and the faint easterly morning breezes die away; the haze thickens, and covers the sky with a white veil, the thermometer rising to 82° at noon, and the west wind succeeding in parching tornados and furious gusts, increasing with the temperature, which attains its maximum in the afternoon, and falling again with its decline at sunset. The evenings are calm; but the earth is so heated, that the thermometer stands at 10 P.M. at 66°, and the minimum at night is not below 55°: great drought accompanies the heat at this season, but not to such a degree as in North-west India, or other parts of this meridian further removed from the hills. In the month of March, and during the prevalence of west winds, the mean temperature was 79°, and the dew-point 22° lower, indicating great drought. The temperature at Calcutta was 7° warmer, and the atmosphere very much damper.

On the second day we arrived at Jeelpigoree, a large straggling village near the banks of the Teesta, a good way south of the forest: here we were detained for several days, waiting for elephants with which to proceed northwards. The natives are Cooches, a Mogul (Mongolian) race, who inhabit the open country of this district, replacing the Mechis of the Terai forest. They are a fine athletic people, not very dark, and formed the once-powerful house of Cooch Behar. Latterly the upper classes have adopted the religion of the Brahmins, and have had caste conferred upon them; while the lower orders have turned Mahomedans: these, chiefly agriculturists, are a timid, oppressed class, who everywhere fled before us, and were with difficulty prevailed upon even to direct us along our road. A rude police is established by the British Government all over the country, and to it the traveller applies for guides and assistance; but the Cooches were so shy and difficult to deal with, that we were generally left to our own resources.

Grass is the prevailing feature of the country, as there are few shrubs, and still fewer trees. Goats and the common Indian cow are plentiful; but it is not swampy enough for the buffalo; and sheep are scarce, on account of the heat of the climate. This uniformity of feature over so immense an area is, however, due to the agency of man, and is of recent introduction; as all concur in affirming, that within the last hundred years the face of the country was covered with the same long jungle-grasses which abound in the Terai forest; and the troops cantoned at Titalya (a central position in these plains) from 1816 to 1828, confirm this statement as far as their immediate neighbourhood is concerned.

These gigantic *Gramineæ* seem to be destroyed by fire with remarkable facility at one season of the year; and it is well that this is the case; for, whether as a retainer of miasma, a shelter for wild beasts, both carnivorous and herbivorous, alike dangerous to man, or from their liability to ignite, and spread destruction far and wide, the grass-jungles are most serious obstacles to civilization. Next to the rapidity with which it can be cleared, the adaptation of a great part of the soil to irrigation during the rains, has greatly aided the bringing of it under cultivation.

By far the greater proportion of this universal short turf grass is formed of *Andropogon acicularis*, *Cynodon Dactylon*,* and in sandy places, *Imperata cylindrica*;

* Called "Dhob." This is the best pasture grass in the plains of India, and the only one to be found over many thousands of square miles.

where the soil is wetter, *Ameletia Indica* is abundant, giving a heather-like colour to the turf, with its pale purple flowers: wherever there is standing water, its surface is reddened by the *Azolla*, and *Salvinia* is also common.

At Jeelpigoree we were waited upon by the Dewan; who governs the district for the Rajah, a boy about ten years old, whose estates are locked up during the trial of an interminable suit for the succession, that has been instituted against him by a natural son of the late Rajah: we found the Dewan to be a man of intelligence, who promised us elephants as soon as the great Hooli festival, now commenced, should be over.

The large village, at the time of our visit, was gay with holiday dresses. It is surrounded by trees, chiefly of banyan, jack, mango, peepul, and tamarind: interminable rice-fields extend on all sides, and except bananas, slender betel-nut palms, and sometimes pawn, or betel-pepper, there is little other extensive cultivation. The rose-apple, orange, and pine-apple are rare, as are cocoa-nuts: there are few date or fan-palms, and only occasionally poor crops of castor-oil and sugar-cane. In the gardens I noticed jasmine, *Justicia Adhatoda*, *Hibiscus*, and others of the very commonest Indian ornamental plants; while for food were cultivated *Chenopodium*, yams, sweet potatos, and more rarely peas, beans, and gourds. Bamboos were planted round the little properties and smaller clusters of houses, in oblong squares, the ridge on which the plants grew being usually bounded by a shallow ditch. The species selected was not the most graceful of its family; the stems, or culms, being densely crowded, erect, as thick at the base as the arm, copiously branching, and very feathery throughout their whole length of sixty feet.

A gay-flowered *Osbeckia* was common along the roadsides, and, with a *Clerodendron*,* whose strong, sweet odour was borne far through the air, formed a low undershrub beneath every tree, generally intermixed with three ferns (a *Polypodium*, *Pteris*, and *Goniopteris*).

The cottages are remarkable, and have a very neat appearance, presenting nothing but a low white-washed platform of clay, and an enormous high, narrow, black, neatly thatched roof, so arched along the ridge, that its eaves nearly touch the ground at each gable; and looking at a distance like a gigantic round-backed elephant. The walls are of neatly-platted bamboo: each window (of which there are two) is crossed by slips of bamboo, and wants only glass to make it look European; they have besides shutters of wattle, that open upwards, projecting during the day like the port-hatches of a ship, and let down at night. Within, the rooms are airy and clean: one end contains the machans (bedsteads), the others some raised clay benches, the fire, frequently an enormous Hookah, round wattled stools, and various implements. The inhabitants appeared more than ordinarily well-dressed; the men in loose flowing robes of fine cotton or muslin, the women in the usual garb of a simple thick cotton cloth, drawn tight immediately above the breast, and thence falling perpendicularly to the knee; the colour of this is a bright blue in stripes, bordered above and below with red.

I anticipated some novelty from a visit to a Durbar (court) so distant from European influence as that of the Rajah of Jeelpigoree. All Eastern courts, subject to the Company, are, however, now shorn of much of their glory;

* *Clerodendron* leaves, bruised, are used to kill vermin, fly-blows, &c., in cattle; and the twigs form toothpicks. The flowers are presented to Mahadeo, as a god of peace; milk, honey, flowers, fruit, amrit (ambrosia), &c., being offered to the pacific gods, as Vishnu, Krishna, &c.; while Mudar (*Asclepias*), Bhang (*Cannabis sativa*), *Datura*, flesh, blood, and spirituous liquors, are offered to Siva, Doorga, Kali, and other demoniacal deities.

and the condition of the upper classes is greatly changed. Under the Mogul rule, the country was farmed out to Zemindars, some of whom assumed the title of Rajah: they collected the revenue for the Sovereign, retaining by law ten per cent. on all that was realized: there was no intermediate class, the peasant paying directly to the Zemindar, and he into the royal treasury. Latterly the Zemindars have become farmers under the Company's rule; and in the adjudication of their claims, Lord Cornwallis (then Governor-General) made great sacrifices in their favour, levying only a small tribute in proportion to their often great revenues, in the hope that they would be induced to devote their energies, and some of their means, to the improvement of the condition of the peasantry. This expectation was not realized: the younger Zemindars especially, subject to no restraint (except from aggressions on their neighbours), fell into slothful habits, and the collecting of the revenue became a trading speculation, entrusted to "middle men." The Zemindar selects a number, who again are at liberty to collect through the medium of several sub-renting classes. Hence the peasant suffers, and except a generally futile appeal to the Rajah, he has no redress. The law secures him tenure as long as he can pay his rent, and to do this he has recourse to the usurer; borrowing in spring (at 50, and oftener 100 per cent.) the seed, plough, and bullocks: he reaps in autumn, and what is then not required for his own use, is sold to pay off part of his original debt, the rest standing over till the next season; and thus it continues to accumulate, till, overwhelmed with difficulties, he is ejected, or flees to a neighbouring district. The Zemindar enjoys the same right of tenure as the peasant: the amount of impost laid on his property was fixed for perpetuity; whatever his revenue be, he must pay so much to the Company, or he forfeits his estates, and they are put up for auction.

One evening we visited the young Rajah at his residence, which has rather a good appearance at a distance, its white walls gleaming through a dark tope of mango, betel, and cocoa-nut. A short rude avenue leads to the entrance gate, under the trees of which a large bazaar was being held; stocked with cloths, simple utensils, ornaments, sweetmeats, five species of fish from the Teesta, and the betel-nut.

We entered through a guard-house, where were some of the Rajah's Sepoys in the European costume, and a few of the Company's troops, lent to the Rajah as a security against some of the turbulent pretenders to his title. Within was a large court-yard, flanked by a range of buildings, some of good stone-work, some of wattle, in all stages of disrepair. A great crowd of people occupied one end of the court, and at the other we were received by the Dewan, and seated on chairs under a canopy supported by slender silvered columns. Some slovenly Natch-girls were dancing before us, kicking up clouds of dust, and singing or rather bawling through their noses, the usual indelicate hymns in honour of the Hooli festival; there were also fiddlers, cutting uncouth capers in rhythm with the dancers. Anything more deplorable than the music, dancing, and accompaniments, cannot well be imagined; yet the people seemed vastly pleased, and extolled the performers.

The arrival of the Rajah and his brothers was announced by a crash of tom-toms and trumpets, while over their heads were carried great gilt canopies. With them came a troop of relations, of all ages; and amongst them a poor little black girl, dressed in honour of us in an old-fashioned

English chintz frock and muslin cap, in which she cut the drollest figure imaginable; she was carried about for our admiration, like a huge Dutch doll, crying lustily all the time.

The festivities of the evening commenced by handing round trays full of pith-balls, the size of a nutmeg, filled with a mixture of flour, sand, and red lac-powder; with these each pelted his neighbour, the thin covering bursting as it struck any object, and powdering it copiously with red dust. A more childish and disagreeable sport cannot well be conceived; and when the balls were expended, the dust itself was resorted to, not only fresh, but that which had already been used was gathered up, with whatever dirt it might have become mixed. One rude fellow, with his hand full, sought to entrap his victims into talking, when he would stuff the nasty mixture into their mouths.

At the end attar of roses was brought, into which little pieces of cotton, fixed on slips of bamboo, were dipped, and given to each person. The heat, dust, stench of the unwashed multitude, noise, and increasing familiarity of the lower orders, warned us to retire, and we effected our retreat with precipitancy.

The Rajah and his brother were very fine boys, lively, frank, unaffected, and well disposed: they have evidently a good guide in the old Dewan; but it is melancholy to think how surely, should they grow up in possession of their present rank, they will lapse into slothful habits, and take their place amongst the imbeciles who now represent the once powerful Rajahs of Bengal.

We rode back to our tents by a bright moonlight, very dusty and tired, and heartily glad to breathe the cool fresh air, after the stifling ordeal we had undergone.

On the following evening the elephants were again in waiting to conduct us to the Rajah. He and his relations were assembled outside the gates, mounted upon elephants, amid a vast concourse of people. The children and Dewan were seated in a sort of cradle; the rest were some in howdahs, and some astride on elephants' backs, six or eight together. All the idols were paraded before them, and powdered with red dust; the people howling, shouting, and sometimes quarrelling. Our elephants took their places amongst those of the Rajah; and when the mob had sufficiently pelted one another with balls and dirty red powder, a torchlight procession was formed, the idols leading the way, to a very large tank, bounded by a high rampart, within which was a broad esplanade round the water.

The effect of the whole was very striking, the glittering cars and barbaric gaud of the idols showing best by torchlight; while the white robes and turbans of the undulating sea of people, and the great black elephants picking their way with matchless care and consideration, contrasted strongly with the quiet moonbeams sleeping on the still broad waters of the tank.

Thence the procession moved to a field, where the idols were placed on the ground, and all dismounted: the Dewan then took the children by the hand, and each worshipped his tutelary deity in a short prayer dictated by the attendant Brahmin, and threw a handful of red dust in its face. After another ordeal of powder, singing, dancing, and suffocation, our share in the Hooli ended; and having been promised elephants for the following morning, we bade a cordial farewell to our engaging little hosts and their staid old governor.

On the 10th of March we were awakened at an early hour by a heavy thunder-storm from the south-west. The sunrise was very fine, through an arch 10° high of bright blue sky, above which the whole firmament was mottled with cirrus. It continued cloudy, with light winds, throughout the day, but clear on the horizon. From this time such storms became frequent, ushering in the equinox; and the less hazy sky and rising hygrometer predicted an accession of moisture in the atmosphere.

We left for Rangamally, a village eight miles distant in a northerly direction, our course lying along the west bank of the Teesta.

The river is here navigated by canoes, thirty to forty feet long, some being rudely cut out of a solid log of Sal, while others are built, the planks, of which there are but few, being sewed together, or clamped with iron, and the seams caulked with the fibres of the root of Dhak (*Butea frondosa*), and afterwards smeared with the gluten of *Diospyros embryopteris*. The bed of the river is here three-quarters of a mile across, of which the stream does not occupy one-third; its banks are sand-cliffs, fourteen feet in height. A few small fish and water-snakes swarm in the pools.

The whole country improved in fertility as we advanced towards the mountains: the grass became greener, and more trees, shrubs, herbs, and birds appeared. In front, the dark boundary-line of the Sal forest loomed on the horizon, and to the east rose the low hills of Bhotan, both backed by the outer ranges of the Himalaya.

Flocks of cranes were abundant over-head, flying in wedges, or breaking up into "open order," preparing for their migration northwards, which takes place in April, their return occurring in October; a small quail was also common on the ground. Tamarisk ("Jhow") grew in the sandy bed of the river; its flexible young branches are used in various parts of India for wattling and basket-making.

In the evening we walked to the skirts of the Sal forest.

The great trunks of the trees were often scored by tigers' claws, this animal indulging in the cat-like propensity of rising and stretching itself against such objects. Two species of *Dillenia* were common in the forest, with long grass, *Symplocos, Emblica*, and *Cassia Fistula*, now covered with long pods. Several parasitical air-plants grew on the dry trees, as *Oberonia, Vanda*, and *Ærides*.

At Rangamally, the height of the sandy banks of the Teesta varies from fifteen to twenty feet. The bed is a mile across, and all sand; * the current much divided, and opaque green, from the glacial origin of most of its head-streams. The west bank was covered with a small Sal forest, mixed with *Acacia Catechu*, and brushwood, growing in a poor vegetable loam, over very dry sand.

The opposite (or Bhotan) bank is much lower, and always flooded during the rains, which is not the case on the western side, where the water rises to ten feet below the top of the bank, or from seven to ten feet above its height in the dry season, and it then fills its whole bed. This information we had from a police Jemadar, who has resided many years on this unhealthy spot, and annually suffers from fever. The Sal forest has been encroached upon from the south, for many miles, within the memory of man, by clearing in patches, and by indiscriminate felling.

About ten miles north of Rangamally, we came to an extensive flat, occupying a recess in the high west bank, the site of the old capital (Bai-kant-pore) of the Jeelpigoree Rajah. Hemmed in as it is on three sides by a dense forest, and on all by many miles of malarious Terai, it appears sufficiently secure from ordinary enemies, during a great part of the year. The soil is sandy, overlying gravel,

* Now covered with *Anthistiria* grass, fifteen feet high, a little *Sissoo*, and *Bombax*.

and covered with a thick stratum of fine mud or silt, which is only deposited on these low flats; on it grew many naturalized plants, as hemp, tobacco, jack, mango, plantain, and orange.

About eight miles on, we left the river-bed, and struck westerly through a dense forest, to a swampy clearance occupied by the village of Rummai, which appeared thoroughly malarious; and we pitched the tent on a narrow, low ridge, above the level of the plain.

It was now cool and pleasant, partly due, no doubt, to a difference in the vegetation, and the proximity of swamp and forest, and partly also to a change in the weather, which was cloudy and threatening; much rain, too, had fallen here on the preceding day.

Brahmins and priests of all kinds are few in this miserable country: near the villages, and under the large trees, are, every here and there, a few miniature thatched cottages, four to six feet high, in which the tutelary deities of the place are kept; they are idols of the very rudest description, of Vishnu as an ascetic (Bai-kant Nath), a wooden doll, gilt and painted, standing, with the hands raised as if in exhortation, and one leg crossed over the other. Again, Kartik, the god of war, is represented sitting astride on a peacock, with the right hand elevated and holding a small flat cup.

Some fine muscular Cooches were here brought for Mr. Hodgson's examination, but we found them unable or unwilling to converse in the Cooch tongue, which appears to be fast giving place to Bengalee.

We walked to a stream, which flows at the base of the retiring sand-cliffs, and nourishes a dense and richly-varied jungle, producing many plants, as beautiful *Acanthaceæ*, Indian horse-chesnut, loaded with white racemes of flowers, gay *Convolvuli*, laurels, terrestrial and parasitic *Orchideæ*, *Dillenia*, casting its enormous flowers as big as two fists, pepper, figs, and, in strange association with these, a hawthorn, and the yellow-flowered Indian strawberry, which ascends 7,500 feet on the mountains, and *Hodgsonia*, a new *Cucurbitaceous* genus, clinging in profusion to the trees, and also found 5000 feet high on the mountains.

In the evening we rode into the forest (which was dry and very unproductive), and thence along the river-banks, through *Acacia Catechu*, belted by *Sissoo*, which often fringes the stream, always occupying the lowest flats. The foliage at this season is brilliantly green; and as the evening advanced, a yellow convolvulus burst into flower like magic, adorning the bushes over which it climbed.

It rained on the following morning; after which we left for the exit of the Teesta, proceeding northwards, sometimes through a dense forest of Sal timber, sometimes dipping into marshy depressions, or riding through grassy savannahs, breast-high. The coolness of the atmosphere was delicious, and the beauty of the jungle seemed to increase the further we penetrated these primæval forests.

Eight miles from Rummai we came on a small river from the mountains, with a Cooch village close by, inhabited during the dry season by timber-cutters from Jeelpigoree: it is situated upon a very rich black soil, covered with *Saccharum* and various gigantic grasses, but no bamboo. These long grasses replace the Sal, of which we did not see one good tree.

We here mounted the elephants, and proceeded several miles through the prairie, till we again struck upon the high Sal forest-bank, continuous with that of Rummai and Rangamally, but much loftier: it formed one of many terraces which stretch along the foot of the hills, from

Punkabaree to the Teesta, but of which none are said to occur for eight miles eastwards along the Bhotan Dooars: if true, this is probably due in part to the alteration of the course of the Teesta, which is gradually working to the westward, and cutting away these lofty banks.

The elephant-drivers appeared to have taken us by mistake to the exit of the Chawa, a small stream which joins the Teesta further to the eastward. The descent to the bed of this rivulet, round the first spur of rock we met with, was fully eighty feet, through a very irregular depression, probably the old bed of the stream; it runs southwards from the hills, and was covered from top to bottom with slate-pebbles. We followed the river to its junction with the Teesta, along a flat, broad gulley, bounded by densely-wooded, steep banks of clay-slate on the north, and the lofty bank on the south: between these the bed was strewed with great boulders of gneiss and other rocks, luxuriantly clothed with long grass, and trees of wild plantain, *Erythrina* and *Bauhinia*, the latter gorgeously in flower.

The Sal bank formed a very fine object: it was quite perpendicular, and beautifully stratified with various coloured sands and gravel: it tailed off abruptly at the junction of the rivers, and then trended away south-west, forming the west bank of the Teesta. The latter river is at its outlet a broad and rapid, but hardly impetuous stream, now fifty yards across, gushing from between two low, forest-clad spurs: it appeared about five feet deep, and was beautifully fringed on both sides with green *Sissoo*.

Some canoes were here waiting for us, formed of hollowed trunks of trees, thirty feet long: two were lashed together with bamboos, and the boatmen sat one at the head and one at the stern of each: we lay along the bottom of the vessels, and in a second we were darting down the river, at the rate of at least ten or fifteen miles an hour, the bright waters leaping up on all sides, and bounding in *jets-d'eau* between prows and sterns of the coupled vessels. Sometimes we glided along without perceptible motion, and at others jolted down bubbling rapids, the steersmen straining every nerve to keep their bark's head to the current, as she impatiently swerved from side to side in the eddies. To our jaded and parched frames, after the hot forenoon's ride on the elephants, the effect was delicious: the fresh breeze blew on our heated foreheads and down our open throats and chests; we dipped our hands into the clear, cool stream, and there was "music in the waters" to our ears. Fresh verdure on the banks, clear pebbles, soft sand, long English river-reaches, forest glades, and deep jungles, followed in rapid succession; and as often as we rounded a bend or shot a rapid, the scene changed from bright to brighter still; so continuing until dusk, when we were slowly paddling along the then torpid current opposite Rangamally.*

The absence of large stones or boulders of rock in the bed of the Teesta is very remarkable, considering the great volume and rapidity of the current, and that it shoots directly from the rocky hills to the gravelly plains. At the

* The following temperatures of the waters of the Teesta were taken at intervals during our passage from its exit to Rangamally, a distance of fifteen linear miles, and thirty miles following the bends:—

		Water.	Air.
Exit 2h.	30m. P.M.	62°	
3		62° 2	74°
3	30	63° 2	
4		64°	
4	30	65°	
5		65° 4	72° 5 opposite Rummai.
5	30	66°	
6		66°	71° 7 opposite Baikant.

embouchure there are boulders as big as the head, and in the stream, four miles below the exit, the boatmen pointed out a stone as large as the body as quite a marvel.

They assured us that the average rise at the mouth of the river, in the rains, was not more than five feet: the mean breadth of the stream is from seventy to ninety yards. From the point where it leaves the mountains, to its junction with the Megna, is at this season thirteen days' voyage, the return occupying from twenty to twenty-five days, with the boats unladen. The name "Teesta" signifies "quiet," this river being so in comparison with other Himalayan torrents further west, the Cosi, Konki, &c., which are devastators of all that bounds their course.

We passed but two crossing-places: at one the river is divided by an island, covered with the rude chaits and flags of the Boodhists. We also saw some Cooch fishermen, who throw the net much as we do: a fine "Mahaser" (a very large carp) was the best fish they had. Of cultivation there was very little, and the only habitations were a few grass-huts of the boatmen or buffalo herdsmen, a rare Cooch village of Catechu and Sal cutters, or the shelter of timber-floaters, who seem to pass the night in nests of long dry grass.

Our servants not having returned with the elephants from Rummai, we spent the following day at Rangamally shooting and botanizing. I collected about 100 species in a couple of hours, and observed perhaps twice that number: the more common I have repeatedly alluded to, and excepting some small terrestrial *Orchids*, I added nothing of particular interest to my collection.*

* The following is a list of the principal genera, most of which are English:—*Polygonum, Quercus, Sonchus, Gnaphalium, Cratægus, Lobelia, Lactuca, Hydrocotyle, Saponaria, Campanula, Bidens, Rubus, Oxalis, Artemisia, Fragaria, Clematis, Dioscorea, Potamogeton, Chara, Veronica, Viola, Smilax.*

On the 14th of March we proceeded west to Siligoree, along the skirts of the ragged Sal forest. Birds are certainly the most conspicuous branch of the natural history of this country, and we saw many species, interesting either from their habits, beauty, or extensive distribution. We noticed no less than sixteen kinds of swimming birds, several of which are migratory and English. The Shoveller, white-eyed and common wild ducks; Merganser, Brahminee, and Indian goose (*Anser Indica*); common and Gargany teal; two kinds of gull; one of Shearwater (*Rhynchops ablacus*); three of tern, and one of cormorant. Besides these there were three egrets, the large crane, stork, green heron, and the demoiselle; the English sand-martin, kingfisher, peregrine-falcon, sparrow-hawk, kestrel, and the European vulture: the wild peacock, and jungle-fowl. There were at least 100 peculiarly Indian birds in addition, of which the more remarkable were several kinds of mina, of starling, vulture, kingfisher, magpie, quail, and lapwing.

The country gradually became quite beautiful, much undulated and diversified by bright green meadows, sloping lawns, and deeply wooded nullahs, which lead from the Sal forest and meander through this varied landscape. More beautiful sites for fine mansions could not well be, and it is difficult to suppose so lovely a country should be so malarious as it is before and after the rains, excessive heat probably diffusing widely the miasma from small stagnant surfaces. We noticed a wild hog, absolutely the first wild beast of any size I saw on the plains, except the hispid hare (*Lepus hispidus*) and the barking deer (*Stylocerus ratna*). The hare we found to be the best game of this part of India, except the teal. The pheasants of Dorjiling are poor, the deer all but uneatable, and the florican, however dressed, I considered a far from excellent bird.

A good many plants grow along the streams, the sandy beds of which are everywhere covered with the marks of tigers' feet. The only safe way of botanizing is by pushing through the jungle on elephants; an uncomfortable method, from the quantity of ants and insects which drop from the foliage above, and from the risk of disturbing pendulous bees' and ants' nests.

A peculiar species of willow (*Salix tetrasperma*) is common here; which is a singular fact, as the genus is characteristic of cold and arctic latitudes, and no species is found below 8000 feet elevation on the Sikkim mountains, where it grows on the inner Himalaya only, some kinds ascending to 16,000 feet.

East of Siligoree the plains are unvaried by tree or shrub, and are barren wastes of short turf or sterile sand, with the dwarf-palm (*Phœnix acaulis*), a sure sign of a most hungry soil.

The latter part of the journey I performed on elephants during the heat of the day, and a more uncomfortable mode of conveyance surely never was adopted; the camel's pace is more fatiguing, but that of the elephant is extremely trying after a few miles, and is so injurious to the human frame that the Mahouts (drivers) never reach an advanced age, and often succumb young to spine-diseases, brought on by the incessant motion of the vertebral column. The broiling heat of the elephant's black back, and the odour of its oily driver, are disagreeable accompaniments, as are its habits of snorting water from its trunk over its parched skin, and the consequences of the great bulk of green food which it consumes.

From Siligoree I made a careful examination of the

gravel beds that occur on the road north to the foot of the hills, and thence over the tertiary sandstone to Punkabaree. At the Rukti river, which flows south-west, the road suddenly rises, and crosses the first considerable hill, about two miles south of any rock *in situ*. This river cuts a cliff from 60 to 100 feet high, composed of stratified sand and water-worn gravel: further south, the spur declines into the plains, its course marked by the Sal that thrives on its gravelly soil. The road then runs north-west over a plain to an isolated hill about 200 feet high, also formed of sand and gravel. We ascended to the top of this, and found it covered with blocks of gneiss, and much angular detritus. Hence the road gradually ascends, and becomes clayey. Argillaceous rocks, and a little ochreous sandstone appeared in highly-inclined strata, dipping north, and covered with great water-worn blocks of gneiss. Above, a flat terrace, flanked to the eastward by a low wooded hill, and another rise of sandstone, lead on to the great Baisarbatti terrace.

Bombax, *Erythrina*, and *Duabanga* (*Lagœrstrœmia grandiflora*), were in full flower, and with the profusion of *Bauhinia*, rendered the tree-jungle gay: the two former are leafless when flowering. The Duabanga is the pride of these forests. Its trunk, from eight to fifteen feet in girth, is generally forked from the base, and the long pendulous branches which clothe the trunk for 100 feet, are thickly leafy, and terminated by racemes of immense white flowers, which, especially when in bud, smell most disagreeably of assafœtida. The magnificent Apocyneous climber, *Beaumontia*, was in full bloom, ascending the loftiest trees, and clothing their trunks with its splendid foliage and festoons of enormous funnel-shaped white flowers.

The report of a bed of iron-stone eight or ten miles west

of Punkabaree determined our visiting the spot; and the locality being in a dense jungle, the elephants were sent on ahead.

We descended to the terraces flanking the Balasun river, and struck west along jungle-paths to a loosely-timbered flat. A sudden descent of 150 feet landed us on a second terrace. Further on, a third dip of about twenty feet (in some places obliterated) flanks the bed of the Balasun; the river itself being split into many channels at this season. The west bank, which is forty feet high, is of stratified sand and gravel, with vast slightly-worn blocks of gneiss: from the top of this we proceeded south-west for three miles to some Mechi villages, the inhabitants of which flocked to meet us, bringing milk and refreshments.

The Lohar-ghur, or "iron hill," lies in a dense dry forest. Its plain-ward flanks are very steep, and covered with scattered weather-worn masses of ochreous and black iron-stone, many of which are several yards long: it fractures with faint metallic lustre, and is very earthy in parts: it does not affect the compass. There are no pebbles of iron-stone, nor water-worn rocks of any kind found with it.

The sandstones, close by, cropped out in thick beds (dip north 70°): they are very soft, and beds of laminated clay, and of a slaty rock, are intercalated with them, also an excessively tough conglomerate, formed of an indurated blue or grey paste, with nodules of harder clay. There are no traces of metal in the rock, and the lumps of ore are wholly superficial.

Below Punkabaree the Baisarbatti stream cuts through banks of gravel overlying the sandstone (dip north 65°). The sandstone is gritty and micaceous, intercalated with beds of indurated shale and clay; in which I found the shaft (apparently) of a bone; there were also beds of the same clay conglomerate which I had seen at Lohar-ghur, and thin seams of brown lignite, with a rhomboidal cleavage. In the bed of the stream were carbonaceous shales, with obscure impressions of fern leaves, of *Trizygia*, and *Vertebraria*: both fossils characteristic of the Burdwan coal-fields (see p. 8), but too imperfect to justify any conclusion as to the relation between these formations.*

Ascending the stream, these shales are seen *in situ*, overlain by the metamorphic clay-slate of the mountains, and dipping inwards (northwards) like them. This is at the foot of the Punkabaree spur, and close to the bungalow, where a stream and land-slip expose good sections. The carbonaceous beds dip north 60° and 70°, and run east and west; much quartz rock is intercalated with them, and soft white and pink micaceous sandstones. The coal-seams are few in number, six to twelve inches thick, very confused and distorted, and full of elliptic nodules, or spheroids of quartzy slate, covered with concentric scaly layers of coal: they overlie the sandstones mentioned above. These scanty notices of superposition being collected in a country clothed with the densest tropical forest, where a geologist pursues his fatiguing investigations under disadvantages that can hardly be realized in England, will I fear long remain unconfirmed.

* These traces of fossils are not sufficient to identify the formation with that of the Sewalik hills of North-west India; but its contents, together with its strike, dip, and position relatively to the mountains, and its mineralogical character, incline me to suppose it may be similar. Its appearance in such small quantities in Sikkim (where it rises but a few hundred feet above the level of the sea, whereas in Kumaon it reaches 4000 feet), may be attributed to the greater amount of wearing which it must have undergone; the plains from which it rises being 1000 feet lower than those of Kumaon, and the sea having consequently retired later, exposing the Sikkim sandstone to the effects of denudation for a much longer period. Hitherto no traces of this rock, or of any belonging to a similar geological epoch, have been found in the valleys of Sikkim; but when the narrowness of these is considered, it will not appear strange that such may have been removed from their surfaces: first, by the action of a tidal ocean; and afterwards, by that of tropical rains.

I may mention, however, that the appearance of inversion of the strata at the foot of great mountain-masses has been observed in the Alleghany chain, and I believe in the Alps.*

A MECH, NATIVE OF THE SIKKIM TERAI.

A poor Mech was fishing in the stream, with a basket curiously formed of a cylinder of bamboo, cleft all round in innumerable strips, held together by the joints above and below; these strips being stretched out as a balloon in the

* Dr. M'Lelland informs me that in the Curruckpore hills, south of the Ganges, the clay-slates are overlain by beds of mica-slate, gneiss, and granite, which pass into one another.

middle, and kept apart by a hoop: a small hole is cut in the cage, and a mouse-trap entrance formed: the cage is placed in the current with the open end upwards, where the fish get in, and though little bigger than minnows, cannot find their way out.

On the 20th we had a change in the weather: a violent storm from the south-west occurred at noon, with hail of a strange form, the stones being sections of hollow spheres, half an inch across and upwards, formed of cones with truncated apices and convex bases; these cones were aggregated together with their bases outwards. The large masses were followed by a shower of the separate conical pieces, and that by heavy rain. On the mountains this storm was most severe: the stones lay at Dorjiling for seven days, congealed into masses of ice several feet long and a foot thick in sheltered places: at Purneah, fifty miles south, stones one and two inches across fell, probably as whole spheres.

Ascending to Khersiong, I found the vegetation very backward by the road-sides. The rain had cleared the atmosphere, and the view over the plains was brilliant. On the top of the Khersiong spur a tremendous gale set in with a cold west wind: the storm cleared off at night, which at 10 P.M. was beautiful, with forked and sheet lightning over the plains far below us. The equinoctial gales had now fairly set in, with violent south-east gales, heavy thunder, lightning, and rain.

Whilst at Khersiong I took advantage of the very fair section afforded by the road from Punkabaree, to examine the structure of the spur, which seems to be composed of very highly inclined contorted beds (dip north) of metamorphic rocks, gneiss, mica-slate, clay-slate, and quartz; the foliation of which beds is parallel to the dip of the strata. Over all reposes a bed of clay, capped with a layer of vegetable mould, nowhere so thick and rich as in the more humid regions of 7000 feet elevation. The rocks appeared in the following succession in descending. Along the top are found great blocks of very compact gneiss buried in clay. Half a mile lower the same rock appears, dipping north-north-east 50°. Below this, beds of saccharine quartz, with seams of mica, dip north-north-west 20°. Some of these quartz beds are folded on themselves, and look like flattened trunks of trees, being composed of concentric layers, each from two to four inches thick: we exposed twenty-seven feet of one fold running along the side of the road, which was cut parallel to the strike. Each layer of quartz was separated from its fellows, by one of mica scales, and was broken up into cubical fragments, whose surfaces are no doubt cleavage and jointing planes. I had previously seen, but not understood, such flexures produced by metamorphic action on masses of quartz when in a pasty state, in the Falkland Islands, where they have been perfectly well described by Mr. Darwin;* in whose views of the formation of these rocks I entirely concur.

The flexures of the gneiss are incomparably more irregular and confused than those of the quartz, and often contain flattened spheres of highly crystalline felspar, that cleave perpendicularly to the shorter axis. These spheres are disposed in layers parallel to the foliation of the gneiss: and are the result of a metamorphic action of great intensity, effecting a complete rearrangement and crystallization of the quartz and mica in parallel planes, whilst the felspar is aggregated in spheres; just as in the rearrangement of the mineral constituents of mica-schists, the alumina is crystallized in the garnets, and in the clay-slates the iron into pyrites.

* Journal of Geological Society for 1846, p. 267, and "Voyage of the Beagle."

The quartz below this dips north-north-west 45° to 50°, and alternates with a very hard slaty schist, dipping north-west 45°, and still lower is a blue-grey clay-slate, dipping north-north-west 30°. These rest on beds of slate, folded like the quartz mentioned above, but with cleavage-planes, forming lines radiating from the axis of each flexure, and running through all the concentric folds. Below this are the plumbago and clay slates of Punkabaree, which alternate with beds of mica-schist with garnets, and appear to repose immediately upon the carboniferous strata and sandstone; but there is much disturbance at the junction.

On re-ascending from Punkabaree, the rocks gradually appear more and more dislocated, the clay-slate less so than the quartz and mica-schist, and that again far less than the gneiss, which is so shattered and bent, that it is impossible to say what is *in situ*, and what not. Vast blocks lie superficially on the ridges; and the tops of all the outer mountains, as of Khersiong spur, of Tonglo, Sinchul, and Dorjiling, appear a pile of such masses. Injected veins of quartz are rare in the lower beds of schist and clay-slate, whilst the gneiss is often full of them; and on the inner and loftier ranges, these quartz veins are replaced by granite with tourmaline.

Lime is only known as a stalactitic deposit from various streams, at elevations from 1000 to 7000 feet; one such stream occurs above Punkabaree, which I have not seen; another within the Sinchul range, on the great Rungeet river, above the exit of the Rummai; a third wholly in the great central Himalayan range, flowing into the Lachen river. The total absence of any calcareous rock in Sikkim, and the appearance of the deposit in isolated streams at such distant localities, probably indicates a very remote origin of the lime-charged waters.

From Khersiong to Dorjiling, gneiss is the only rock, and is often decomposed into clay-beds, 20 feet deep, in which the narrow, often zigzag folia of quartz remain quite entire and undisturbed, whilst every trace of the foliation of the softer mineral is lost.

At Pacheem, Dorjiling weather, with fog and drizzle, commenced, and continued for two days: we reached Dorjiling on the 24th of March, and found that the hail which had fallen on the 20th was still lying in great masses of crumbling ice in sheltered spots. The fall had done great damage to the gardens, and Dr. Campbell's tea-plants were cut to pieces.

POCKET-COMB USED BY THE MECH TRIBES.

END OF VOL. I.

BRADBURY AND EVANS, PRINTERS, WHITEFRIARS.

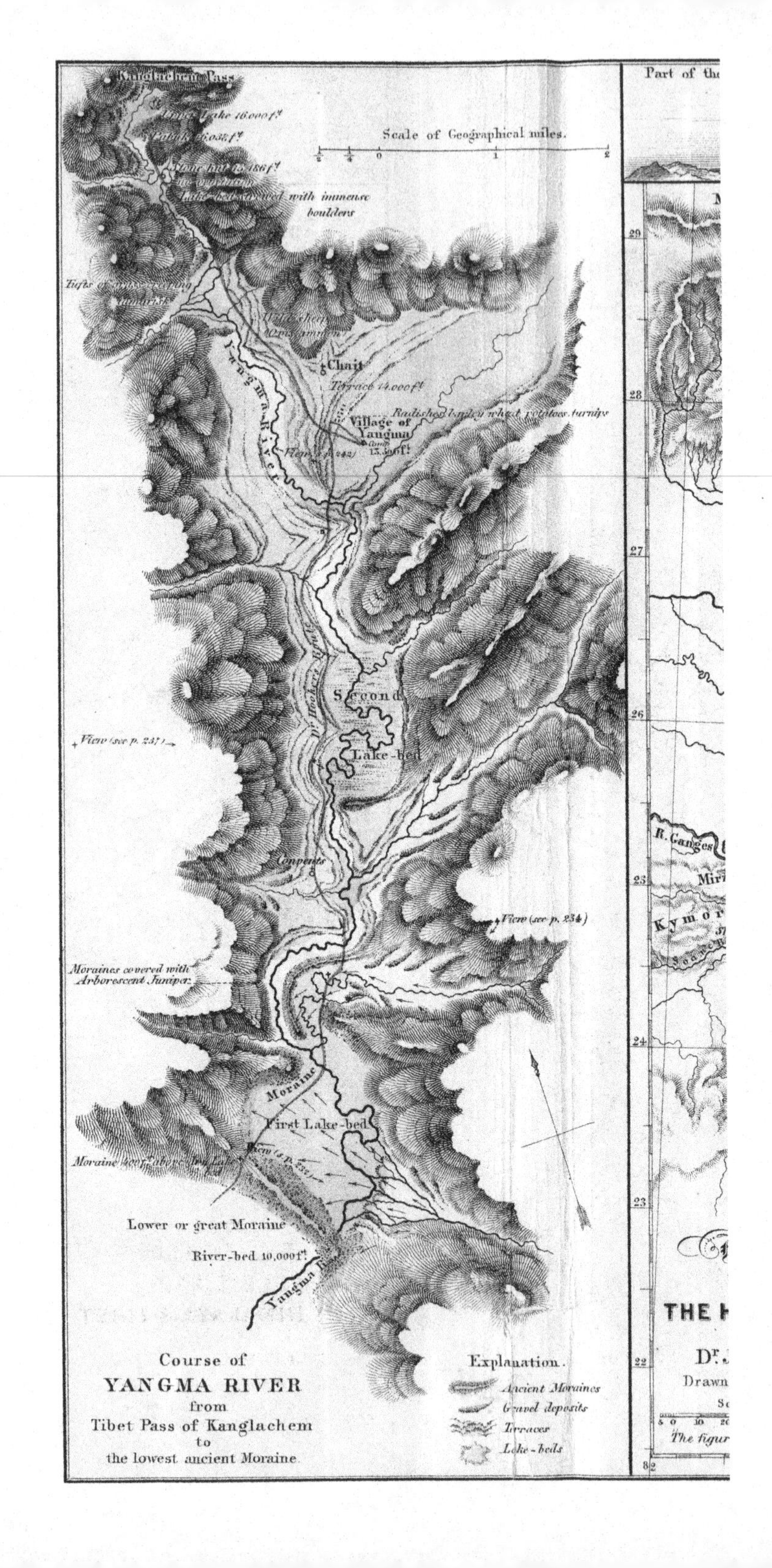
Kanglachem Pass
Scale of Geographical miles.
Chait
Terrace 14,000 ft.
Village of Yangma
Yangma River
Second Lake-bed
View (see p. 237)
Convents
View (see p. 234)
Moraines covered with Arborescent Juniper
Moraine
First Lake-bed
Lower or great Moraine
River-bed 10,000 ft.
Yangma R.
Course of
YANGMA RIVER
from
Tibet Pass of Kanglachem
to
the lowest ancient Moraine.
Explanation.
Ancient Moraines
Gravel deposits
Terraces
Lake-beds
Part of the
R. Ganges
THE H
Dr. J
Drawn

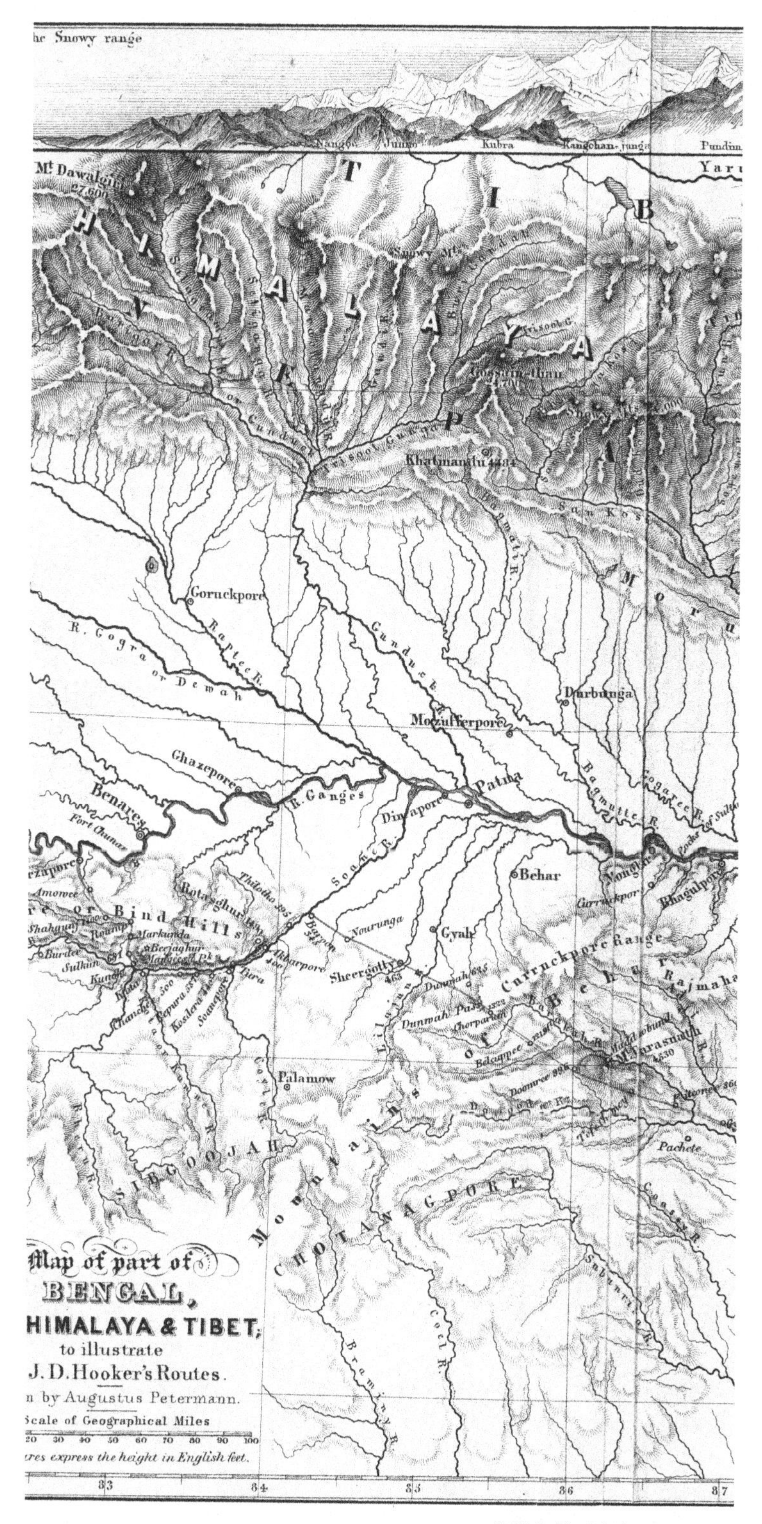
Map of part of
BENGAL,
HIMALAYA & TIBET;
to illustrate
J. D. Hooker's Routes.
by Augustus Petermann.
Scale of Geographical Miles
Goruckpore
Ghazepore
Benares
Patna
Dinapore
Behar
Gyah
Palamow
CHOTANAGPORE
Khatmandu
Durbunga
Mozufferpore
R. Ganges
Published by John Murray, Albem

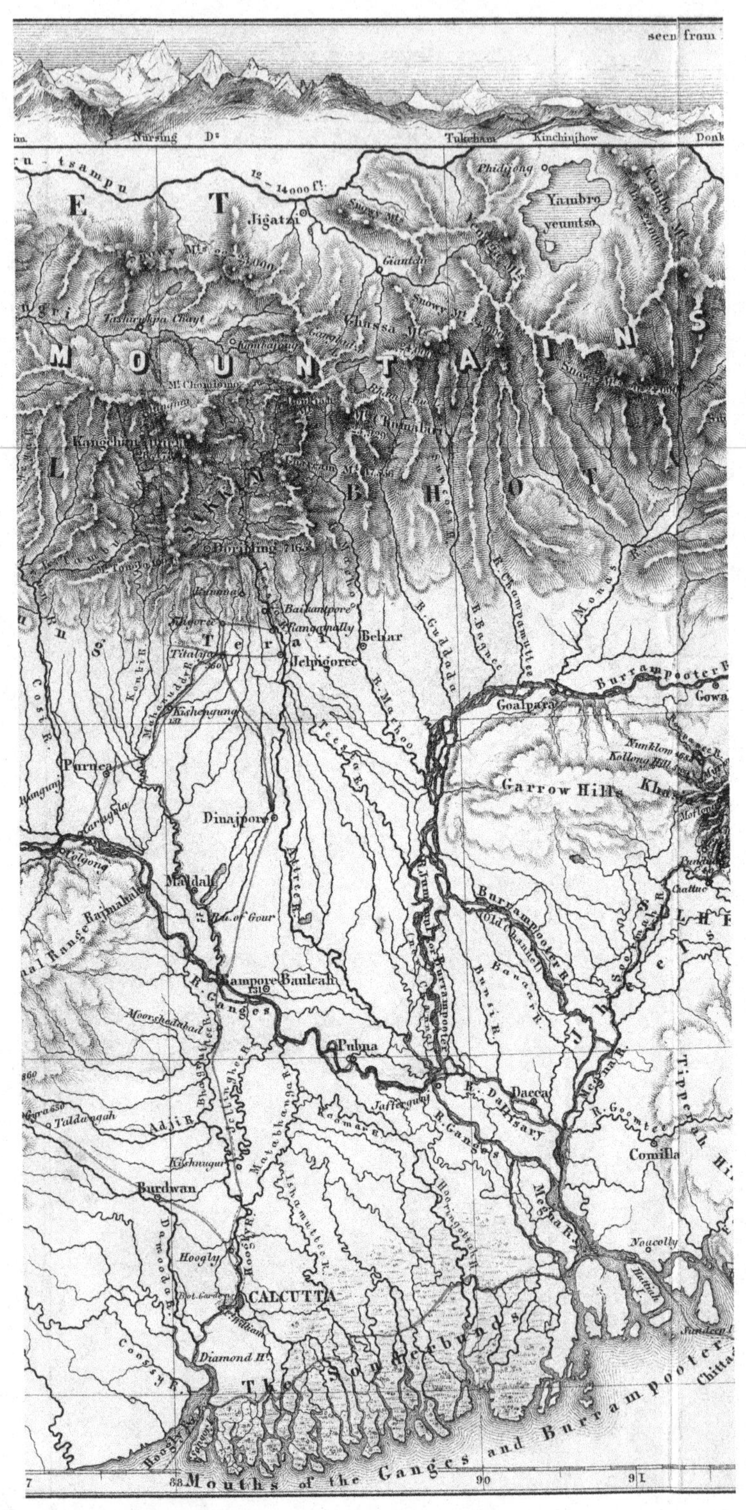

emarle Street , London 1854 .

Chumalari 23,929 ft. dist. 84 miles.
Chola Peak, 17,319 ft. dist. 51 miles.

Chumalari Mt., seen over the Chola range, from Tonglo (10,079 ft.)
(See Vol. I. p. 185)

Kangchan-junga 28,178 ft. bearing N.37° 30' E. dist. 25 miles

View in East Nepal, looking over the Singalelah Mts
(See Vol I. p. 276.)

Kangchan-junga, bearing S.60° W., dist. 37 miles, from Tibet, elev. 18,600 ft.
(See Vol II. p. 165.)

Ghassa Mountains, bearing N.84° E. dist. 90 miles and N.87° E. dist. 60 miles, from Tibet, elev. 18,600 ft.

Lith. by A. Petermann, 9, Charing Cross.

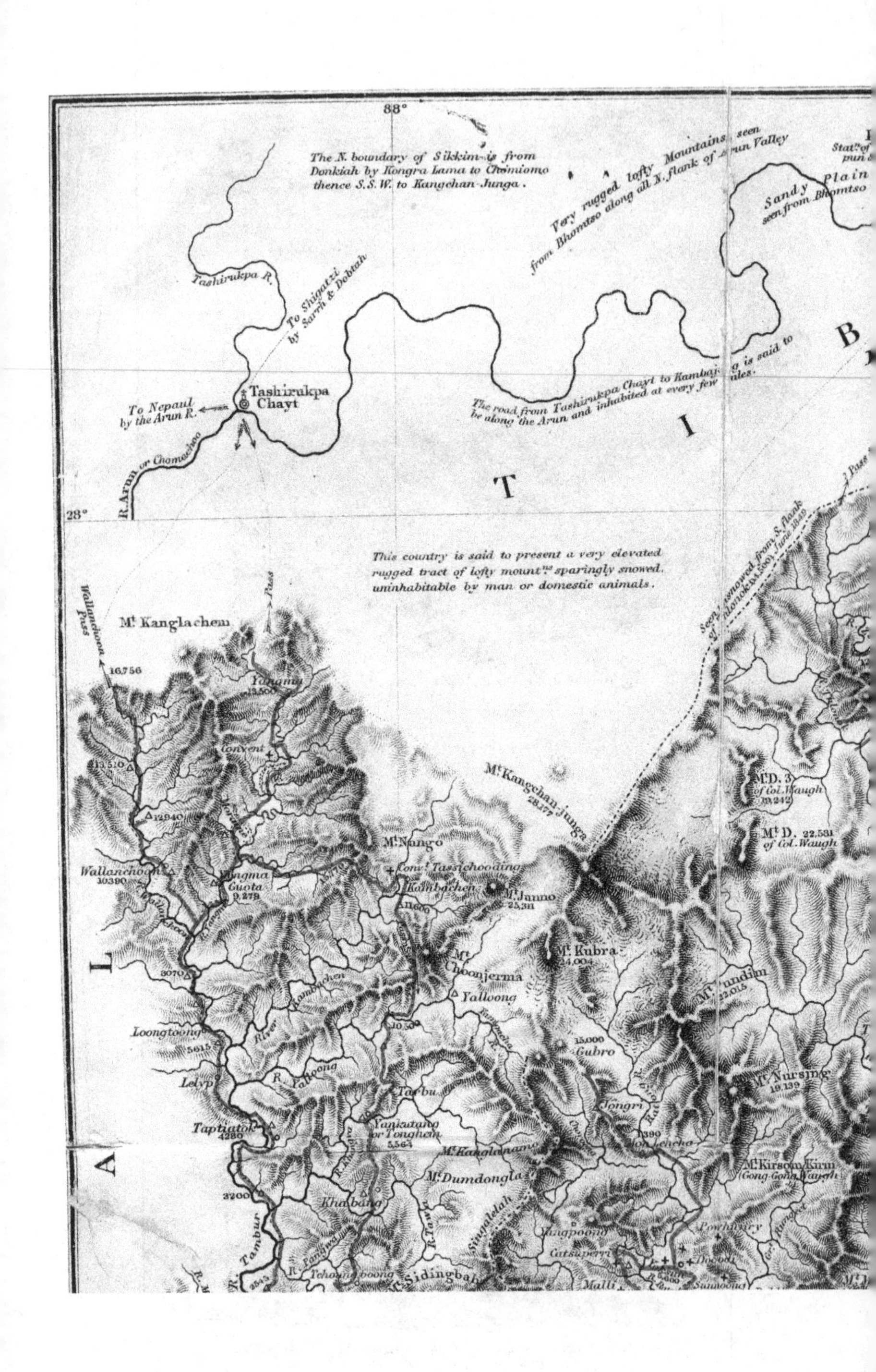
88°
The N. boundary of Sikkim is from Donkiah by Kongra Lama to Chomiomo thence S.S.W. to Kangchan-Junga.
Very rugged lofty Mountains seen from Bhomtso along all N. flank of Arun Valley
Sandy Plain seen from Bhomtso
Tashirukpa R.
To Shigatzi by Sarrh & Dobtah
Tashirukpa Chayt
To Nepaul by the Arun R.
R. Arun or Chomochoo
The road from Tashirukpa Chayt to Kambajong is said to be along the Arun and inhabited at every few miles.
T
I
B
28°
This country is said to present a very elevated rugged tract of lofty mount.ns sparingly snowed, uninhabitable by man or domestic animals.
Wallanchoon Pass
Pass
Mt. Kanglachem
16,756
Yangma
13,500
Convent
12,940
Mt. Kangchan-Junga
28,177
Mt. D. 3 of Col. Waugh
Mt. D. 22,531 of Col. Waugh
Mt. Nango
Wallanchoon
10,390
Yangma Guota
9,279
Conv.t Tassichooding
Kambachen
Mt. Junno
25,311
Mt. Kubra
24,004
Mt. Choonjerma
Yalloong
3070
Loongtoong
5615
Lelyp
R. Yalloong
Tambur
Taptiatok
4280
Tarbu
Yankutang or Tonghean
5,564
15,000
Gubro
Jongri
Mt. Nursing
19,139
Mt. Kanglanamo
Mt. Dumdongla
Mt. Kirsom Kiru (Gong-Gong Waugh)
3200
Khabang
Singalelah
Yangpoong
Catsuperri
Sidingbah
Malli
L
A

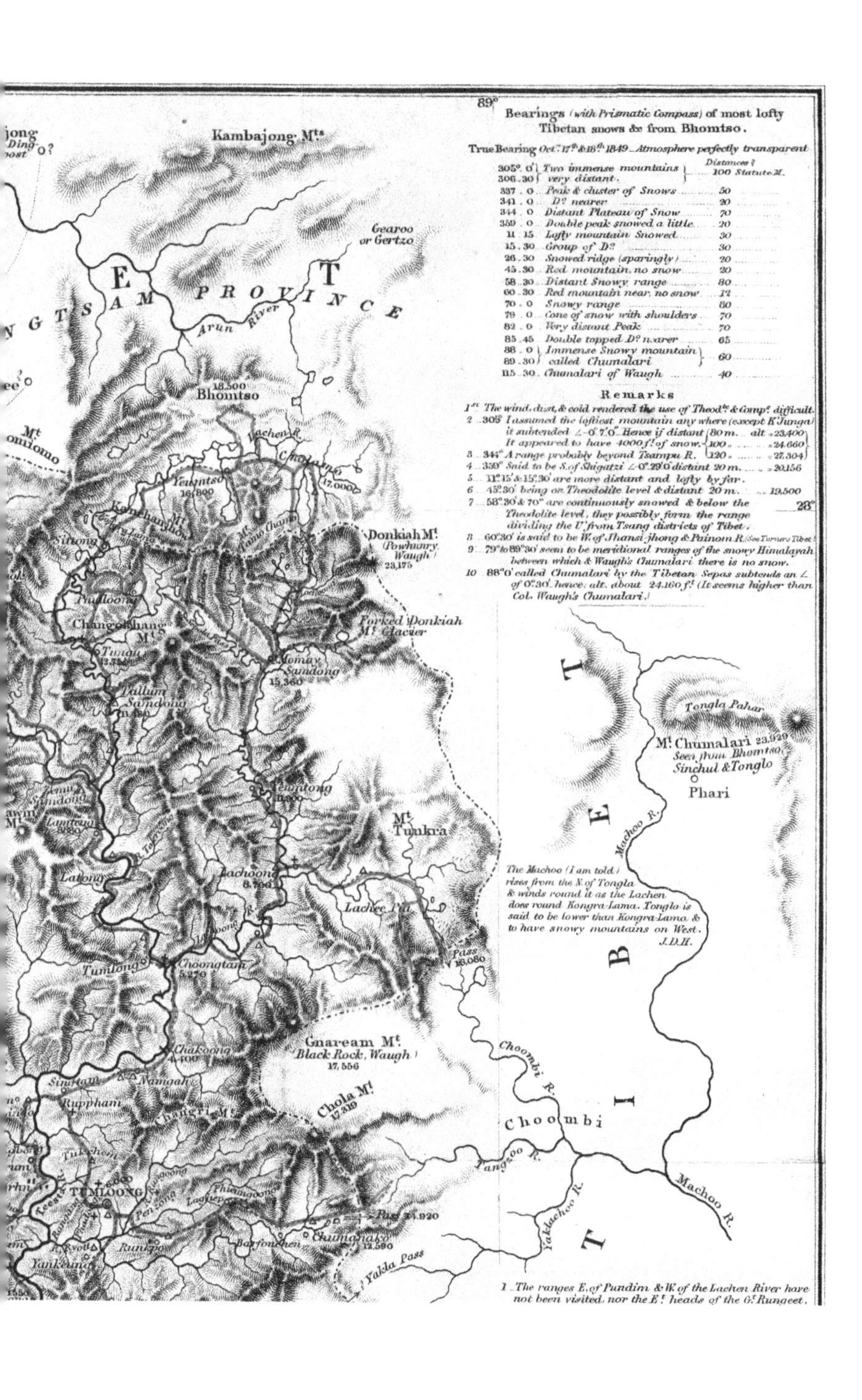
89°
Bearings (with Prismatic Compass) of most lofty Tibetan snows &c from Bhomtso.
True Bearing Oct.r 17th & 18th 1849 _ Atmosphere perfectly transparent
Distances?
305°. 0′ 306.30 Two immense mountains very distant. 100 Statute M.
337 . 0 Peak & cluster of Snows 50
341 . 0 D° nearer 20
344 . 0 Distant Plateau of Snow 70
359 . 0 Double peak snowed a little 20
11 . 15 Lofty mountain Snowed 30
15 . 30 Group of D° 30
26 . 30 Snowed ridge (sparingly) 20
45 . 30 Red mountain, no snow 20
58 . 30 Distant Snowy range 80
60 . 30 Red mountain near, no snow 12
70 . 0 Snowy range 80
79 . 0 Cone of snow with shoulders 70
82 . 0 Very distant Peak 70
85 . 45 Double topped D° nearer 65
88 . 0 89 . 30 Immense Snowy mountain called Chumalari 60
115 . 30 Chumalari of Waugh 40
Remarks
1st The wind, dust, & cold rendered the use of Theod.te & Comp.s difficult.
2 305° I assumed the loftiest mountain any where (except K. Junga) it subtended ∠-0°7′0″. Hence if distant 80 m. alt =23,400; 100 " =24.660; 120 " =27.304. It appeared to have 4000 f.t of snow.
3 344° A range probably beyond Tsampu R.
4 359° Said to be S. of Shigatzi ∠-0°29′0″ distant 20 m. =20.156
5 11°15′ & 15°30′ are more distant and lofty by far.
6 45°30′ being on Theodolite level & distant 20 m. 19,500
7 58°30′ & 70° are continuously snowed & below the Theodolite level, they possibly form the range dividing the U. from Tsang districts of Tibet.
8 60°30′ is said to be W. of Jhansi-jhong & Painom R. (See Turner's Tibet)
9 79° to 89°30′ seem to be meridional ranges of the snowy Himalayah between which & Waugh's Chumalari there is no snow.
10 88°0′ called Chumalari by the Tibetan Sepas subtends an ∠ of 0°30′ hence: alt. about 24.160 f.t (It seems higher than Col. Waugh's Chumalari.)
28°
Kambajong M.ts
Gearoo or Gertzo
T E
PROVINCE
Arun River
18,500
Bhomtso
Lachen R.
Cholamo
17,000
Yeumtso
16,800
Kinchanjhow
Sittong
Donkiah M.t
(Powhunry Waugh)
23,175
Phalloong
Changachang M.t
Tungu
Forked Donkiah M.t Glacier
Momay Samdong
15,360
Tallum Samdong
Tongla Pahar
M.t Chumalari 23,929
Seen from Bhomtso, Sinchul & Tonglo
Phari
Zemu Samdong
Lamteng
8880
Lachoong
M.t Tunkra
Lachoong
8,796
Latong
Lachee Pia
Machoo R.
The Machoo (I am told) rises from the N. of Tongla & winds round it as the Lachen does round Kongra-Lama. Tongla is said to be lower than Kongra-Lama & to have snowy mountains on West.
J.D.H.
T I B E T
Pass
16,080
Tumlong
Choongtam
5,270
Gnaream M.t
Black Rock, Waugh
17,556
Choombi R.
Chakoong
4,400
Singtam
Namgah
Rupphum
Chola M.t
17,319
Choombi
TUMLOONG
Tangzoo R.
Machoo R.
Yakchoo R.
Phiamgoong
Pass 14,920
Chumanako
12,590
Yakla Pass
Runipo
Yankeum
1 _ The ranges E. of Pundim & W. of the Lachen River have not been visited, nor the E.t heads of the G.t Rungeet.

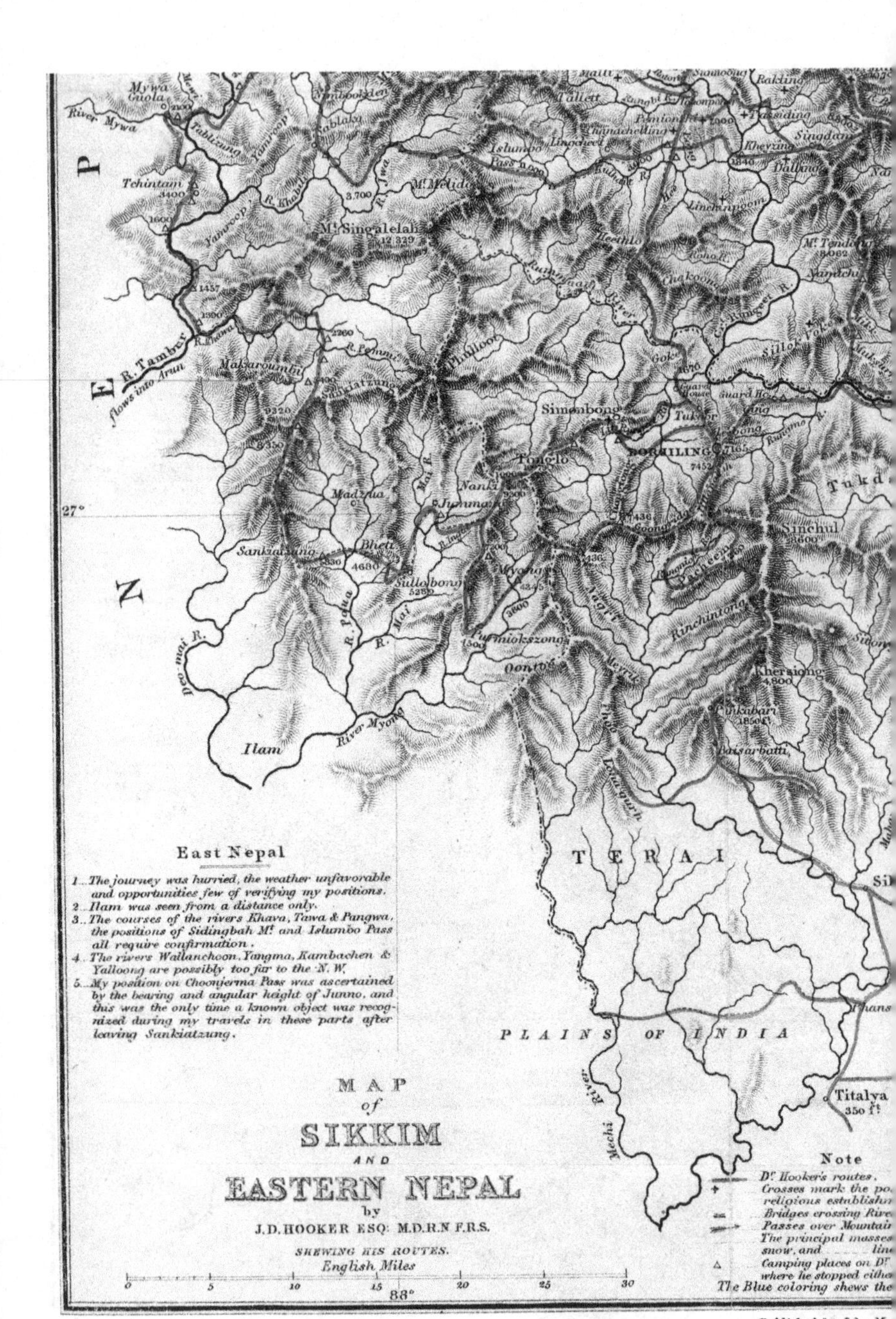

Published by John Mu

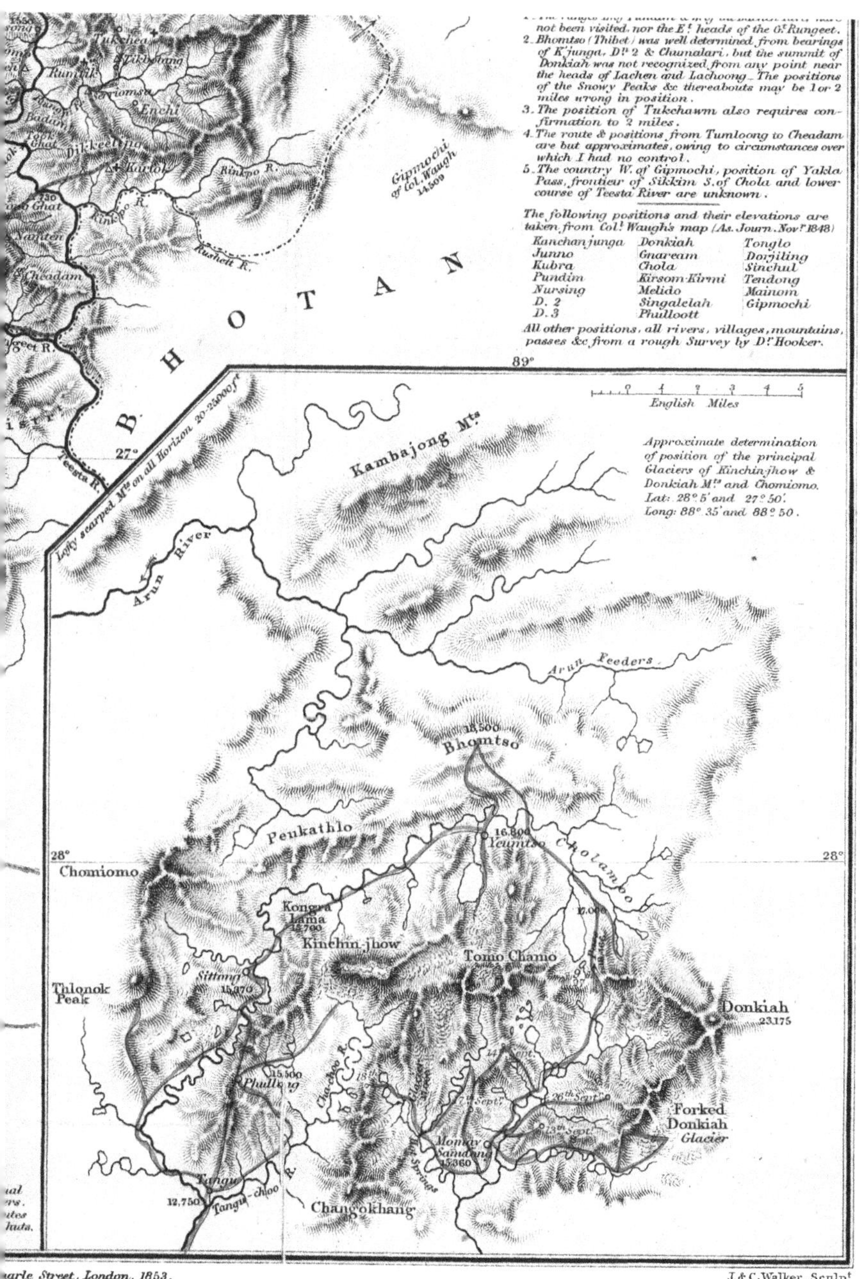
not been visited, nor the E.t heads of the G.t Rungeet.
2. Bhomtso (Thibet) was well determined from bearings of K'junga, D.o 2 & Chumalari, but the summit of Donkiah was not recognized from any point near the heads of Lachen and Lachoong. The positions of the Snowy Peaks &c thereabouts may be 1 or 2 miles wrong in position.
3. The position of Tukchawm also requires confirmation to 2 miles.
4. The route & positions from Tumloong to Cheadam are but approximates, owing to circumstances over which I had no control.
5. The country W. of Gipmochi, position of Yakla Pass, frontier of Sikkim S. of Chola and lower course of Teesta River are unknown.
The following positions and their elevations are taken from Col.l Waugh's map (As. Journ. Nov.r 1848)
Kanchanjunga
Junno
Kubra
Pundim
Nursing
D. 2
D. 3
Donkiah
Gnaream
Chola
Kirsom-Kirmi
Melido
Singalelah
Phulloott
Tonglo
Dorjiling
Sinchul
Tendong
Mainom
Gipmochi
All other positions, all rivers, villages, mountains, passes &c from a rough Survey by D.r Hooker.
89°
0 1 2 3 4 5
English Miles
Approximate determination of position of the principal Glaciers of Kinchin-jhow & Donkiah M.ts and Chomiomo.
Lat: 28° 5' and 27° 50'.
Long: 88° 35' and 88° 50.
Gipmochi of Col. Waugh 14,509
Rinkpo R.
Rusheit R.
Enchi
Rumtik
Tukchea
Kartol
Cheadam
B H O T A N
Teesta R.
27°
Lofty scarped M.ts on all Horizon 20-25000 f.t
Kambajong M.ts
Arun River
Arun Feeders
Bhomtso
18,500
Peukathlo
Yeumtso
16,800
Cholamoo
28°
Chomiomo
Kongra Lama
15,700
Kinchin-jhow
Tomo Chamo
Sittong
15,370
Thlonok Peak
Donkiah
23,175
Phullung
15,500
Forked Donkiah Glacier
Momay Samdong
15,360
Tangu
12,750
Tangu-choo R.
Changokhang
...arle Street, London, 1853.
J. & C. Walker, Sculp.t

Printed in the United States
By Bookmasters